Rethinking Rural

Series Editors
Philomena de Lima
Centre for Remote and Rural Studies
University of the Highlands and Islands
Inverness, UK

Belinda Leach
Guelph, ON, Canada

This series will foreground rural places and communities as diverse, mutually constitutive and intrinsic to contemporary Sociology scholarship, deeply imbricated in globalisation and colonisation processes stretched across national spaces. This is in contrast to an urban focus which is the implicit norm where (urban) place often can appear as the sole backdrop to social life.

Rather than rural places being marginal to Sociology, this series emphasises these places as embodying plural visions, voices and experiences which are fundamental to making sense of places as sites of solidarity, contestation and disruption in different national contexts.

Sharada Srinivasan
Editor

Becoming A Young Farmer

Young People's Pathways Into Farming: Canada, China, India and Indonesia

Editor
Sharada Srinivasan
Department of Sociology & Anthropology
University of Guelph
Guelph, ON, Canada

ISSN 2730-7123 ISSN 2730-7131 (electronic)
Rethinking Rural
ISBN 978-3-031-15232-0 ISBN 978-3-031-15233-7 (eBook)
https://doi.org/10.1007/978-3-031-15233-7

The original version of the book has been revised. A correction to this book can be found at https://doi.org/10.1007/978-3-031-15233-7_16

To the generations of farmers young and not-so-young who have shared with us their time and stories of becoming farmers and who are creating farming futures

Acknowledgements

The research for this book has been funded by the Social Sciences and Humanities Research Council (Insight 435-2016-0307).

Funding support for this book to be published open access has been provided by the College of Social and Applied Human Sciences, University of Guelph, the Social Sciences and Humanities Research Council (Insight 435-2016-0307), the International Institute of Social Studies at the Erasmus University Rotterdam, the University of Manitoba, and the College of Humanities and Development Studies at China Agricultural University, Beijing.

Alicia Filipowich provided incredible editorial support on several drafts.

Contents

Notes on Contributors

A. Haroon Akram-Lodhi is a professor in the Department of International Development Studies at Trent University. His research interests include agrarian political economy, feminist development economics, peasant economics, political ecology and sustainable rural livelihoods, and food systems analysis.

Aprilia Ambarwati is doing an MA programme in Anthropology from Gadjah Mada University in Yogyakarta, Indonesia. Her research interests include rural youth in agriculture, rural youth employment, and the dynamics of village governance.

Hannah Jess Bihun has an MA in Geography from the University of Manitoba. Her graduate work focused on analysing the social impacts of changing land tenure patterns in Manitoba, Canada.

Charina Chazali has an MA in Development Studies from the International Institute of Social Studies in The Hague. Her research interests focus on rural youth in agriculture and evaluation of poverty alleviation programmes.

Annette Aurélie Desmarais main areas of research are agrarian change, the land question, rural social movements, and food sovereignty. Prior to obtaining her doctorate in geography, Annette was a grain farmer in Canada for over a decade.

Roy Huijsmans is an associate professor at the International Institute of Social Studies in The Hague, Netherlands. His research concentrates on young people's situated encounters with development and change.

Travis Jansen is a University of Guelph alumnus, with an MSc in Food, Agriculture and Resource Economics (FARE), and a farmer who was involved in the data collection and writing for the research in Ontario.

Dong Liang research interest includes rural governance, agrarian change, and rural development.

Sudha Narayanan Research fellow, International Food Policy Research Institute (IFPRI), New Delhi.

Joshua Nasielski holds the MacSon Professorship at the University of Guelph. He is an alumnus of the PhD programme in Plant Agriculture and International Development.

Lu Pan research interests encompass left-behind populations and the sociology of agricultural and agrarian change in China. She is actively involved in action research in rural China to promote social support and social awareness for left-behind populations and rural vitalizations.

Isono Sadoko Senior researcher, AKATIGA Center for Social Analysis, Bandung, Indonesia.

Sharada Srinivasan Associate Professor of Development Studies and Canada Research Chair in Gender, Justice and Development, Department of Sociology & Anthropology, University of Guelph.

Radha Varadarajan Doctoral fellow, Department of Politics and Public Administration, University of Madras.

M. Vijayabaskar Professor, Madras Institute of Development Studies, Chennai, India.

Ben White has been involved in research on agrarian change and the anthropology and history of childhood and youth since the early 1970s, mainly in Indonesia.

Hanny Wijaya has an MA in Anthropology from Gadjah Mada University in Yogyakarta, Indonesia. Her research interests focus on agrarian change, gender, and labour relations.

List of Figures

List of Maps

List of Tables

1

Introduction: Young People's Pathways into Farming

Sharada Srinivasan and Ben White

Who Wants to Be a Farmer?

Will there be a new generation of farmers to take the place of today's age-ing farmers? What are the experiences of young people who are establish-ing themselves as farmers, and how are these pathways gendered? How can young farmers be supported to feed the world's growing population?

These are the questions that stimulated us and our colleagues in Canada, China, India, and Indonesia to join together in the multi-country research project *Becoming a Young Farmer: Young People's Pathways*

S. Srinivasan
Department of Sociology & Anthropology, University of Guelph,
Guelph, ON, Canada
e-mail: sharada@uoguelph.ca

B. White (✉)
International Institute of Social Studies, The Hague, The Netherlands
e-mail: white@iss.nl

© The Author(s) 2024
S. Srinivasan (ed.), *Becoming A Young Farmer*, Rethinking Rural,
https://doi.org/10.1007/978-3-031-15233-7_1

into Farming in Four Countries (2016–2021).[1] Each team used multi-sited case study research to bring to life the experiences of young farmers and would-be farmers, the various challenges they face, and important differences in their experiences both within and between the countries and study sites.

Thinking about young people and farming raises fundamental questions about the future, both of rural young women and men and of agriculture itself (Ní Laoire 2002; White 2012). The future shape of rural communities, and of the world's agriculture, will depend to a large extent on these and future generations of young rural people and their interest in—and their ability to acquire the needed resources for—farming careers and livelihoods. In recent years, numerous panels of international experts have highlighted the ecological, economic, and social advantages of small-scale "family farming" compared to large-scale industrial agriculture. While "family farms" (including both small and large-scale farms) occupy around 70–80 per cent of all the world's farmland and produce about 80 per cent of the world's food in value terms, "small family farms" (of less than 2.0 hectares), which account for more than 80 per cent of all farm units, produce 35 per cent of the world's food on only 12 per cent of all agricultural land (Lowder et al. 2021). Smallholder farms of this type thus generally achieve higher yields per unit of land—and higher yields per unit of non-human energy input, but less per unit of human labour—than industrial farming. They also provide better outcomes in terms of food security, environmental sustainability, employment, and community cohesion and development (FAO-IFAD 2019; Committee on World Food Security 2019, 2020; IAASTD 2009; Herren et al. 2020; Ricciardi et al. 2021).

If visions of sustainable agricultural futures based on "family farm" units are to be realized—as envisaged in the current (2019–2028) "United Nations' Decade of Family Farming" (FAO-IFAD 2019)—the problems that young people face in establishing themselves as farmers have to be taken seriously and given much more attention than has been the case in recent research and policy debates (FAO 2014; White 2020; Committee on World Food Security 2021).

---

[1] The project was funded by the Social Sciences and Humanities Research Council of Canada.

The world's crisis-ridden agriculture and food systems, besides huge environmental challenges, are facing a looming problem of generational renewal. Farming populations are ageing, many farmers appear to have no successor, and it is widely claimed that young people are not interested in farming; smallholder farming in its present state appears to be so unattractive to young people that they are turning away from agricultural futures. While this generalization is to some extent a myth—as this book also hopes to show—there is growing and justified concern about the problem of generational renewal in agriculture, and it is finally getting some attention internationally. This has been recently reflected, for example, in the United Nations' Decade of Family Farming (2019–2028) with its Action Pillar underlining the need to "ensure the generational sustainability of family farming" (FAO-IFAD 2019), and in various policy reports on the broader problems of rural youth livelihoods (IFAD 2019) and young people's engagement in agri-food systems (Committee on World Food Security 2021).

Meanwhile, youth unemployment is increasing worldwide. Open unemployment rates for youth are three times higher than for adults in all world regions. In most regions, youth unemployment was rising even before the COVID-19 crisis (ILO 2020). Large-scale youth unemployment has long been a matter of concern in national and international policy discourse and is flagged in the United Nations' Sustainable Development Goals,[2] but it is hard to find realistic proposals to address it.

Close to half of the world's unemployed are youth (World Bank 2006, 2015), and many others are underemployed, having insufficient work, and/or precarious informal sector employment. In some regions, rural and urban youth employment rates are the same; in some others, the rural rate is slightly lower, but it must be remembered that significant numbers of young urban unemployed themselves originate from rural areas. Agriculture remains the majority world's largest employer. As a sector that faces growing demand for food, fibres, and other products, if

---

[2] Goal 8: "Promote sustained, inclusive and sustainable economic growth, full and productive employment and decent work for all" includes as targets 8.5 and 8.6: "By 2030, achieve full and productive employment and decent work for all women and men, including for young people and persons with disabilities, and equal pay for work of equal value" and "By 2020, substantially reduce the proportion of youth not in employment, education or training" (United Nations n.d.)

supported, it has the potential to provide decent livelihoods for many more. But various surveys in different parts of the world have all found that young rural people, including the children of farmers, do not aspire to farming futures; even in regions where agriculture is the most important contribution to rural livelihoods, agriculture is not the first option (Committee on World Food Security 2021, 26; White 2021).

Our study and this book come at a strategic moment, responding directly and concretely to these concerns whilst also challenging some widely held assumptions. We decided not to study young rural people's apparent turn-away from farming, but instead to focus on young men and women who are—or are trying to become—farmers, the many constraints they face, and their efforts to overcome them. This also meant diverging from the conventional focus in youth studies on young people in their late teens and early twenties (or as in United Nations' definitions, between ages 15 and 24). Many of the young generation of farmers featuring in this book did not become farmers during their youth as defined in this way; they embarked on the long-drawn-out process of becoming a farmer, sometimes in their youth or in later years after a period of migratory non-farm work or education.

The process of farm transmission and the intergenerational tensions that often surround it have been studied in detail in Europe, North America, and other late capitalist regions (Cassidy et al. 2019). Among other world regions, it is only in sub-Saharan Africa that we find a rich tradition of research on intergenerational tensions and conflict surrounding farmland, and young people's difficult pathways into farming (for example, Skinner 1961; Quan 2007; Amanor 2010; Kouamé 2010; Peters 2011; Berckmoes and White 2016). On rural young people and farming generally, it is again in sub-Saharan Africa that research is relatively well-developed compared to Asia and Latin America (see, for example, Flynn and Sumberg 2021; Chamberlain et al. 2021). This was one reason for our decision to devote three of our four country studies to Asia.

Our study provides a comparative analysis of young men's and women's pathways into farming in four countries. As explained further in the concluding chapter by Haroon Akram-Lodhi and Roy Huijsmans, one (Canada) is a major agro-exporting western country characterized by modern, highly capital-intensive farming and fewer farms of increasing size (now down to around 272,000 farms and up to an average 274 hectares), which employ only a small fraction of the labour force. Farming in the other three countries in Asia (China, India, and Indonesia) is dominated by smallholder farmers numbering in the (tens of) millions, with farms of much smaller size (in India averaging 1.3 hectares (ha), in Indonesia 0.8 ha, and in China only 0.4 ha), and much greater proportions of the labour force depend wholly or partly on agriculture for their livelihoods. Despite industrialization and growing service sectors, agriculture is still the largest single source of employment in these three countries. But these countries also share some trends with Canada: smallholder farming is losing out to agribusiness (except in India),[3] the average age of farmers is rising, farmland prices are rising, and young people in rural areas apparently do not want to become farmers. Smallholder farmers themselves often say that they hope their children will find better work than farming (see for Southeast Asia, Hall et al. 2011, 118). At the same time, for those young people who are interested in farming, access to farmland—for both young "newcomers" to farming and would-be "continuers" from farming families—appears to be a major problem, with the possible exception of China. And for those who have some form of access to land, as a recent Prindex report shows, in nearly all regions of Asia, the perception of tenure insecurity is much higher among younger people (aged 18–45) than among older generations[4] (Prindex 2020, 12, 25).

Our Approach, Guiding Concepts, and Methods

There are several key features worth noting. First, we agreed to focus on those who have become, are in process of becoming, or are aspiring to become, farmers. In contrast to the more common focus in youth studies, we did not limit the research to young people in the age range 15–24,

---

[3] In India, businesses cannot own land and agribusiness is only in the form of agricultural inputs or in post-production.
[4] China and Vietnam being important exceptions in Asia.

which often leads only to studies on why young people leave the country-side and, for many of those who are on the way to becoming farmers, stops before their trajectory into independent farming is completed. Second, the life-history approach in our interviews with young farmers has enabled us to document the—perhaps obvious, but not always recog-nized—fact that "becoming a farmer" is not an event but a process, which in the majority of cases involves some period of non-farm work, most often through migration to urban centres. Capturing people's typical mobilities between places and sectors during the first decades of life, the study contributes to rethinking conventional rural-urban and farmer-nonfarmer divides and does so from a life course perspective. Third, we have mainstreamed gender distinctions and also differences between "continuer" young farmers (those who have grown up in farming fami-lies) and "newcomer" young farmers. By concentrating on women and men who have managed, or are trying, to set up their own farming liveli-hoods at a relatively early stage in their lives, we aimed to contribute both to theory by clarifying the generational dimension in the social reproduc-tion of agrarian communities, and to policy by clarifying the barriers that young rural men and women confront in accessing land and other resources as well as the role of policies, institutions, and young people's own (individual and collective) efforts in overcoming these barriers.

Guiding Concepts

Our study in the four countries has engaged with the term "youth" in the chronological (age) sense as well as in the social (generational) meaning of youth, as a relative stage in the lifecycle, depending on the context. At the time of data collection, although the average age of a farmer across our sample of 378 farmers was 34 years, most had their first experience of farm work as young as 13 years of age and began farming independently when they were 23 years old (Table 1.1). The farmers that we interviewed are established as farmers or are well set on the path to becoming a farmer. This may not be true for the vast majority of youth outside of our study who aspire to or are trying to become farmers. Most youth do not become farmers when they are chronologically young. For many the *process* of

Table 1.1 Summary characteristics of young farmers in our research

	Canada	China	India	Indonesia	Summary
Farmers interviewed	95	76	98	109	378
Female farmers	43	34	25	49	150
Male farmers	52	42	73	60	228
Age started farming	16	12	13	12	13
Age farming independently	25	22	21	24	23
Mean age	33	36	34	33	34
% under 35	66%	45%	55%	67%	59%
% married	60%	93%	76%	82%	77%
% with >12 years education	89%	0%	33%	3%	32%
% working full-time	93%	99%	97%	77%	90%
% full-time, primary income farming	79%	87%	85%	50%	74%
% primary income—animal farmer	24%	0%	2%	1%	7%
% primary income—plant farmer	54%	87%	81%	49%	66%
% primary income—farmer—unspecified	1%	0%	2%	0%	1%
% full-time, primary income—not farming	14%	12%	12%	28%	17%
% farmers owning land	67%	76%	66%	83%	73%
Minimum acres owned	1.00	0.08	0.00	0.00	0.00
Average acres owned	527.81	0.53	5.92	0.89	134.55
Maximum acres owned	4300.00	2.47	56.83	77.83	4300.00
% farmers that have inherited land	7%	16%	60%	50%	35%
Minimum acres inherited	62.50	0.16	0.00	0.00	0.00
Average acres inherited	1511.79	0.15	4.79	0.49	381.36
Maximum acres inherited	5500.00	2.47	20.00	3.93	5500.00
% farmers likely to inherit land	57%	34%	36%	28%	38%
Minimum acres likely to be inherited	0.00	0.12	0.00	0.00	0.00
Average acres likely to be inherited	805.34	0.23	2.09	0.12	203.02
Maximum acres likely to be inherited	3400.00	1.4	10.50	2.47	3400.00
% farmers renting in land	60%	9%	16%	24%	28%
Minimum acres rented in	1.00	0.33	0.00	0.00	0.00

(continued)

Table 1.1 (continued)

	Canada	China	India	Indonesia	Summary
Average acres rented in	719.47	0.07	2.31	0.11	181.46
Maximum acres rented in	4060.00	1.65	20.00	1.42	4060.00
% farmers sharing land	33%	1%	37%	28%	26%
Minimum acres shared	0.25	0.49	0.00	0.00	0.00
Average acres shared	967.40	0.01	2.65	0.18	243.87
Maximum acres shared	4800.00	0.49	12.36	3.71	4800.00
% with access to community land		17%	6%	5%	6%
Minimum acres of community land farmed		0.12	0.00	0.00	0.00
Average acres of community land farmed		26.38	0.05	0.04	5.33
Maximum acres of community land farmed		49.4	1.00	1.36	49.40

becoming a farmer or aspiring to be a farmer begins when they are chronologically identified as "youth," but only culminates in an independent farming existence many years later. Many are unable to get into farming right away for generational reasons, waiting to inherit land and other farm resources. Others may migrate or work in the non-farm/urban sectors to diversify their experiences, livelihoods, and to bring an investment into farming. For many women (especially in China, India, and Indonesia), farming as a vocation becomes possible only after marriage.

Combining chronological and social age allowed us to focus on the relatively young women and men who are (trying to become) farmers in specific contexts. We have not included farmers in their later adulthood, who might be 60 years old or more and are trying their hands at farming for the first time, notwithstanding the importance of such experiences in some countries. Thus, for example, when the two authors of this Introduction visited Sanggang village in Hebei province in 2016 and asked to be introduced to the youngest farmer in the village, they found themselves on the organic apple farm of 43-year-old Zhang Changchun, who had started farming at the age of 37 after a long period of migratory non-farm work (White 2020, 199–120). While Mr Zhang is not chronologically a youth, he is indeed a young farmer in the demographic context of his village. This has important implications for policy to support young farmers and the "generational sustainability of family farming" (FAO-IFAD 2019).

Our guiding research framework combines core concepts from the interdisciplinary fields of agrarian studies, youth studies, generation studies, and gender and development. Agrarian political economy allows us to depict and compare "the social relations and dynamics of production and reproduction, property and power in agrarian formations and their processes of change, both historical and contemporary" (Bernstein 2010, 1). It also helps us to better understand and compare the structure of the rural communities that we study, the possible future trajectories of the agri-food sector, and in particular, the underlying and continuing debate on large versus small-scale agricultural futures as well as the special characteristics of smallholder farming (van der Ploeg 2013).

Youth studies help us to move away from adult-centric perspectives that continue to dominate agrarian studies and development studies more broadly. It foregrounds young people's own perspectives and priorities, thereby shedding new light on their paradoxical (apparent) turn away from farming in this era of mass rural un(der)employment (Cuervo and Wyn 2012). It also provides an important reminder of the need and the right of young people to be properly researched, not as objects, but as subjects and where possible as participants in research (Beazley et al. 2009). Key concepts that we draw from the "new" youth studies are the ideas of youth as actors in social and economic renewal, youth as identity, and youth as generation (Jones 2009; Huijsmans 2016). We complement this with a generational perspective that helps illuminate the "generational ordering" (Punch 2020) of agrarian structures. Viewing generation as a relationship is critical for understanding the various intergenerational relations, practices, and life course events frustrating and/or facilitating agrarian generational renewal. This includes issues of inheritance and the generational transfer of agricultural knowledge, but also young people's often marginal position in farmers' organizations and rural movements. It also directs attention to how the process of becoming a farmer intersects with other key life course events, such as education, marriage, and family formation.

For many or most young would-be farmers, becoming a farmer depends on the transmission of agrarian resources, particularly land, between the generations. Farm transmission is normally but not always between parents and their children; "extra-familial" transmission is increasingly common in late capitalist countries where many ageing farmers have no successor (see, for example, McGreevy et al. 2019 for Japan and Korzensky 2019 for Austria).

These ideas support a relational approach to studying young people's experiences with farming, the dynamics of relations between younger and older generations, within the same generation, and their role in the social reproduction of agrarian communities (Archambault 2014; Berckmoes and White 2014). At the same time, young people are neither homogeneous nor exist in a vacuum; generation intersects with other important social categories such as social-economic class and gender (Hajdu et al. 2013; Jones 2009; Nayak and Kehily 2013; Wyn and White 1997). As already noted, our study incorporates a systematic focus on young women (would-be) farmers. Traditional agrarian societies are typically sites of heteronormative patriarchy in both gender and generational relations (Ní Laoire 2002; Stearns 2006). Young people are not passive victims within these patriarchal structures but exercise a "constrained" agency. This is most evident in the gender and generation-neutral term "family farm," which hides the imbalances within families in ownership, access, and decision-making around resources and in the division of labour. Yet, as our research reveals, young (would-be) farmers, and especially young women farmers, confront or subvert these structures. At the same time, we hope that future research will focus on the impacts of sexualities and gender plurality on traditional agrarian contexts. Bringing these perspectives together has helped us to understand the intergenerational tensions that we see almost everywhere in rural communities, particularly young people's problems in gaining access to farmland and other agriculture-related opportunities.

As already mentioned, another key distinction that we have adopted and further explored in this study is that between "continuer" young farmers (those who take over their parents' farm) and "newcomer" young farmers (those not from farming backgrounds). There is a strong supposition that "newcomer" farmers are likely to be more critical of mainstream farming practices and innovators. By "family farming" and "smallholder farming" in this study, we refer not to the size (acreage) of the farm unit, but to the manner and "scale" of its operation, where owners or tenants themselves manage and work on the farm, often with the help of family members but not ruling out the use of hired workers. It can thus encompass both farms of half or one hectare in one of our Asian sample countries and farms of 100 hectares or more in Canada, depending on the manner in which they are owned, managed, and worked.[5]

Our Methods

We aimed to look for commonalities, comparisons, and contrasts in the experiences and trajectories of young farmers between countries, between regions within countries, and within localities. In each country, the teams selected two sites (three in Indonesia), offering contrasts in forms of smallholder farming. In methodology workshops while preparing the project, the four teams developed together a set of general research questions, summarized below.[6]

Agrarian context: What are the general patterns and trends of farmland ownership and access, farm sizes, and labour use? How have farmland prices changed? Who gets what in smallholder agriculture? What are the trends in the age structure of the farming population?

Becoming a young farmer: How do young people become farmers? What are the resources that they need in this process? What are the typical modes of transfer of farmland and property between generations? How are resources divided among sons and daughters? How do young people access land and credit? How do they acquire and develop farming knowledge and skills? What kind of social networks do they rely on and what support do they receive from these networks? How do young women farmers fare? What challenges do they encounter and how do they deal with them? How do they deal with social, economic, and other barriers to becoming farmers in their own right?

---

[5] For the distinction between "size" and "scale" in farming, see van der Ploeg (2013).
[6] These questions are adapted from the original "Becoming a young farmer: Young people's pathways into farming" project proposal. See also White (2015).

Young farmers and innovation: What are young farmers' attitudes to conventional farming practices? Are young farmers in general, and newcomers in particular, more flexible and innovative with regard to farming compared to older farmers and continuers? What role do relatively new technologies such as mobile phones, the internet, and social media play in the innovation process and dissemination?

Young farmers in policy and agenda setting: How do agrarian and rural policies affect young people engaged in farming? Which policies make it more or less easy for young people to enter into farming? What specific kinds of support are available for young farmers? How do young people attempt to influence the level and contents of such support? Are young farmers organized? How are they involved in existing farmer unions, associations, and/or political parties, in dedicated young-farmer organizations, and in new modes of networking among young farmers (with particular attention here to social media)? How do they influence political parties and policymakers, and with what degree of success?

Qualitative, in-depth methods of inquiry are best suited to study young (would-be) farmers' lived experiences and trajectories. Identification and selection of young men and women (would-be) farmers, both "continuers" and "newcomers," was facilitated by the research team members' existing local contacts and relationships, with farmers, (youth) farmers' organizations, women's organizations, and non-governmental organizations.

In each country, we aimed for 100 young farmer interviews. Our final sample consists of a total of 378 young farmers, covering newcomer and continuer farmers, established farmers and young women farmers. The primary informants were young men and women who (a) were already farmers in their own right, (b) managed farms that they may not yet own or control completely, but with a degree of independence (that is, not merely working or helping in their parents' fields), or (c) were in process of trying to establish themselves as farmers (which may include "apprenticeship"-like stints on established farms), either by choice or lack of it.

Interviews were guided by a common set of questions while leaving research teams in the individual countries to address further country-specific issues and questions essential for the completeness of their own case studies. The interviews included a life-history component, starting from the respondents' first childhood experience of helping on the farm and continuing with their work and livelihood trajectories after leaving school or college. This explains our decision to use the age range 18–45 years in selecting our "young farmer" samples. To avoid misunderstanding, it should be underlined that this was not because we want to expand the definition of "youth" or consider a 45-year-old farmer to be "young" but because—as we have already explained—becoming a farmer is often a long-drawn-out process, and we miss important parts of that process if we only interview those who have recently started farming. As the Canadian research team puts it: "we believed that it was important to include some 'older' young farmers in our study; farmers who are still young enough to remember their own experiences as young farmers but who might be further along in their journey toward becoming a successful farmer" (Nasielksi et al. this book, page).

We also interviewed some older farmers in the same locations to capture and compare their earlier experiences as young farmers and some young people who were keen to exit farming. Also included were (depending on the location and their relevance) non-governmental and community-based organization activists, farmers' union members, staff of agricultural universities and other training centres, and officials of regional agriculture departments.

Table 1.1 provides a summary profile of the farmers that we interviewed. The average age of the farmer respondents is 34 years. Most farmer respondents in China, India, and Indonesia are married, with Canada having the lowest proportion of young farmers who reported being married. Nearly 90 per cent of Canadian farmers in our sample have more than 12 years of education compared to zero in China, 33 per cent in India, and 3 per cent in Indonesia.

Most respondents work full-time in farming in all four countries, but many combine farming with other income sources. For the majority in Canada, China, and India, farming is the primary source of income, while it is the primary source of income for only for 50 per cent of Indonesian respondents.

Nearly three-fourths of our sample already owns some land, although there is a huge variation in the acreage owned. In Canada it is more than 500 acres, while in China and Indonesia it is less than one acre. About 28 per cent rent land for farming, with the highest share of renters in Canada as well as sharing land or engaged in sharecropping. "Land sharing" could be as in community-supported agriculture in Canada or amongst family members as in India.

That not all of the land owned is obtained through inheritance is evident in that only 35 per cent of the 378 farmers in our sample report having inherited land. The lowest number of farmers to inherit land is in Canada (7 per cent), while in India it is 60 per cent and in Indonesia, 50 per cent. About 38 per cent of young farmer respondents expect to inherit land, with the highest number being in Canada.

All of the country teams have incorporated an explicit focus on the interactions of generational and gender relations. Researchers prioritized interviewing young women farmers, even in the case of couples where the male partner was (self) identified as the "farmer," in order to identify the special problems of resource access and recognition that young women farmers often face. And finally, the selection of two contrasting research locations in each country (in Indonesia, three locations) has allowed the authors to draw comparisons and contrasts within as well as between countries and thus avoid country-based stereotypes of "the young (Canadian, Chinese, Indian, or Indonesian) farmer."

As already explained, our field research aimed to privilege young people's own perspectives on and experiences with farming, the challenges that they encounter, the ways that they deal with them as well as the impact that young farmers' practices have on farming. This approach is consistent with young people's right to be properly researched on their own terms and in their own perspectives (Huijsmans et al. 2014; Naafs and White 2012; Srinivasan 2014).

Organization of This Book

The book is organized by country into four sections. Each section begins with a country overview chapter, which serves to contextualize the case studies and summarize what is known about young people and farming from available studies and secondary data. These overview chapters also explain the choice of research sites in each country. The remaining two chapters in each section (three for Indonesia) present the results of our local case studies.

Each country team, while following a general set of shared guidelines, has chosen their own distinct style of presentation and writing, which also reflects different scholarly orientations and practices between the teams. We have consciously opted to preserve this variety, rather than squeezing it all into a standard format and style.

Findings

In the concluding chapter, Haroon Akram-Lodhi and Roy Huijsmans, members of the Canada and Indonesia research teams, respectively, have provided some inter-country and inter-locality comparisons and concluding reflections, based on the country overviews and case-study chapters. In this section, we highlight just a few general findings of our study that readers may need to bear in mind when reading the chapters.

Perhaps the most important contribution of our study is the way it has demonstrated that becoming a (young) farmer is a process rather than an "event." While most respondents begun farm work as young as 13 years of age, the great majority of farmers in all four countries did not start farming independently until much later. The average age when respondents began farming independently was 23 years (Table 1.1), but for many this milestone came much later. After leaving school or college, young people typically go through a period of non-farm employment, frequently migrating to urban centres for work. This applies as much to the "continuers" (those from farming families) as to the "newcomer" farmers and as much to female as to male young farmers. It also applies to young graduates of vocational agricultural schools. This has many policy implications for the kinds, and the timing, of the needed support for young farmers, including education, land allocations, and subsidies. Most rural youth, including those enrolled in agricultural schools, are very unlikely to make a start in farming immediately after leaving school. This suggests that support for entry into farming—training, internships, help with accessing land, and other resources—may be needed much later in life.

Our study also provides a reminder that young rural people are generally landless, even if their parents have land. The only young people who may obtain access to parental land while still young are the children of land-rich farmers or those whose landowning parents die early. Most young farmers, therefore, don't have access to parental land when they start. Even if there is access to land, gaining more control over farming, farm-related decisions, and earnings poses a challenge. Large numbers of young farmers start their farming life on rented land; sometimes, as we have seen in Canada and Indonesia, on land rented or sharecropped from their parents. This has clear implications for policy. In countries or regions where significant amounts of farmland are not privately but state or community owned (China, parts of Indonesia, Canada's Crown Lands), there are many possibilities for allocation of use-rights on this land, at low rental rates, to young farmers. There are many examples around the world of good practice in government facilitation of young farmers' access to land at reasonable rates (see, for example, Committee on World Food Security 2021, 56–61).

While the degree of gender discrimination (whether in law or social practice) differs between countries and also in some cases within countries, young women farmers face problems of resource access and recognition as farmers in the majority of our research sites. National and local efforts to counter these biases are important, but unlikely to emerge unless (young) women farmers emerge as a political force.

Young farmers' pluriactivity—combining farm and non-farm incomes, with or without migration—is the norm in most cases, often at individual and certainly at the household level. This provides another insight for policies to support young people in farming: they need to recognize the reality that for today's young farmers, engagement with farming is seen—as it was for many of the previous generations—as both a part life-course and a part livelihood activity. "Farmer" is an identity as well as an occupation; sometimes self-identified as in the case of Canada, and sometimes ascribed as the China situational analysis makes clear.

Another insight from the case studies is that "smallholder" farms are not necessarily "family farms." In all countries, we find some young farmer couples sharing the burden of farm work and sharing farming decisions, sometimes also—but much less than in previous generations—helped by their children; but in all countries too, it is also common to find that only one household member has any significant engagement with the farm.

Are young farmers by nature innovative, or simply more innovative than older farmers? Our study provides reason to question this common assumption, as some previous studies have done (Sumberg and Hunt 2019; Chamberlain and Sumberg 2021). Young continuer farmers may find themselves locked into crop choices and farm practices that their parents established, as noted in the Canadian studies. Newcomer farmers, on the other hand, may come to farming with an intention to do things differently. The Canadian studies have a good number of respondents who identify as newcomers, while they are absent or rare in the other country samples. This is an area for future research.

Most government policies towards agriculture continue their long-standing focus on scale enlargement and promotion of industrial agriculture. To date, there is little evidence of any commitment—beyond lip service—to the support of alternative, more earth-friendly modes of farming.

Finally, our study found little evidence of young farmers emerging as a significant political force, whether locally, regionally, or nationally. Smallholder farmer movements, despite their long history, are generally weak, and young farmers' movements in most cases are non-existent. Sustainable futures of the world's farming and agri-food systems, and their role in providing the planet's growing population with healthy, safe, and nutritious food while combating climate breakdown, are unlikely to be ensured by state policies and actions from above unless these are forced to act by large-scale and sustained pressures from below. In all of these matters, today's young people are at the front line and it is important that their voices are heard.

References

Amanor, K. 2010. Family values, land sales and agricultural commodification in South-eastern Ghana. *Africa* 80 (1): 104–125.

Archambault, C. 2014. Young perspectives on pastoral rangeland privatization: Intimate exclusions at the intersection of youth identities. *European Journal of Development Research* 26 (2): 204–218.

Beazley, H., S. Bessell, J. Ennew, and R. Waterson. 2009. The right to be properly researched: Research with children in a messy, real world. *Children's Geographies* 7 (4): 365–378.

Berckmoes, L., and B. White. 2014. Youth, farming and precarity in rural Burundi. *European Journal of Development Research* 26: 190–203.

———. 2016. Youth, farming and precarity in rural Burundi. In *Generationing development: A relational approach to children, youth and development*, ed. R. Huijsmans, 291–312. London: Palgrave Macmillan.

Bernstein, H. 2010. *Class analysis of agrarian change*. Agrarian Change and Peasant Studies Series. Halifax: Fernwood Press.

Cassidy, A., S. Srinivasan, and B. White, eds. 2019. Generational transmission of smallholder farms in late capitalism. *Canadian Journal of Development Studies* 40 (2): 220–308.

Chamberlain, J., and J. Sumberg. 2021. Are young people transforming the rural economy? In *African youth and the rural economy: Points of departure*, ed. J. Sumberg, 58–97, 92–124. Boston, MA, Wallingford: CABI International. https://www.cabi.org/cabebooks/FullTextPDF/2021/20210138603.pdf.

Chamberlain, J., F. Yeboah, and J. Sumberg. 2021. Young people and land. In *Youth and the rural economy in Africa: Hard work and hazard*, ed. J. Sumberg, 58–77. Boston, MA: CABI International.

Committee on World Food Security. 2019. *Agroecological and other innovative approaches for sustainable agriculture and food systems that enhance food security and nutrition: A report by the high level panel of experts on food security and nutrition of the committee on world food security*. Rome: Committee on World Food Security. www.fao.org/3/ca5602en/ca5602en.pdf.

———. 2020. *Food security and nutrition: Building a global narrative towards 2030. A report by the high level panel of experts on food security and nutrition of the committee on world food security*. Rome: Committee on World Food Security. www.fao.org/3/ca9731en/ca9731en.pdf.

————. 2021. *Promoting youth engagement and employment in agriculture and food systems. A report by the high level panel of experts on food security and nutrition of the committee on world food security*. Rome: Committee on World Food Security.

Cuervo, H., and J. Wyn. 2012. *Young people making it work: Continuity and change in rural places*. Carlton, VIC: Melbourne University Publishing.

Flynn, J., and J. Sumberg. 2021. Are Africa's rural youth abandoning agriculture? In *Youth and the rural economy in Africa: Hard work and hazard*, ed. J. Sumberg, 43–57. Boston, MA: CABI International.

Food and Agriculture Organization (FAO). 2014. *Youth and agriculture: Key challenges and concrete solutions*. Rome, Food and Agriculture Organization of the United Nations (FAO) in collaboration with the Technical Centre for Agricultural and Rural Cooperation (CTA) and the International Fund for Agricultural Development (IFAD).

Food and Agriculture Organization (FAO)-International Fund for Agricultural Development (IFAD). 2019. *Global action plan: United Nations decade of family farming 2019–2028*. Rome: Food and Agriculture Organization of the United Nations (FAO).

Hajdu, F., N. Ansell, E. Robson, and L. van Blerk. 2013. Rural young people's opportunities for employment and entrepreneurship in globalized Southern Africa: The limitations of targeting policies. *International Development Planning Review* 35 (2): 155–174.

Hall, D., P. Hirsch, and T.M. Li. 2011. *Powers of exclusion: Land dilemmas in Southeast Asia*. Singapore: National University of Singapore Press.

Herren, H., B. Haerlin, and IAASTD+10 Advisory Group, eds. 2020. *Transformation of our food systems*. Berlin/Zurich: Zukunftsstiftung Landwirtschaft/Biovision.

Huijsmans, R. 2016. *Generationing development: A relational approach to children, youth and development*. London: Palgrave Macmillan.

Huijsmans, R., S. George, R. Gigengack, and S.J.T.M. Evers. 2014. Introduction. Theorizing age and generation in development: A relational approach. *European Journal of Development Research* 26 (2): 163–174.

International Assessment of Agricultural Science and Technology for Development (IAASTD). 2009. *Agriculture at a crossroads: International assessment of agricultural science and technology for development*. Synthesis Report. Washington: Island Press.

International Fund for Agricultural Development (IFAD). 2019. *Rural development report 2019: Creating opportunities for rural youth.* Rome: International Fund for Agricultural Development.

International Labour Office (ILO). 2020. *Global employment trends for youth 2020: Technology and the future of jobs.* Geneva: International Labour Office. www.ilo.org/global/publications/books/ WCMS_737648/lang--en/ index.htm.

Jones, G. 2009. *Youth.* Cambridge: Polity Press.

Korzensky, A. 2019. Extrafamilial farm succession: An adaptive strategy contributing to the renewal of peasantries in Austria. *Canadian Journal of Development Studies* 40 (2): 291–308.

Kouamé, G. 2010. Intra-family and socio-political dimensions of land markets and land conflicts: The case of the Abure, Côte d'Ivoire. *Africa: The Journal of the International African Institute* 80 (1): 126–146.

Lowder, S., M. Sanchez, and R. Bertini. 2021. Which farms feed the world and has farmland become more concentrated? *World Development* 142.

McGreevy, S., M. Kobayashi, and K. Tanaka. 2019. Agrarian pathways for the next generation of Japanese farmers. *Canadian Journal of Development Studies* 40 (2): 271–290.

Naafs, S., and B. White. 2012. Intermediate generations: Reflection on Indonesia youth studies. *The Asia Pacific Journal of Anthropology* 13 (1): 3–20.

Nayak, A., and M.J. Kehily. 2013. *Gender, youth and culture: Global masculinities and femininities.* 2nd ed. Basingstoke: Palgrave Macmillan.

Ní Laoire, C. 2002. Young farmers, masculinities and change in rural Ireland. *Irish Geography* 35 (1): 16–27.

Peters, K. 2011. *War and the crisis of youth in Sierra Leone.* Cambridge: Cambridge University Press.

Prindex. 2020. *Prindex comparative report 2020.* London: Prindex. https://www. prindex.net/reports/prindex-comparative-report-july-2020/.

Punch, Samantha. 2020. Why have generational orderings been marginalised in the social sciences including childhood studies? *Children's Geographies* 18 (2): 128–140. https://doi.org/10.1080/14733285.2019.1630716.

Quan, J. 2007. Changes in intra-family land relations. In *Changes in "customary land tenure systems in Africa",* ed. L. Cotula, 51–63. London: International Institute for Environment and Development (IIED).

Ricciardi, V., Z. Mehrabi, H. Wittman, D. James, and N. Ramankutty. 2021. Higher yields and more biodiversity on smaller farms. *Nature Sustainability* 4: 651–657. https://doi.org/10.1038/s41893-021-00699-2.

Skinner, E. 1961. Intergenerational conflict among the Mossi: father and son. *The Journal of Conflict Resolution* 5 (1): 55–60.

Srinivasan, S. 2014. Growing up unwanted: Girls' experiences of gender discrimination and violence in Tamil Nadu, India. *European Journal of Development Research* 26 (2): 233–246.

Stearns, P. 2006. *Childhood in world history*. London: Routledge.

Sumberg, J., and S. Hunt. 2019. Are African rural youth innovative? Claims, evidence and implications. *Journal of Rural Studies* 69: 130–136. https://doi.org/10.1016/j.jrurstud.2019.05.004.

United Nations. n.d. *"The 17 goals." Sustainable development, department of economic and social affairs*. https://sdgs.un.org/goals.

Van der Ploeg, J.D. 2013. *Peasants and the art of peasant farming: A chayanovian manifesto*. Halifax: Fernwood Press.

White, B. 2012. Agriculture and the generation problem: Rural youth, employment and the future of farming. *IDS Bulletin* 43 (6): 9–19.

———. 2015. Generational dynamics in agriculture: Reflections on rural youth and farming futures. *Cahiers Agricultures* 24 (6): 330–334. https://doi.org/10.1684/agr.2015.0787.

———. 2020. *Agriculture and the Generation Problem*. Halifax and Rugby: Fernwood Publishing and Practical Action Publishing.

———. 2021. Human capital theory and the defectology of aspirations in policy research on rural youth. *European Journal of Development Research* 33 (1): 54–70.

World Bank. 2006. *Development and the next generation: World development report 2007*. Washington, DC: The World Bank.

———. 2015. *Country data*. http://data.worldbank.org/indicator/SL.UEM.1524.ZS.

Wyn, J., and R. White. 1997. *Rethinking youth*. London: Sage Publications.

2

"Passion Alone Is Not Sufficient": What Do We Know About Young Farmers in Canada?

Joshua Nasielski, Sharada Srinivasan, Travis Jansen, and A. Haroon Akram-Lodhi

Introduction

In 2016, Canada's 271,935 farm operators represented less than 0.8 per cent of the Canadian population (Statistics Canada 2017a). This reflects a loss of close to 120,000 farmers over the past 25 years as Canadian livelihoods continue to shift away from agriculture (about 1.4 per cent of the population farmed in 1991). Considering that less than 10 per cent of Canadian farmers are under the age of 35, it is hard to imagine these

J. Nasielski • T. Jansen
University of Guelph, Guelph, ON, Canada
e-mail: nasielsk@uoguelph.ca; traviswjansen@gmail.com

S. Srinivasan (✉)
Department of Sociology & Anthropology, University of Guelph, Guelph, ON, Canada
e-mail: sharada@uoguelph.ca

A. H. Akram-Lodhi
Trent University, Peterborough, ON, Canada
e-mail: haroonakramlodhi@trentu.ca

© The Author(s) 2024
S. Srinivasan (ed.), *Becoming A Young Farmer*, Rethinking Rural,
https://doi.org/10.1007/978-3-031-15233-7_2

23

numbers rebounding anytime in the near future (Statistics Canada 2017a). Clearly, Canadian farming faces a generational challenge (Qualman et al. 2018). However, despite these generational challenges, there has been little research that focuses specifically on young farmers in Canada and their experiences in becoming "successful" farmers. Therefore, the purpose of this chapter is to provide an overview of the information available on Canadian young farmers. This overview references existing research from scholarly literature and government statistics. This is done to offer an understanding of the context within which Canada's young farmers are embedded. A young person's desire to farm is partly shaped by but also shapes their experiences in becoming and being a young farmer. This overview helps inform the discussion in the next two chapters that are based upon interviews with young farmers in our two case study provinces: Ontario and Manitoba.

In this chapter, we argue that if there is to be a generational renewal on Canada's farms, challenges facing young farmers will need to be overcome. Such an undertaking requires listening to the voices of young farmers and providing public support that addresses the different needs of both newcomer and returning young farmers (defined below). We begin by describing the agrarian context in Canada in Section "Canada's Agrarian Context," focusing specifically on farming in Ontario and Manitoba, the two provinces selected for our case studies. Section "The Constraints to Becoming a Young Farmer" focuses on the constraints facing young farmers entering farming. It does this by examining whether there are common characteristics among those young people who enter farming before exploring their aspirations for farming, their pathways into farming, and some of the specific challenges that they face. Section "Being a Young Farmer" examines the experiences of those who have become farmers. The concluding section provides a segue into the next two chapters, offering description and analysis of in-depth interviews with nearly 100 young farmers in Ontario and Manitoba.

We start by defining a farmer, a farm operator, a young farmer, a newcomer farmer, and a returning farmer. These terms are described differently across regions and literature; providing clear definitions improves the ease of comparison between existing work and the research presented here and in the next two chapters.

Farmers and Farm Operators

The terms "farmer" and "farm operator" can be used interchangeably. Statistics Canada uses the term "farm operator," as "those persons responsible for the management decisions in operating an agricultural operation. These can be owners, tenants or hired managers of the agricultural operation, including those responsible for management decisions pertinent to particular aspects of the farm—planting, harvesting, raising animals, marketing and sales, and making capital purchases and other financial decisions." A given farm can have up to three farm operators and these can be women or men, including youth (Statistics Canada 2019a).

Statistics Canada changed its definition of farm operators prior to the 1991 Census of Agriculture. These changes make farm operator data from previous Censuses incomparable. The agency explains:

> Prior to the 1991 Census, the farm operator referred to only one person who was responsible for the day-to-day decisions made in the operation of an agricultural holding. Because only one operator was listed for each census farm, the number of operators was the same as the number of census farms. Beginning in 1991, up to three operators per operation could be listed on the questionnaire. (Statistics Canada 2019a)

Young Farmer

The age of young farmers interviewed for this study ranged from 18 to 45 years old. This age range is inconsistent with Statistics Canada data on farm operators, which is often reported using three broad age segments: 15–34; 35–54; and 55 and older. We believe that it is important to include some "older" young farmers in our study—farmers who are still young enough to remember their experiences as young farmers but who might be further along in their journey towards becoming a successful farmer. However, where this chapter uses Statistics Canada data, it refers exclusively to the 15–34 age range.

Newcomer and Returning Farmers

Drawing on the work of Monllor (2012), we separated our young farmers into two broad categories: newcomers and returning farmers. A "newcomer" is a young farmer who is farming now but did not grow up on a farm or benefit from having a family farm from which to launch their career. This is in contrast to a "returning" farmer who grew up on a farm and whose farming career was in some way tied to their family's farm. These two types of young farmers are not distinct dichotomies; for example, a returning farmer may exit from farming for a significant period of time before returning and may not be involved with their family's farm. Nonetheless, there are enough similarities within each group as to make these categories useful.

Canada's Agrarian Context

According to the Food and Agricultural Organization of the United Nations, Canada is the seventh largest country in the world by arable land area, with just under 3 per cent of the world's total arable land. In 2018, it produced 1.1 per cent of the gross value of the world's agricultural output, measured in constant terms (FAOSTAT 2021). Following the 2016 census, the Canadian agri-food system accounted for 6.7 per cent of the country's Gross Domestic Product (GDP). The agri-food system represented about 12.5 per cent of Canadian employment, and about 1.6 per cent of Canadians are employed as farm operators (Statistics Canada 2017d). Canada produces a range of agricultural products across its many diverse landscapes. With ports on the Pacific and Atlantic Oceans, a rail system across the country, and a shared land border with one of the largest economies in the world, Canada is well positioned to engage in global agri-food markets.

As a result, it is not surprising that for more than 25 years, the foundation of Canadian government agricultural policy has been the export maximization of agri-food products produced for global markets. The narrative that both federal and provincial governments have advanced is one of Canada as a top-tier food exporter—the country must contribute to feeding the world, and if it succeeds in increasing exports, everyone benefits. While Canadian farmers have helped the government to meet ambitious export expansion targets, this period has also been associated with fewer farmers, fewer local processors, and increased consolidation along national supply chains. Some 17 per cent of all cash receipts in Canada's agricultural sector in 2017 were accounted for by the supply-managed sector, in which farmers must hold a quota, a form of permit that provides a per unit licence to produce and sell a specific level of output and which is set by national agencies (Heminthavong 2018). Dairy, chicken and turkey products, broiler hatching eggs, and table eggs are all subject to supply management. As production for global markets requires that commodities be relatively homogeneous across farms and countries, agriculture tends to focus on efficiency, consistency, and standardization. For farmers, this encourages expansion in those commodities where there are significant economies of scale, in order to be price competitive. Globally competitive markets in many agricultural products result in thin margins between farm gate prices and cost of production. This is in stark contrast to a more local food supply chain, where producers have a strong relationship with their processors and the people who eat the food that they produce. Over time, Canada's focus on efficiency, consistency, and the global market has marginalized local food supply chains. This has continued despite a renewed interest in and demand from customers for rebuilding local food supply chains.

Just as the agri-food markets for Canadian farmers have changed, so too have demographics. Between 1991 and 2016, the average age of a Canadian farmer increased from 47.5 to 55, and more than 54 per cent of Canadian farmers are now older than 55 (Statistics Canada 2017a). Improvements in technology have reduced the sector's dependence on physical labour, which may be enabling some farmers to extend the length of their careers. However, delayed retirement may also be due to the financial challenges that have emerged out of their participation in a stringently price-competitive global market (Qualman et al. 2018) as well as the challenges that farmers face in developing plans for the future of their farm when the value of farm properties has dramatically increased (discussed later in this chapter). Not only are farmers trying to construct a way to finance their retirement, they also want to treat their non-farming children fairly while leaving enough money for the farm to remain viable for future generations. This is not easy to do as farming businesses tend to have little liquid cash (Qualman et al. 2018); most of the total net worth of a farm is tied up in land and non-land assets that are needed for the farm to operate and cannot be easily sold to support retirement or non-farming children.

The traditions associated with farm management and intergenerational succession are also changing. Many women raised on a farm who want to continue farming seek the same opportunities to farm as their brothers and their brothers' peers. This is in contrast to previous generations where the farm was conventionally passed on to the oldest son. Notwithstanding this, however, the proportion of women operators is still significantly lower than men and has only marginally increased, from 25 per cent in 1991 to 28 per cent in 2016 (Ainley 2013; Statistics Canada 2017a). In addition to women being underrepresented within the overall farm operator population, there are also disproportionately fewer young farmers in Canada within the overall farm operator population. In 2016, less than 10 per cent of Canadian farmers were under the age of 35 (Statistics Canada 2017a). Figure 2.1 presents a beehive graph that offers a detailed picture of the ages and genders of Canadian farmers. The length of each horizontal bar represents the number of farmers in a specific one-year age category. Farmers aged 15 are at the bottom and those aged 90 are at the top. Note the shift upward on both sides of the graph—the dramatic shift upward in the age profiles of both women and men farmers in Canada. The loss of young farmers is plainly visible (Figs. 2.2 and 2.3).

Canada is a settler-colonial country and throughout its history has relied on immigration to increase its population. While historically a significant share of immigrants entered farming, particularly from the United Kingdom and the Netherlands, that is less in evidence contemporarily. In the 2016 Census of Agriculture immigrants comprised 8.7 per cent of Canadian farm operators, or 23,440 people (Statistics Canada 2019b). The main source of immigrant farmers in Canada were the United States and China. Interestingly, in 2016 nearly a quarter of American immigrants to Canada who entered farming did so from an urban background and were thus newcomers to farming. Some 53.8 per cent of US farm immigrants to Canada were women, and the average age of American farm immigrants was 47.3. Typically, American farm immigrants are involved in beef cattle farming or in activities to support beef cattle farming, both of which influence the size of farm of American farm immigrants, which was, on average, 628.6 acres. More than 40 per cent of American farm immigrants relied on an off-farm income. By way of contrast, almost three-quarters of Chinese farm immigrants were men, the average age of Chinese farm immigrants was 45.1, and more than half were located in Ontario. Typically, Chinese farm immigrants are involved in greenhouse, nursery, and floriculture production, their farms average 360.2 acres, and no Chinese farm immigrant reported an off-farm income. Having said that, for both American and Chinese farm immigrants, average total income was at $33,321 and $13,627, respectively, low. However, perhaps most importantly, immigrants have comprised a declining share of the Canadian farm operator population since 1996, and only 1.7 per cent of Canadian farm operators arrived in the country between 2011 and 2016 (Statistics Canada 2019b), suggesting that immigration is no longer a driver of Canadian farm demographics. This decline has occurred alongside the sector continuing to rely on temporary foreign workers through the official Seasonal Agricultural Workers Program, which grew from recruiting 264 Jamaicans to Ontario in 1966 to more than 40,000 workers from Mexico and English-speaking island nations in the Caribbean in ten Canadian provinces by 2018 (Leigh Binford 2019).

While these trends are consistent across all of Canada's provinces, Ontario and Manitoba were chosen for our research because they each have their own unique agrarian context. Ontario is Canada's most populated province and accounts for 26 per cent of Canada's farmers and 15.5 per cent of Canada's agricultural land (Connell et al. 2016). Most of Ontario's agriculture is located in the southern part of the province, near the Great Lakes. A temperate, humid climate combined with quality soil, access to major American markets, and proximity to urban centres make Ontario a desirable place to farm. Compared to Manitoba, Ontario also has significantly more poultry, swine, and dairy farmers. Livestock can allow farmers to successfully operate on significantly smaller land bases given that they are not solely dependent on cropping income. All of these traits are reflected in higher average land prices (US$8459[1] per acre) than found in other provinces and smaller average farm sizes (249 acres) than those found in the prairie provinces (Statistics Canada 2019c). While these geospatial characteristics make farming favourable in southern Ontario, the region's comparatively higher population density also creates more opportunities for farmers' off-farm employment. In Ontario as a whole, off-farm income accounted for 78.6 per cent of total income among farmers in 2013 (Statistics Canada 2014). This income stream may slow down the loss of farmers as smaller operations are able to supplement their farm revenue with income from other jobs. Furthermore, proximity to urban centres and large populations may make southern Ontario a good candidate for new and existing farmers who are looking to avoid conventional commodity-based supply chains and instead market their food directly to local customers.

Manitoba is immediately west of Ontario and is one of Canada's three prairie provinces. With about 1.4 million people compared to Ontario's 14.6 million, Manitoba is much smaller than southern Ontario. Moreover, 62 per cent of the population live in Manitoba's capital city of Winnipeg (City of Winnipeg 2020). Notwithstanding this urbanized population distribution, farmers make up only about 3 per cent of Manitoba's workforce, while Manitoba has 11.4 per cent of Canada's total agricultural land (Connell et al. 2016). The average farm size in Manitoba is close to 1200 acres—significantly larger than Ontario (Government of Manitoba 2017). Although larger, Manitoba's farms are worth about one-fifth of the price of an Ontario farm, with farmland having an average value of C$2193 per acre (Statistics Canada 2019c). Most of Manitoba's farmland is seeded with wheat or canola (Statistics Canada 2022)—commodity crops for global agri-food markets. As in Ontario, Manitoba farmers rely on off-farm income, earning about 70 per cent of their total income from off-farm work, which is less than Ontario or the national average (Statistics Canada 2014). Far fewer urban areas and larger farm sizes make Manitoba less amenable to production for local food markets, although when farms have proximity to Winnipeg, this does present opportunities for some farmers to direct market.

---

[1] Using a US dollar to Canadian dollar exchange rate of 1.3269.

The Constraints to Becoming a Young Farmer

Most young farmers fit into one of two broad groups: returning farmers and newcomers. Then there are, of course, those who never leave the farm. These groups can have different aspirations and motivations to farm, may be exposed to different opportunities to farm, could have different goals for their farms, might deploy different resources as they follow different pathways into farming, and face different challenges as farm operators. Cumulatively, the experiences of returning and newcomer young farmers can be profoundly different. In what follows, aspects of these different experiences are explored through a discussion of farming aspirations, pathways into farming, and constraints on farming.

Aspirations

Martz and Brueckner (2003) found that 120 of the 200 youth that they interviewed across Canada who grew up on a farm would continue to farm if given the opportunity, even as most acknowledged the challenges that can result from a farming career. More recent reports provide glimpses into a continued interest in farming (Xiong 2017). Indeed, it is not difficult to find testimonies from young people who grew up on farms and who say that, since they could remember, they wanted to be farmers, providing the basis of what is defined above as returning farmers:

> There's a lot of us that it's a born and bred passion. It's the reason I get up in the morning. We just love what we do ... There's a lot of dignity that goes with farming. I think it's an honest way to make a living.—Young male farmer (RealAgriculture 2014)

> We wouldn't work 16 and 18 hour days in the spring if we didn't love what we do.—Young female dairy farmer from Ontario (Farm and Food Care 2014a)

For those who grew up around farming, it is seen as a career that is stimulating and satisfying (CBC News: The National 2017)—a profession that comes with intrinsic rewards for successfully navigating the vagaries that accompany a life as a farmer. This includes building and honing a unique blend of physical and intellectual skills, at different points during the year being not only a farmer but also an accountant, a heavy-machinery operator, a mechanic, a botanist, a scientist, a chemist, a labourer, a marketer, and a manager of people (National Farmers Union 2007, 17).

The motivations for newcomer farmers differ in that they often compare the enjoyment that they receive from farming with the ennui that their former profession generated. Leonard (2015) and Haalboom (2013) studied young first-generation farmers who founded small-scale alternative farms (e.g. organic and biodynamic). Leonard (2015) studied four farms in Manitoba, and Haalboom (2013) interviewed eight farmers in Nova Scotia. Participating young farmers often contrasted farming with their old urban occupations:

> I wanted to do something different …. I was tired of getting dressed up in the morning to go sit in an office, and I just felt like I was part of a rat race heading towards a finish line I had not consciously chosen. I mean, there's the expression, 'Even if you win the rat race, you're still a rat.'—Young sheep farmer from Nova Scotia (Haalboom 2013, 29)

> I've contrasted this [farming] from cubicle life and screen-oriented work. I've learned the satisfaction of growing and doing. With manual labour I get to integrate mind, body and spirit. It's important to have dirt under your fingernails.—Young small-scale organic farmer from Manitoba (Leonard 2015, 72)

> I really enjoy not having a boss of any kind, basically having a completely open schedule and all our time is our own, not dictated by anyone else …. There is no separation between life and work, it's all the same thing, and it all happens here at the farm.—Young organic vegetable farmer from Nova Scotia (Haalboom 2013, 35)

> I hated being inside with my job in summer. I couldn't do it.—Young small-scale organic farmer from Manitoba (Leonard 2015, 89)

While farming is a difficult profession, these testimonies from newcomer young farmers suggest that there are likely to be significant numbers of young Canadians who aspire to farm. Nonetheless, despite this aspiration, the number of young farmers in Canada has been decreasing over time. To understand this decline, it is helpful to examine the experiences of current young farmers, specifically in understanding the pathways into how young people become successful farmers.

Education and Training

The first pathway into becoming successful at anything is learning how to do it. Young farmers, by their own admission, benefit tremendously from agricultural education and training (Robinson 2003; Agriculture and Agri-Food Canada 2010; Roessler 2005). In focus interviews with seven young farmers in Manitoba, Durnin (2010, 111) notes that "learning and the importance of keeping oneself up-to-date was mentioned by all participants" when asked to identify the factors critical to their success as farmers (see also House of Commons Canada 2010). In their Canada-wide survey, Martz and Brueckner (2003) found that, on average, farmers access eight sources of agricultural information: talking to others, attending meetings, readings newsletters, attending agricultural fairs, reading on the internet, television, reading newspapers and books, and meeting with or attending training sessions given by government extension officers and industry representatives. In terms of education and training modalities, young farmers prefer hands-on and practical training, learning from other farmers and in situations where they can observe different practices (Roessler 2005; House of Commons Canada 2010; Agriculture and Agri-Food Canada 2010; Laforge 2017). Examples of these types of education and training include intensive short courses or workshops focused on aspects of farm management such as crop production or financial management as well as farmer field days and hands-on training or internships in specific aspects of farm operations. In general, most studies find that access to farmer training and educational materials is not a major barrier for young farmers: the opportunities to access training and education are, in general, available to those who want it (Agriculture and Agri-Food Canada 2010; Dennis 2015; Ekers et al. 2016).

Nonetheless, it is clear from the literature that young farmers with a farming background—returning farmers and those who never left—and those without a farming background—newcomers—have different education and training needs. For those from a farming background, growing up on a farm and helping parents with day-to-day decisions and farm chores provided tacit and practical agricultural knowledge (Martz and Brueckner 2003; Lobley et al. 2010) as well as agro-ecological knowledge that is specific to the farm (National Farmers Union 2007; Uchiyama et al. 2008). Yet practical "learning by doing" does not take place only on the farm. Many young returning farmers have a favourable view of working off-farm for a couple of years in the agricultural sector (e.g. in marketing or agronomy roles), viewing it as an opportunity to gain valuable experience in the broader sector while also saving to support future farming plans (Robinson 2003; Ahearn 2016). Young returning farmers also see the value in post-secondary agricultural education, although it is not normally ranked as highly as practical, hands-on training (Robinson 2003; Durnin 2010; Monllor 2012).

Newcomers who did not grow up on a farm have somewhat different training and educational requirements, at least initially. This is especially so if they want to use alternative farming practices (Ekers et al. 2016; Laforge 2017). As Knibb (2012, 12) concludes after surveying 436 new entrants to farming in Ontario, 61 per cent of whom were under the age of 40, "Entry-level practitioners often have little, if any, farming experience." Lacking the hands-on training and accumulation of tacit knowledge that those who grow up on a farm normally receive during their youth, and lacking access to family members who could act as farm mentors, newcomers must purposefully invest in practical farmer training and consciously seek out mentors (Epps 2017). Young newcomers who do not follow conventional, commodity-focused farming practices emphasize hands-on learning through farmer-to-farmer education, such as internships and fields days, rather than formal education or government extension services, which are in general not geared to the needs of small-scale farming (Knibb 2012; Laforge 2017). Indeed, the popularity of unpaid internships on alternative farms in Canada may be a response to the need for practical hands-on training and a result of the lack of formal programmes specializing in alternative agriculture at educational institutions (Ekers and Levoke 2016; Ekers et al. 2016). In research undertaken in 1991–1992 with 203 organic and conventional farmers in southern Ontario and British Columbia, Egri (1999) found that organic farmers, who tend to be younger than farmers producing for commodity markets, are much less likely to use the traditional sources of information that more conventional farmers relied upon, such as government publications, extension agents, and input suppliers such as seed representatives and company agronomists. Instead, they tended to rely upon farm organizations and farmer-to-farmer learning opportunities to a much greater degree. Roessler (2005), in a survey of 14 alternative farmers in British Columbia, found that organic farmers strongly prefer training resources or educational materials tailored to their specific farming region, perceiving these kinds of materials as most applicable to their own farm. Testimonials from the literature reinforce the point:

> Farmers learn best from other farmers. We were learning from other farmers in Manitoba; some who were just getting into agriculture, and some who were grain farmers but were sympathetic to what we were doing.— Small-scale organic farmer from Manitoba (Leonard 2015, 77)

> [Sharing knowledge has] always been a strong value because that's how I learned. I think that the mentors in our community, the farmers we learned from, are very into sharing knowledge and encouraging new farmers to get into it.—Small-scale organic farmer from British Columbia (Roessler 2005, 54)

> … the vast majority of the resources that I look at [on the internet] are American university extension services … I'm glad that it's there, but it just may not be at all applicable! It's a long way away, different ecology, different climate …—Organic farmer from British Columbia (Roessler 2005, 54)

These quotations demonstrate the importance of the local agricultural community as a source of information, education, and training for young farmers engaged in alternative agricultural practices.

Land

To become a farmer inevitably requires access to farmland. Both Dennis (2015) and Leonard (2015) report that the desire to start a farm becomes more difficult to sustain when there is no clear route to accessing farmland. This becomes the most binding constraint on establishing a pathway into farming (Qualman et al. 2018). In interviews with 35 young Albertan farmers (35 years of age or younger or those with less than 10 years of experience farming), Robinson (2003) found that the most common farming challenge identified was a small land base and an inability to expand due to high farmland prices, which have been increasing across Canada since 1993 (Farm Credit Canada 2016b). Access to land has become the most important barrier facing young farmers, particularly for new entrants to farming who cannot access land through family succession (Robinson 2003; Agriculture and Agri-Food Canada 2010; Dennis 2015). Thus, in a May 2017 report, Statistics Canada noted, "In 2016 the average value of land and buildings was C$2,696 per acre, which is an increase of 38.8% from 2011 (in 2016 constant dollars)." In part, well-documented increased investor ownership of farmland has driven rising prices (Holtslander 2015; Desmarais et al. 2017), in addition to farmland concentration among fewer farmers (Qualman et al. 2020). At present, Canadian farmland affordability, which can be defined by the ratio of land price to the net agricultural returns that the land can generate, is so low that a young farmer cannot hope to service the debt that they would be required to take on to purchase even a small farm (Qualman et al. 2018).

With farmland ownership being out of the reach of many young farmers, both new entrants and those taking over the family farm are increasingly relying on rented land or alternative land use arrangements such as cooperative ownership, land trusts, and farming public lands (Dennis 2015). Thus, Statistics Canada notes that "young farmers [are] more likely to rent land" than to own it: "Of agricultural operations where all operators were under the age of 35, 50.6% rented land from others, compared with 35.1% of all agricultural operations. On agricultural operations that used only rented land, the average operator age was 46.0 years, 9 years younger than the national average" (Statistics Canada 2017a).

However, while renting might be a good option for returning farmers, finding rented land can be more challenging for new farmers who do not have extensive networks in the agricultural community. This is particularly true for young farmers from urban centres or those who need only a fraction of the land that is often being offered for rent; for example, those who only need between 1 and 5 acres of land when parcels are typically a minimum of 50 acres. In this light, it is perhaps not surprising that in an online survey of 430 Ontario farmers, of whom 61 per cent were under the age of 40, Knibb (2012) found that 70 per cent of respondents wanted to learn more about alternative land tenure options because conventional pathways to accessing farmland through purchase or rental were no longer perceived to be accessible. By way of contrast, certain ecological practices such as conservation tillage, fallowing, and crop rotation are most practically applied on larger land bases. Young farmers who can access enough land to operate a smaller farm can then be disadvantaged because unaffordable land means that they may not be able to get to the economies of scale necessary to implement the kinds of agricultural practices that they want to use (Davey and Furtan 2008; Monllor 2012).

Farm Succession

For young people who grew up on a family farm, the challenges involved in accessing land could potentially be overcome by acquiring the existing family farm. However, the process of one generation transferring ownership of the farm to the next is rarely a straightforward affair. Many studies find that farm succession can be emotionally and financially difficult and, therefore, farm succession planning is often avoided (Uchiyama et al. 2008; Pitts et al. 2009). Although informal dinner table conversations with family members about the future of the farm occur on many Canadian farm operations (Taylor et al. 1998), successfully transferring the farm to the next generation requires forethought and planning (Durnin 2010; Brown 2011; Kirkpatrick 2013). Statistics Canada (2017e) reports that only 8 per cent of Canadian farms have a formal succession plan[2] in place. In a survey of 33 farmers in Haldimand Country, Ontario, Earls (2017) found that 45 per cent of surveyed farmers had no succession plan at all. The widespread inattention to this important step has resulted in much more difficult intergenerational farm transfers in Canada. In part, the avoidance of succession planning can reflect the interpersonal challenges and financial challenges that complicate such planning. An older farmer from Ontario shared: "As a mom of four children, I want a succession plan that comes away with the two older children that are not involved [in the farm] still feeling valued, still feeling they are an integral part of the family. And yet, not doing so at the expense of the farm business [taken over by two younger children]. In other words, fair is not always equal" (AMI 2013a).

In this light, it is not surprising that most Canadian studies conclude that farm succession processes can be stressful for both retiring and succeeding generations. For many farmers, farming is not a job but a vocation. This makes it difficult for farmers to think about, let alone decide, to retire. As it is a vocation, even when the decision to retire has been made, most Canadian farmers never intend to fully retire from farming. This creates a (potentially long) period of time when both older and younger generations work side-by-side. While this has the potential for great outcomes (Ferrier et al. 2013), "the prolonged period of intergenerational involvement may pose problems for family relationships" (Taylor et al. 1998, 554). An older grain and oilseed farmer from Ontario explained:

> The most difficult thing for me has been to give up control. And I understand that several of my peers going through the same process have the same problem. But if the business is going to continue to succeed, someone else has to be at the helm and take charge and control. I believe I still play a significant part in the [farm] business, but I am not the leader of the charge anymore. (AMI Ontario 2013b)

---

[2] Statistics Canada (2017a) defines a "formal" succession plan as a document that "lays out how the operation will be transferred to the next generation of farmers." A formal plan is also often defined as a legally binding document, crafted with the assistance of professionals such as lawyers and accountants, that outlines exactly when and how farm transfer will take place. An informal succession plan can consist of a verbal agreement or handshake.

A farm is a system where one decision can impact the entire direction that the farm is headed. The challenge comes when two farmers have different ideas about the future of the farm.

> [Son] came home from college knowing it all and with the attitude that I knew nothing. And I looked at him as wet behind the ears, knowing nothing. It took a number of years of mellowing for us both in order to work it out.—Older Canadian farmer (Taylor et al. 1998, 564)

> We argued a lot more the first five to six years than we do now. It was a power struggle because who's going to be boss?—Older Canadian farmer (Taylor et al. 1998, 563)

Taylor et al. (1998) found that these kinds of intergenerational bargaining struggles last several years and are resolved once the older generation relinquishes management responsibility to the younger generation. In other words, once control is given to the successor, both parents and children report that intergenerational conflicts are alleviated. Yet even after management control is relinquished, the older generation normally stays involved on the farm, helping out with non-physical tasks such as driving tractors and combines or keeping financial records.

Given how easily family conflicts can arise in succession planning, it may be psychologically easier for older generations to delay tough decisions, maintaining the status quo by making no formal plans for succession. As a result, however, their children remain uncertain about their future (Pitts et al. 2009; Kirkpatrick 2013). It is likely then that many young people, especially from farm families, may be delaying their entry into full-time farming, a situation that perpetuates the idea of a "generational crisis" in Canadian agriculture. Anecdotally, many young people who grew up on "smallish-medium" sized farms opt out of trying to farm right after their schooling in order to avoid some of the conflict and stress (that comes with wanting to be financially dependent on the farm) and allow themselves to build some personal stability before trying to navigate the process of incorporating the family farm into their "actual" workload. That is not to say that they do not help, but at this point they would not consider themselves farmers either.

Another reason that farm succession processes can be stressful is that retiring farmers prioritize varied and potentially conflicting goals (Wasney 1992; Uchiyama et al. 2008; Brown 2011). In a survey of 225 Manitoba farmers, Wasney (1992) found that the retiring generation normally has four goals in succession planning: they would like enough wealth or income to support their desired lifestyle during retirement; they want to ensure the continuation of the family farm; they want to maintain good family relationships; and they want to provide financial assistance to both farming and non-farming children. Some of these goals are not always compatible, and the dramatic rise in the price of farmland may have created further incompatibilities across these goals (Su 2017).

Succession planning challenges are not insurmountable (Brown 2011; Su 2017), but they certainly reinforce the need for a formal succession plan, tailored to the specific needs of the family and created with the help of professionals so that it is in place years or decades before the farm transfer occurs (Kirkpatrick 2013). Formal succession planning gives older farmers the best chance to achieve their retirement goals and gives younger farmers the stability that they need when they take responsibility for running the family farm on their own. Yet the potential for intra-family conflicts and the difficulties that farmers may have in achieving both a financially secure retirement and a farm transition that does not overburden the next generation can be a major source of stress on Canadian farms.

Financial Viability

In their testimony, young farmers showed their love for farming; they appreciate the lifestyle. Nevertheless, this passion alone is not sufficient to convince them to set up in agriculture. They consider it first and foremost as a business that must be profitable.—Federal Parliamentary Committee, which interviewed 132 farmers in British Columbia, Alberta, Saskatchewan, and Manitoba. (House of Commons Canada 2010, 7)

In this light, the well-documented problems of low net farm incomes and significant inequalities in the distribution of net farm incomes

(Qualman et al. 2018) can act as a major constraint on establishing a pathway into farming for young farmers. Moreover, insufficient net farm income can force the exit of young farmers. Statistics Canada provides income information for Canadian farmers broken down by age, using combined data from the Census of Agriculture and the National Household Survey. The Census of Agriculture is released every five years and targets all farms in Canada. The National Household Survey was released only in 2011, sampling about 30 per cent of Canadian households drawn from a random subsample of those completing the 2011 Canadian census. As such, 2011 is the only year where farm income is broken down by age, using Statistics Canada's definition of a young farmer as those under 35. A summary of the relevant figures is provided in Table 2.1. According to Table 2.1, farms operated by young farmers tend to generate less net income than those operated by older farmers (Statistics Canada 2011a). For Canadian farm operators under 35 years old, 25 per cent had net farm incomes of C$50,000 or greater compared to 35 per cent of operators between 35 and 55 years of age and 27 per cent of operators 55 or older (Statistics Canada 2011a). Data for Ontario and Manitoba are similar to the rest of Canada. In Ontario, 24 per cent of operators under 35 earned at least C$50,000 versus 36 per cent of farm operators between 35 and 55. In Manitoba, these figures are 27 per cent and 36 per cent, respectively (Statistics Canada 2011a).

Table 2.2 presents relevant figures from Statistics Canada (2011b) on the major sources of income for farm operators by age. While most farmers earn the majority of their income from off-farm or non-farm sources regardless of age, younger farmers are more reliant on non-farm income sources (Statistics Canada 2011b). In 2011, about 21 per cent of farmers under 35 years old earned the majority of their income from their farm operation, compared to 24 per cent for farmers between 35 and 55 years old (Statistics Canada 2011b). Wages and salaries were the most common main source of income for farm operators under 35 or between 35 and 55. Over half of farm operators derived most of their income from off-farm or non-farm work in these age categories (Statistics Canada 2011b). For farm operators 55 years or older, sources of retirement income such as pensions or investment income from registered retirement savings plans is the main income source for 40 per cent of farm operators. Only about 20 per cent of farmers 55 and over earn most of their income from their farm operation. In Ontario and Manitoba, the picture is similar to the rest of Canada. In these provinces, 24 per cent and 22 per cent of young farmers, respectively, earn the majority of their net income from farming. This compares to provincial averages of 24 per cent and 30 per cent for farmers aged 35–55 in Ontario and Manitoba (Statistics Canada 2011b). One of the problems in generating sufficient net farm income is that farming requires access to assets (such as land, buildings, and equipment) to grow farm products. Typically, acquiring these assets and the associated inputs (seed, fertilizer, feed, etc.) require access to working capital and credit (Robinson 2003; Agriculture and Agri-Food Canada 2010; Pouliot 2011; Junior Farmers of Ontario 2013; Food Secure Canada 2015; Laforge 2016). Conducting interviews with six key informants, both young farmers and researchers, in British Columbia, Gichungwa (2015) reported that after access to land, lack of capital and the difficulties associated with accessing credit were the most important barriers to beginning to farm. A similar finding is reported from a Canada-wide survey with 1326 new farmers, the majority of who were in the age group 26–35 (Laforge 2016). Most young farmers do not have access to the financing available through banks or credit unions, who prefer to deal with more experienced farmers with credit histories and access to collateral such as land (Murphy 2012; Epps 2017). A Junior Farmers of Ontario survey (2013) of 250 young farmers found that only 27 per cent of respondents accessed lines of credit through traditional financial institutions, and most of those farmers had farm experience, suggesting that this is a specific challenge that newcomers can face. Most young farmers relied on personal savings (53 per cent) and financial support from their extended family (29 per cent) to fund asset purchases (Junior Farmers of Ontario 2013). Having said that, in many provinces, small seed grants are available to young and beginning farmers (FarmStart 2016; Epps 2017), and young farmers generally express support for Farm Credit Canada (FCC) (Agriculture and Agri-Food Canada 2010), a federal crown corporation that has a specific loan programme for young farmers, the FCC Young Farmer Loan (FCC 2016a). Conversely, as Leonard (2015) mentions, some young farmers engaged in alternative agriculture to eschew traditional farm entry that relies on debt, preferring options that grant greater autonomy, such as community-supported agriculture, direct support from customers, and personal savings from off-farm work.

The supply-managed sub-sectors in Canadian agriculture (dairy, poultry, and eggs) provide a unique example of how capital and net farm incomes are intertwined for Canada's young farmers. By managing supply in the market, these sectors are designed to support sufficient farm-gate prices for producers. However, the resulting income security from these sectors has significantly inflated the prices of the production quotas that are needed to produce these agricultural products. A production quota provides a per unit licence to produce, and these quotas have become very expensive. The cost of quotas presents a major barrier to farm expansion, but also to new entrants. For some perspective, a 2012 study by University of Calgary researchers (Findlay 2012) found that to own a single cow, a Canadian farmer would need $28,000 worth of quota. While they support strong operating incomes (income before taxes, interest, etc.), accessing the financing to acquire these quotas poses a challenge for young farmers. While most supply management organizations have special programmes to help beginning farmers obtain quotas through low interest loans, there are limitations to the quotas available and the number of entrants accepted to this programme annually (Mitchell 2015; Dairy Farmers of Ontario 2021; Chicken Farmers of Ontario 2017).

Creating Social Networks

For returning farmers, the succession process can provide a major means through which they enter into and create the social networks necessary to sustain their farm operation. As the young farmer begins to take over various responsibilities and decision making, they start to develop their own relationships with input suppliers, purchasers, and advisors, and they find their own way to fit into the local community, building upon the social networks that their parents established.

For newcomers, however, the creation of social networks can be more challenging. When young people move to a new rural community to start farming, they face the task of building such support networks of like-minded farmers for guidance and mentorship from scratch (Roessler 2005; Ngo and Brklacich 2014; Haalboom 2013). In a study of 1480 Quebec farmers who were less than 38 years old, Parent (2012) found that 58 per cent of respondents considered themselves socially isolated and that young single farmers, who were more at-risk of social isolation than partnered farmers, felt that farming made it more difficult to find a spouse. Conversely, Ngo and Brklacich (2014), surveying nine new farmers in Ontario from urban backgrounds, of whom six were young, found that while most respondents indicated that developing a sense of community in their new rural locale was challenging, most felt they were making progress in establishing their social lives and developing a sense of place.

Being a Young Farmer

Newcomer Farmers: Farming as a Political Act

For newcomers, their social and environmental views provide a strong motivation to start farming. They also provide a strong reason to continue to farm even when the farm operation itself is only marginally profitable or if "profit" is not the primary goal of their farm operation (Wilson 2015; Ekers et al. 2016). Small qualitative studies from across Canada consistently find that young first-generation farmers see agriculture as a way of building an alternative agricultural economy, one that promotes social justice, environmental stewardship, healthy food, and prioritizes meeting local and regional food needs (Mills 2013; Wilson 2015; Laforge 2017). Young farmers seek to live out environmental and social values that they find important, gaining a sense of place and of purpose from agriculture that they would not receive from another occupation. The kinds of social and environmental values that motivate young newcomer farmers reflect how they see themselves fitting into—or rather not fitting into—the dominant agricultural paradigm of high-input, capital-intensive, and export-oriented farming (Laforge 2017).

In this light, it is perhaps not surprising that most of Haalboom's eight research subjects had university degrees, but in non-agricultural subjects (Haalboom 2013). Laforge (2017, 218) notes that of 1326 respondents, who had an average age of 37, "many … had an education in environmental studies or work experience with social justice organizations." They saw their farming activities as a political act, as a way of challenging the dominant agricultural and indeed social paradigm by living as an example. So, it comes as little surprise that according to one farmer in Leonard's study, "We think about social justice as our lifestyle, not as being organizers of a movement" (2015, 83).

Interestingly, while newcomer young farmers place a great deal of intrinsic value on the land that they farm, the fact that they often lack secure tenure on the farmland that they operate because of its unaffordability means that for some, their personal relationship to land is not to a specific parcel of land. Thus, young farmers in Dennis' study indicate that their insecure tenure meant that they could not form deep attachments to the land that they operated because there was an ever-present threat of losing their lease (2015, 63). Having said that, studying 10 small-scale organic farmers in eastern Ontario and western Quebec, Wilson (2015) found that some participants talked about the tangible and intangible benefits they derived from owning a farm:

> Every day and every year I love more and more this land that we live on. I can't really even explain it but it just grows and grows often time; the relationship with the land. Having grown up here, I just love being here …— Young Canadian organic farmer (Wilson 2015, 119)

> I felt that was important to me in terms of long-term stewardship and also good investment of my own personal resources [to buy land for an organic farm] …. And investing in land felt like a great way to have some sort of financial security but more importantly, have a connection to a location where I see it as a responsibility [to] me personally.—Young Canadian organic farmer (Wilson 2015, 121)

When newcomers struggle with profitability, as many do, Leonard (2015) found that young alternative farmers in his sample used off-farm labour to supplement on-farm income, concluding that this strategy may be a way for these farmers to remain autonomous in the face of a corporate-controlled food system. Young farmers were also generating additional income through the value-added processing of agricultural products grown on-farm (Ahearn 2016). The majority of the young alternative farmers who Leonard (2015) and Laforge (2017) surveyed sold value-added processed products (e.g. sausages, salsa, and sauces) made from their own produce. In Manitoba, Durnin (2010) found that value-added processing of on-farm products is a common side business of young farmers, and in remote rural areas, one of the few avenues available to earn additional income.

As they wish to challenge the prevailing parameters of the food system through the way that they live, newcomer alternative young farmers have argued that current Canadian agricultural policies do not support those who want to start small-scale farms. In country-wide roundtable discussions with Agriculture and Agri-Food Canada, many young farmers have opined that Canadian agricultural policy disproportionally favours one "model" of agriculture (Agriculture and Agri-food Canada 2010). Several participants suggested that most government programmes that are geared towards young farmers are targeted at those who are starting or taking over large-scale farms producing for commodity markets, with a lack of programming that supports different models of production. This finding was echoed in separate consultations with young farmers that the Canadian Parliamentary Standing Committee on Agriculture and Agri-Food conducted (House of Commons Canada 2010). One young farmer recounts:

> The government's rules and regulations do not always have the small-scale producer in mind and discriminate against them in many ways. Our guarantee to you is that our food is safe and we eat it as well. My parents eat our food, my children eat our food. We have more and stronger reasons for food safety than any regulation could possibly capture. (Leonard 2015, 44)

It is these personal guarantees of relationships with eaters, safety, and hence quality that set newcomers and local food systems apart from commodity-based agriculture. For the most part, it is very rarely the case that newcomers produce for commodity markets. In part because of their disdain of production for commodity markets, newcomer farmers in alternative agriculture often feel somewhat marginalized by their wider rural community:

I find that even though I've operated a successful CSA [community supported agriculture] for 15 years, I am still not considered to be a 'real farmer' by my conventional neighbours.—Young biodynamic farmer (mixed crop-livestock) from Ontario (Laforge 2017, 219)

Not having community [with conventional neighbours], but then having it at the farmers market was great. People would say 'you changed my life' or 'that's the best sausage I've had.' It was great affirmation.—Young small-scale organic farmer from Manitoba (Leonard 2015, 71)

I do feel a disconnect with the people out here a little bit because it's white, very white. It's small town. It's a lot of industrial agriculture out here so organic agriculture out here is laughable or something.—Organic farmer from Ontario (Ngo and Brklacich 2014, 58)

I mean, even people who have done tractor work for me, will say it outright, like you're not really a farmer, or you're not a true Canadian farmer, backyard gardener, these sorts of things, hobby farmer ...—Organic farmer from Ontario (Ngo and Brklacich 2014, 63)

These young farmers challenge the traditional notions of what it means to be a "good farmer" (e.g. large land base, weed-free fields) and swim against the cultural current in most rural communities. Being surrounded by neighbours that do not really understand them, having little support from government, and struggling with the mundane challenges of small-scale farming, Laforge (2017, 227) concludes that to sustain their motivation, new farmers engaging in alternative agriculture require "a stable framework of supportive customers and peers to reinforce their alternative self-identity in a culture dominated by productivist ideals."

Returning Farmers: Farming as a Way of Life

Once returning farmers take over the management of the farm, it can be challenging for them to take the farm enterprise in the business direction to which they aspire. Monllor's (2012) study of 50 young Ontario farmers found that those taking over the family farm are often locked into particular production practices, irrespective of their actual agricultural and social attitudes. Similarly, in a survey of 57 farmers, Khaledi et al. (2010) found that older Saskatchewan farmers were more likely to convert all or part of their farmland to organic production than younger farmers. This is because a radical change in agricultural production practices is more difficult when one has just taken over a farm operation and is concerned with managing the day-to-day operations of a farm, meeting any outstanding debt obligations, and maintaining arrangements with other family members that hold different views as to the business direction of the farm (Monllor 2012). Moreover, as opposed to older established farmers, many young returning farmers, struggling to stay profitable in the first few years after starting to farm (when most learning is by trial and error), encounter difficulties implementing ecological practices that may be profitable in the long run, but are difficult to sustain in the short run when cash flow, can be a major concern (Smithers and Furman 2003).

Notwithstanding these challenges, returning young farmers have broader motivations sustaining their decision to return to the farm than simply the business direction of the enterprise. In an in-depth study of three intergenerational family farms in Ontario, Ainley (2013) found that younger farmers returning to take over the family farm do so in part because they want to give their own children the upbringing that they themselves experienced growing up. Thus, for returning farmers, farming is not only seen as a career but as an intrinsically good "way of life" that has a range of non-pecuniary benefits that they want to provide to their children as they are being raised (Lobley et al. 2010).

> When we started this family, I wanted the girls to have the same life that I was given ... and that is being on a farm. The country life. Everything that I got to have they get to have.—Young dairy farmer from Ontario (Food and Farm Care 2015b)

The thing I love most about farming is that we can work from home, and that it is a family business ... I know the work I'm doing now will affect my kids.—Poultry farmer from Ontario (Farm and Food Care 2015a)

Well I grew up here [the farm] and I didn't really think I would be a farmer ... I imagined that I'd go off to university and have some sort of professional career ... It was only when I went away that I started to miss the way that I had grown up.—Young Canadian organic farmer (Wilson 2015, 147)

It is important to note though that while the "farming lifestyle" and "way of life" are often cited by farmers as a principal reason to farm, these terms are never clearly defined by study participants or authors. Indeed, it may not be possible to authoritatively define these concepts (Ikerd 2016). Research reports and small studies consistently find that farmers and their families want to live in vibrant rural communities, where amenities such as schools, health care, and childcare are available (OECD 2006; AGree 2010; Agriculture and Agri-Food Canada 2010;). So, for example:

My favorite part of being a farmer [beef and cash crops] is community. Living in an agricultural area, you can get so involved ... It's just a different way of living. It's great, the kids are raised on the farm.—Beef and cash crop farmer from Ontario (Farm and Food Care 2014c)

Dennis (2015, 42) reports that the young farmers in British Columbia who she interviewed are reluctant to move to more remote rural locations where land is considerably cheaper. In part, this may be because of a lack of vibrant rural communities with amenities, which also suggests that in part, this may be because they are concerned about the wider opportunities available in these more remote locations, such as support networks, access to markets, and the availability of off-farm employment. Ongoing farm consolidation in Canada (Qualman et al. 2018, 3) contributes to these issues as it brings an attendant decline in farm families living in rural communities and in so doing can exacerbate the declining availability of rural services. For example, in their study of four small Saskatchewan farming communities, Bacon and Brewin (2008, 1–16) found that these rural populations continue to decline as amenities such as schools, hospitals, and restaurants become less likely to remain open, which in turn makes these areas less likely to attract new migrants, including farmers. While returning farmers are familiar with the communities that surround their farm, this does not mean that they are immune to the challenges of depleting services as rural populations decline.

As just noted, smaller communities also make it more challenging for farmers to find off-farm employment, which is increasingly important as a source of income that sustains the financial viability of Canadian farms (Qualman et al. 2018). This is suggestive of the demands facing Canadian farmers and their families as they try to maintain a "farming lifestyle." A report by Statistics Canada (Alasia and Bollman 2009) found that the probability of a farmer working off-farm peaks at around the age of 35 and declines gradually over time. In qualitative surveys, both Robinson (2003), who interviewed 35 young Albertan farmers under the age of 40, and Haalboom (2013) found that younger farmers report high rates of participation in off-farm income-generating activities and found that such income is used to subsidize the farming enterprise, support living costs, and qualify for agricultural loans.

Martz and Brueckner (2003) found that returning farmers often cite an "attachment to the land" that comprises the farm, or a responsibility towards the land ("stewardship"), as a motivation to continue farming. Given these views, it is not surprising that returning farmers are aware of the problems in the food system and as a general rule seek more responsive policies that protect them against unfair trade practices and, more broadly, the corporate control of the agri-food system. In interviews with 105 farm women across Canada, 25 of whom were under 35 years of age, Roppel et al. (2006) found that participants viewed current agricultural policies as unaccountable to most farmers and unfriendly to family farms. These participants voiced support for reorienting agricultural policies to centre on fair trade (not free trade), farmer needs (not industry needs), public control of research and food safety (not corporate control), quality food (not cheap food), and a policy environment that favours farms that produce for domestic markets and local consumption (not exports). Roundtables with young farmers find similar results: young farmers support "buy local" or "buy Canadian" food campaigns that promote domestic food consumption, demand policies that buffer farmers against the impacts of corporate monopolies in the agri-food system, and ask for fairer trade agreements and adjustments in current trade practices that place Canadian farmers at a disadvantage. This includes the ability of food processors to import agricultural products that do not meet the same standards required of Canadian producers (House of Commons Canada 2010, 31–32).

Conclusion

This chapter has provided glimpses into the Canadian agrarian context facing young farmers, using government statistics and scholarly literature. Generally, studies on Canadian young farmers have been qualitative, with small sample sizes and restricted geographical coverage. The lack of nationally representative comprehensive data on young Canadian farmers is a major gap. As such, we are unable to generalize about "young farmers in Canada" based on results from small-scale studies. Nonetheless, as this chapter demonstrates, these studies are rich in insights that alert us to issues confronting young farmers across Canada.

At the start of the chapter, we set out that young farmers can be subdivided into returning farmers or newcomer farmers. Across the chapter, it has been demonstrated that farmers within these groups face different challenges, opportunities, and would benefit from different forms of government support. Indeed, there can be significant divergence within a group as diverse as "young Canadian farmers." For example, some young farmers would like to expand supply management, while other young farmers would do away with supply management (Agriculture and Agri-Food Canada 2010, 15). This reflects the fact that some young farmers seek greater intervention from the government, while others eschew government regulations and policies as a matter of principle (Laforge 2017, 220). In this chapter, the secondary literature suggests that if there is to be a generational renewal on Canada's farms, challenges facing young farmers will have to be overcome, and that overcoming these challenges requires nuanced supports that address the different needs of newcomer and returning farmers. What is also clear, however, is that across both returning and newcomer farmers' farming capacity to provide a viable livelihood is a specific challenge when the policy environment targets the farmers who already exist. If we are to support new farmers in entering the sector, Canada needs to rethink policy from the ground up. For years, all farmers have been bringing the same value to the market (growing a commodity). Given the lack of differentiation, they can only compete on efficiency and for decades, the result has been expansion as farmers chase economies of scale. Attracting new farmers requires a business environment where new ideas and value propositions are supported. Farmers who bring unique value to the market have more sustainable businesses and their businesses have greater potential, adequate and reliable incomes for farm families. This is necessary for sustaining and rebuilding rural communities.

The next two chapters, based on in-depth interviews with about 100 young farmers in Manitoba and Ontario, tease out many of these issues further, insights from which will be immediately relevant to research and practice.

Appendix

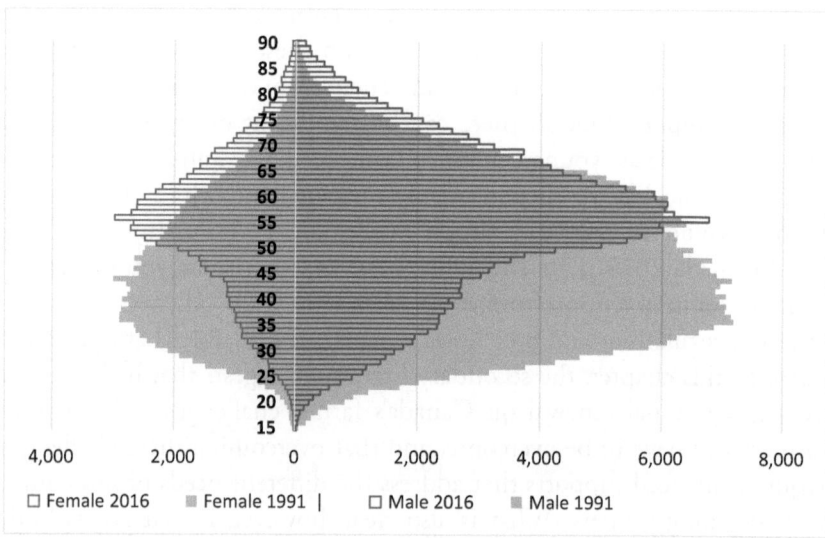

Fig. 2.1 "Beehive" graph showing number of Canadian farmers by age and gender, 1991 versus 2016. (Sources: Statistics Canada 2011 custom tabulation; Beaulieu 2012)

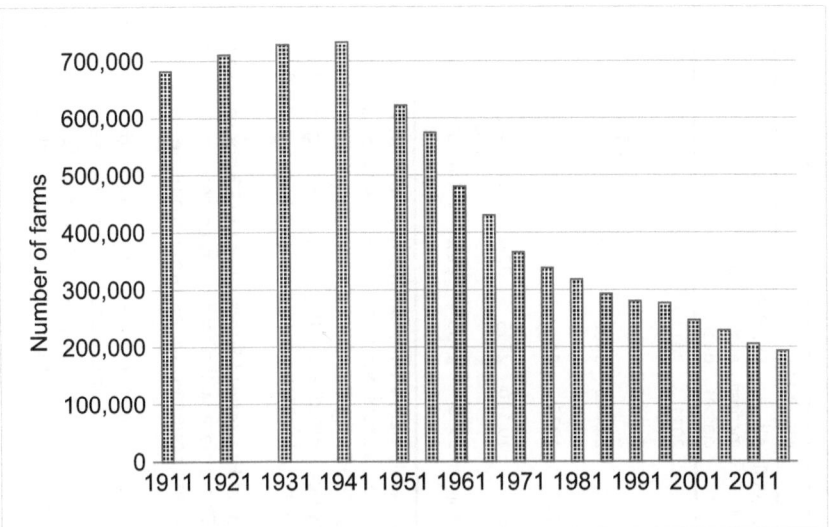

Fig. 2.2 Number of farms ("farm operations") in Canada, Census years 1911–2016. (Sources: Leacy et al. 1983; Statistics Canada 2017b)

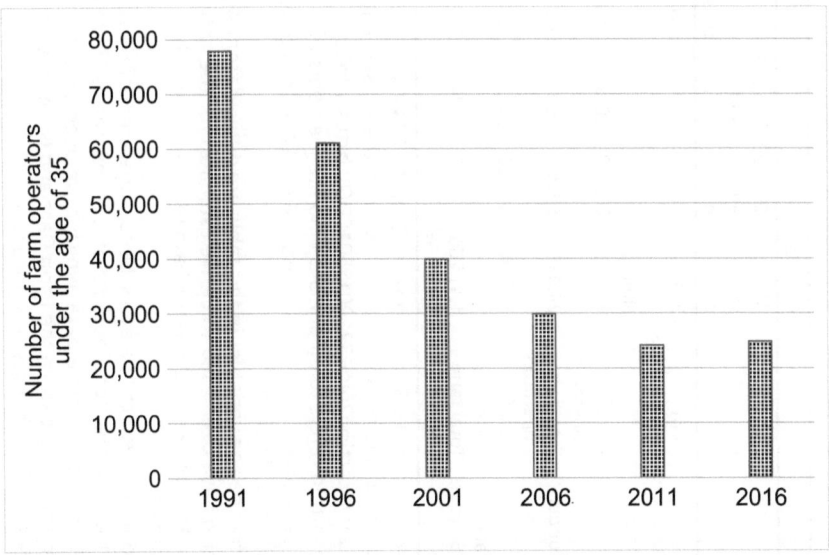

Fig. 2.3 The number of young farmers (farm operators aged 15–34, in Canada). (Source: Statistics Canada 2017c)

Table 2.1 Number of farm operators classified by income class and age in the year prior to the census

	Canada			Ontario			Manitoba			
	All income classes	$50,000 and over		All income classes	$50,000 and over		All income classes	$50,000 and over		
Age 2	2011	2011		2011	2011		2011	2011		
All ages										All ages
Farm operators 3	Number									
Total farm operators	2,92,790	87,710	30%	74,760	23,970	32%	22,040	5300	24%	
Operators under 35 years	24,055	5965	25%	6135	1470	24%	1935	440	23%	Operators under 35 years
Operators 35–54 years	1,27,465	43,920	34%	31,795	11,550	36%	10,110	2715	27%	Operators 35–54 years
Operators 55 years and over	1,41,260	37,830	27%	36,825	10,940	30%	9995	2155	22%	Operators 55 years and over

Source: Table: 32-10-0185-01 (formerly CANSIM 004–0115). Statistics Canada. Table 32-10-0185-01 Number of farm operators classified by income class in the year prior to the census, age and number of operators per farm https://www150.statcan.gc.ca/t1/tbl1/en/tv.action?pid=3210018501; DOI: https://doi.org/10.25318/3210018501-eng

[a]Farm population refers to all persons who are members of a farm operator's household excluding those residing in Canada's three territories or in http://www12.statcan.gc.ca/census-recensement/2011/ref/dict/dwelling-logements002-eng.cfm">collective dwellings

[b]The data for the age variables in this table are derived from Agriculture-National Household Survey linkage 33 per cent sample data, which have been used to derive estimates for the entire population. These data will differ from corresponding data in the Farm Data and Farm Operator Data release tables, which were based on data collected for all farms from the 2011 Census of Agriculture

[c]This is a count of distinct operators; hence, operators of two or more separate farms were included only once in the total. The data for the variable "total number of farm operators" in this table are derived from Agriculture-National Household Survey linkage 33 per cent sample data, which have been used to derive estimates for the entire population. These data will differ from corresponding data in the Farm Data and Farm Operator Data release tables, which were based on data collected for all farms from the 2011 Census of Agriculture

[d]The Canada total excludes the Yukon, Northwest Territories, and Nunavut

Table 2.2 Number of farm operators classified by major source of income and age in the year prior to the census

	Canada						
Farm operators	All income sources	Net farm income 5	Wages and salaries	Non-farm self-employment income	Investment income 6	Other sources of income 7	Zero income or negative income
Age 2	2011	2011	2011	2011	2011	2011	2011
All ages	Number						
Total farm operators	2,92,790	63,175	1,21,070	13,665	21,740	71,585	1560
Operators under 35 years Total farm operators	24,060	4940	14,040	1175	1060	2665	190
Operators 35–54 years Total farm operators	1,27,465	30,465	69,525	7345	8620	10,745	775
Operators 55 years and over Total farm operators	1,41,265	27,770	37,515	5145	12,065	58,175	590

Source: Table: 32-10-0188-01 (formerly CANSIM 004–0118). Statistics Canada. Table 32-10-0188-01 Number of farm operators classified by major source of income in the year prior to the census, age and number of operators per farm https://www150.statcan.gc.ca/t1/tbl1/en/tv.action?pid=3210018801; DOI: https://doi.org/10.25318/3210018801-eng
aFarm population refers to all persons who are members of a farm operator's household excluding those residing in Canada's three territories or in <A HREF="http://www12.statcan.gc.ca/census-recensement/2011/ref/dict/dwelling-logements002-eng.cfm">collective dwellings</A>
bThe data for the age variables in this table are derived from Agriculture-National Household Survey linkage 33 per cent sample data, which have been used to derive estimates for the entire population. These data will differ from corresponding data in the Farm Data and Farm Operator Data release tables, which were based on data collected for all farms from the 2011 Census of Agriculture

(continued)

Table 2.2 (continued)

[c]This is a count of distinct operators; hence, operators of two or more separate farms were included only once in the total. The data for the variable "total number of farm operators" in this table are derived from Agriculture-National Household Survey linkage 33 per cent sample data, which have been used to derive estimates for the entire population. These data will differ from corresponding data in the Farm Data and Farm Operator Data release tables, which were based on data collected for all farms from the 2011 Census of Agriculture

[d]The Canada total excludes the Yukon, Northwest Territories, and Nunavut

[e]Net farm income refers to the profit or loss of the farm operation measured by total farm operating revenues minus total farm operating expenses and capital cost allowance reported on the tax return for the farm. Operating revenues include revenues from the sale of agricultural products and services such as cash advances, rebates, agricultural custom work, and machine rental, plus payments from federal, provincial, and regional agricultural programmes, and insurance proceeds (e.g. income stabilization or crop insurance payments). Operating expenses include business costs in the production of agricultural products plus wages and salaries paid to children and spouses for unincorporated farms. For incorporated farms, operating expenses may include wages and salaries or rent paid to shareholders

[f]Investment income includes payments of dividends including those from farm corporations, interest on deposits and bonds, and other investment income excluding retirement investments (Registered Retirement Savings Plans [RRSPs] and Registered Retirement Income Funds [RRIFs])

References

AGree. 2010. *AGree strategies to attract young people to food and agriculture.* AGree Priority Strategy Report.

Agriculture and Agri-Food Canada. 2010. *2009 dialogue tour on young farmers and farm transfers.* What We Heard: Report from November 2009 Roundtables. https://foodsecurecanada.org/sites/foodsecurecanada.org/files/2009_dialogue_on_young_farmers1.pdf.

Agri-Food Management Institute (AMI Ontario). 2013a. *AMI—Success—Foster.* YouTube, 11:27. Agri-Food Management Institute. https://www.youtube.com/watch?v=o62LPl9EwtQ.

———. 2013b. *Farm succession—The jack family.* YouTube, 08:43. Agri-Food Management Institute. https://www.youtube.com/watch?v=F9O-vsAHUyQ&t=35s.

Ahearn, M.C. 2016. Theme overview: Addressing the challenges of entry into farming. *Choices* 31 (4): 1–4.

Ainley, S.E. 2013. *The experience of Ontario farm families engaged in agritourism.* MA thesis, University of Waterloo. https://uwspace.uwaterloo.ca/handle/10012/7246.

Alasia, A., and R.D. Bollman. 2009. Off-farm work by farmers: The importance of rural labour markets. *Rural and Small Town Canada Analysis Bulletin (Statistics Canada)* 8 (1): 2–31. http://www.statcan.gc.ca/pub/21-006-x/21-006-x2008001-eng.pdf.

Bacon, Brenda, and Derek Brewin. 2008. *Rural community viability: Lessons from 4 communities.* Working Paper. University of Manitoba.

Beaulieu, Martin. 2012. *Demographic changes in Canadian agriculture.* Statistics Canada: Canadian Agriculture at a Glance. https://www150.statcan.gc.ca/n1/pub/96-325-x/2014001/article/11905-eng.htm#a6.

Brown, J. 2011. *Problems with farm succession: The case of Saskatchewan.* Presented at the 18th International Farm Management Conference, Canterbury, March 20–25. http://ifmaonline.org/wp-content/uploads/2014/08/11_NPR_Brown_P164-169.pdf.

CBC News: The National. 2017. *New crop of young farmers emerging.* YouTube. Video, 02:00. CBC News: The National. https://www.youtube.com/watch?v=dpgTBS6aBJw&t=55s.

Chicken Farmers of Ontario. 2017. *New entrant chicken farmer program.* https://www.ontariochicken.ca/Programs/New-Entrant-Chicken-Farmer-Program.aspx. Accessed 15 May 2020.

City of Winnipeg. 2020. *Population of Winnipeg*. https://winnipeg.ca/cao/pdfs/population.pdf. Accessed 12 June 2020.

Connell, David J., et al. 2016. *Farmland: A prerequisite for farmers, food and agri-food policy*. Policy Brief. http://www.aglup.org/uploads/1/1/6/6/1166741771/aglup_policy_brief_603.pdf.

Dairy Farmers of Ontario. 2021. *New entrant program*. https://new.milk.org/Industry/Programs-and-Policies/New-Entrant-Quota-Assistance-Program#:~:text=Dairy%20Farmers%20of%20Ontario's%20(DFO,for%20the%20program%20in%20Ontario. Accessed 13 Feb 2022.

Davey, K.A., and W.H. Furtan. 2008. Factors that affect the adoption decision of conservation tillage in the prairie region of Canada. *Canadian Journal of Agricultural Economics/Revue Canadienne d'agroeconomie* 56 (3): 257–275.

Dennis, J.E. 2015. *Emerging farmer movements and alternative land access initiatives in British Columbia, Canada*. MA thesis, University of British Columbia. https://open.library.ubc.ca/cIRcle/collections/ubctheses/24/items/1.0166546.

Desmarais, A.A., D. Qualman, A. Magnan, and N. Wiebe. 2017. Investor ownership or social investment? Changing farmland ownership in Saskatchewan, Canada. *Agriculture and Human Values* 34 (1): 149–166.

Durnin, D.D. 2010. *Young Manitoba farmer literacy for long term farm viability*. MA thesis, University of Manitoba. https://mspace.lib.umanitoba.ca/bitstream/handle/1993/4208/durnin_debora.pdf?sequence=1.

Earls, A. 2017. *Succession Planning in Haldimand County: A Review of Succession Plans and Available Resources*. Cayuga, ON: Economic Development and Tourism Division, Haldimand County.

Egri, C.P. 1999. Attitudes, backgrounds and information preferences of Canadian organic and conventional farmers: Implications for organic farming advocacy and extension. *Journal of Sustainable Agriculture* 13 (3): 45–72.

Ekers, M., and C.Z. Levoke. 2016. Transformations in agricultural non-waged work: From kinship to intern and volunteer labor: A research brief. *Journal of Agriculture, Food Systems, and Community Development* 6 (2): 179–183.

Ekers, M., C.Z. Levkoe, S. Walker, and B. Dale. 2016. Will work for food: Agricultural interns, apprentices, volunteers, and the agrarian question. *Agriculture and Human Values* 33 (3): 705–720.

Epps, S. 2017. *Barriers to new and young farmers*. Presented at the 2017 Bring Food Home Conference, Ottawa, October 27–28. http://bringfoodhome.com/custom/uploads/2017/10/New-and-Young-Farmers-Network-Research-2017-.pdf.

Farm and Food Care. 2014a. *2015 faces of farming calendar—May—Amanda and Jason.*" YouTube, 04:12. Farm and Food Care. https://www.youtube.com/watch?v=BsbrlvINS5g.

———. 2014c. *2015 faces of farming calendar—August—Sheila.* YouTube, 03:16. Farm and Food Care. https://www.youtube.com/watch?v=VC8x0ltZViw.

———. 2015a. *2016 faces of farming calendar—November—Siblings Joseph and Andrea.* YouTube, 02:21. Farm and Food Care. https://www.youtube.com/watch?v=iaAUm3hFQU0.

———. 2015b. *2016 faces of farming calendar—November—Rob, Julie, Presley, Brinkleigh and Rilee.* YouTube, 03:01. Farm and Food Care. https://www.youtube.com/watch?v=v3iJM5VRNbg.

Farm Credit Canada. 2016a. *FCC young farmer loan program overview.* https://www.fcc-fac.ca/en/about-fcc/media-newsroom/news-releases/2016/fcc-expands-support-for-young-farmers.html.

———. 2016b. *2016 FCC farmland values report.* https://www.fcc-fac.ca/fcc/about-fcc/corporate-profile/reports/farmlandvalues/farmland-values-report-2016.pdf.

FarmStart. 2016. *National farm renewal initiative: Policy recommendations.* http://www.farmstart.ca/wp-content/uploads/PROV-FARM-RENEWAL_Oct-2016.pdf.

Ferrier, H., J. Hardy-Parr, N. Pirie, and H. Watson. 2013. *Transferring knowledge and experience to strengthen the agricultural industry: Step-up—A mentorship program for Canada's future farmers.* Presented at the 19th International Farm Management Conference, Warsaw, Poland, July 24.

Findlay, Martha Hall. 2012. Supply management: Problems, politics and possibilities. *The School of Public Policy Publications* 5 (19): 1–33. https://doi.org/10.11575/sppp.v5i0.42391.

Food and Agriculture Organization of the United Nations. 2021. *FAOSTAT statistical database.* Rome: Food and Agriculture Organization of the United Nations.

Food Secure Canada. 2015. *Home grown crisis—Growing our farmer population.* Research brief. https://foodsecurecanada.org/sites/foodsecurecanada.org/files/support_for_new_farmers.pdf.

Gichungwa, H. 2015. *Young farmers in the lower Mainland.* Vancouver: Vancouver Village Transition Society. https://open.library.ubc.ca/cIRcle/collections/undergraduateresearch/52966/items/1.0075690.

Government of Manitoba. 2017. *Manitoba: Agricultural profile, 2016 census.* https://www.gov.mb.ca/agriculture/markets-and-statistics/statistics-tables/pubs/census-of-agriculture-mb-profile.pdf.

Haalboom, S. 2013. *Young Agrarian culture in Nova Scotia: The initial and ongoing motivations for young farmers from non-agricultural backgrounds*. B.Sc Honours thesis from Dalhousie University. http://dalspace.library.dal. ca/bitstream/handle/10222/21740/Thesis%20S.%20Haalboom.pdf?sequen ce=1&isAllowed=y.

Heminthavong, K. 2018. *Canada's supply management system*. Library of Parliament *Background Paper* 2018-42-E. https://lop.parl.ca/sites/ PublicWebsite/default/en_CA/ResearchPublications/201842E.

Holtslander, C. 2015. *Losing our grip: How corporate farmland buy-up, rising farm debt and agribusiness financing of inputs threaten family farms and food sovereignty—2015 update*. Saskatoon: National Farmers Union (Canada).

House of Commons Canada. 2010. *Young farmers: The future of agriculture*. Report of the Parliamentary Standing Committee on Agriculture and Agri-Food. http://publications.gc.ca/collections/collection_2010/parl/XC12-403-1-1-02-eng.pdf.

Ikerd, John. 2016. *Family farms of North America*. The International Policy Centre for Inclusive Growth Working Paper No. 152. Brasilia: International Policy Centre for Inclusive Growth. http://www.fao.org/3/a-i6354e.pdf.

Junior Farmers Association of Ontario. 2013. *Data summary of 2013 new farmer survey*. http://slideplayer.com/slide/11686578/.

Khaledi, M., S. Weseen, E. Sawyer, S. Ferguson, and R. Gray. 2010. Factors influencing partial and complete adoption of organic farming practices in Saskatchewan, Canada. *Canadian Journal of Agricultural Economics/Revue canadienne d'agroeconomie* 58 (1): 37–56.

Kirkpatrick, J. 2013. Retired farmer—An elusive concept. *Choices* 28 (2): 1–5.

Knibb, H. 2012. *Learning to become a farmer: Findings from a FarmON alliance survey of new farmers in Ontario*. https://foodsecurecanada.org/sites/foodse-curecanada.org/files/learning-to-become-a-farmer-2012-1.pdf.

Laforge, Julia M.L. 2016. *Key findings–new farmers in Canada: A baseline report*. Saskatoon: National Farmers Union Youth Caucus. https://foodsecurecanada. org/sites/foodsecurecanada.org/files/key findings_nnfc_national_survey.pdf.

———. 2017. *Farmer knowledge in alternative agriculture: Community learning and the politics of knowledge*. PhD thesis, University of Manitoba. https:// mspace.lib.umanitoba.ca/xmlui/handle/1993/32667.

Leacy, F.H., M.C. Urquhart, and K.A.H. Buckley, eds. 1983. *Historical statistics of Canada*. 2 ed. Ottawa: Statistics Canada and Social Science Federation of Canada. http://www.statcan.gc.ca/pub/11-516-x/3000140-eng.htm.

Leigh Binford, Arthur. 2019. Assessing temporary foreign worker programs through the prism of Canada's Seasonal Agricultural Worker Program: can they be reformed or should they be eliminated? *Dialectical Anthropology* 43: 347–366. https://doi.org/10.1007/s10624-019-09553-6.

Leonard, D. 2015. *Back-to-the-landers and the emergence of a peasant paradigm in Manitoba.* MA thesis, University of Manitoba. https://mspace.lib.umanitoba.ca/handle/1993/30675.

Lobley, M., J.R. Baker, and I. Whitehead. 2010. Farm succession and retirement: Some international comparisons. *Journal of Agriculture, Food Systems, and Community Development* 1 (1): 49–64.

Martz, D.J.F., and I.S. Brueckner. 2003. *The Canadian farm family at work: Exploring gender and generation.* https://www.grain.org/article/entries/3704-the-canadian.

Mills, Elyse Noble. 2013. *The political economy of young prospective farmers' access to farmland: Insights from industrialised agriculture in Canada.* Major Research Paper, International Institute of Social Studies, Erasmum University Rotterdam. https://thesis.eur.nl/pub/15215/.

Mitchell, B. 2015. *Encouraging new egg farmers.* Egg Farmers of Ontario. https://www.getcracking.ca/members/article/encouraging-new-egg-farmers-0.

Monllor, N. 2012. *Farm entry: A comparative analysis of young farmers, their pathways, attitudes, and practices in Ontario (Canada) and Catalunya (Spain).* http://www.accesstoland.eu/IMG/pdf/monllor_farm_entry_report.

Murphy, Damian. 2012. *Young farmer finance schemes.* Nuffield Australia Farming Scholars. Nuffield Australia Project No. 1203. https://www.gardinerfoundation.com.au/wp-content/uploads/2019/07/1366339342DamienMurphy-YoungFarmerFinanceSchemes.pdf.

National Farmers Union. 2007. *A Brief to the Ontario ombudsman from the national farmers union regarding the Ontario ministry of agriculture, food, and rural affairs and its violation of its public trust.* National Farmers Union (Canada). http://www.nfu.ca/sites/www.nfu.ca/files/Ombudsman-OMAFRA.pdf.

Ngo, M., and M. Brklacich. 2014. New farmers' efforts to create a sense of place in rural communities: Insights from southern Ontario, Canada. *Agriculture and Human Values* 31 (1): 53–67.

Organisation for Economic Co-operation and Development (OECD). 2006. *Investment priorities for rural development—Key messages.* Summary Paper based on the Investment Priorities for Rural Development Conference, Edinburgh, Scotland, October 19–20. https://www.oecd.org/cfe/regionalpolicy/Investment-Priorities-for-Rural-Development.pdf.

Parent, Diane. 2012. *Social isolation among young farmers in Quebec, Canada.* Presented at Producing and Reproducing Farming Systems: New Modes of Organisation for Sustainable Food Systems of Tomorrow, 10th European International Farming System Symposium, Aarhus, Denmark, July 1–4. http://ifsa.boku.ac.at/cms/fileadmin/Proceeding2012/IFSA2012_WS3.2_Parent.pdf.

Pitts, M.J., C. Fowler, M.S. Kaplan, J. Nussbaum, and J.C. Becker. 2009. Dialectical tensions underpinning family farm succession planning. *Journal of Applied Communication Research* 37 (1): 59–79.

Pouliot, Sébastien. 2011. *The beginning farmers' problem in Canada.* Structure and Performance of Agriculture and Agri-products industry (SPAA) Network Working paper #2011-9. Québec: SPAA Network. https://ageconsearch.umn.edu/bitstream/118019/2/Beginning%20farmers%20-%20Pouliot%20Nov%202011.pdf.

Qualman, D., H. Akram-Lodhi, A.A. Desmarais, and S. Srinivasan. 2018. Forever young? The crisis of generational renewal on Canada's farms. *Canadian Food Studies* 5 (3): 100–127.

Qualman, D., A.A. Desmarais, A. Magnan, and M. Wendimu. 2020. *Concentration matters: Farmland inequality on the prairies.* Winnipeg: Canadian Centre for Policy Alternatives. https://www.policyalternatives.ca/publications/reports/concentration-matters?mc_cid=e2d43d368e&mc_eid=43fc8613db.

RealAgriculture. 2014. *Why I love farming—Corey MacQuarrie of New Brunswick young farmers forum.* YouTube, 02:51. RealAgriculture. https://www.youtube.com/watch?v=2MqrHsGvtkU

Robinson, D. 2003. *Prepared for success: A look at the next generation of Alberta farmers.* Presented at the 14th International Farm Management Association Congress, Perth, Australia, August 10–15. http://ifmaonline.org/contents/prepared-for-success-a-look-at-the-next-generation-of-alberta-farmers/.

Roessler, H. M. 2005. *Innovation shared is resilience built: Farmer to farmer knowledge sharing and adapting to climatic change.* MA thesis, University of Victoria. https://dspace.library.uvic.ca:8443/bitstream/handle/1828/4420/Roessler_HannahMaia_MA_2013.pdf?sequence=1&isAllowed=y.

Roppel, C., A.A. Desmarais, and D. Martz. 2006. *Farm women and Canadian agricultural policy.* Research report for Status of Women Canada. Saskatoon: National Farmers Union. https://www.nfu.ca/publications/farm-women-and-canadian-agricultural-policy/.

Smithers, J., and M. Furman. 2003. Environmental farm planning in Ontario: Exploring participation and the endurance of change. *Land Use Policy* 20 (4): 343–356.

Statistics Canada. 2011a. *Number of farm operators classified by income class in the year prior to the census, age and number of operators per farm.* Table 32-10-0185-01. https://www150.statcan.gc.ca/t1/tbl1/en/tv.action?pid=3210018501. Accessed 26 May 2021.

———. 2011b. *Number of farm operators classified by major source of income in the year prior to the census, age and number of operators per farm.* Table 32-10-0188-01. https://www150.statcan.gc.ca/t1/tbl1/en/cv.action?pid=3210018801. Accessed 26 May 2021.

———. 2014. *Total and average off-farm income by source.* Table 32-10-0057-01. https://www150.statcan.gc.ca/t1/tbl1/en/tv.action?pid=3210005701. Accessed 19 June 2020.

———. 2017a. *Number of farm operators by sex, age and paid non-farm work, historical data.* Table 32-10-0169-01. https://www150.statcan.gc.ca/t1/tbl1/en/cv.action?pid=3210016901. Accessed 15 June 2020.

———. 2017b. *Number and area of farms and farmland area by tenure, historical data farm work, historical data.* Table 32-10-0152-01. https://www150.statcan.gc.ca/t1/tbl1/en/cv.action?pid=3210016901. Accessed 15 June 2020.

———. 2017c. *Farm operators classified by number of operators per farm and age.* Table 32-10-0442-01. https://www150.statcan.gc.ca/t1/tbl1/en/cv.action?pid=3210016901. Accessed 15 June 2020.

———. 2017d. *An Overview of the Canadian Agriculture and Agri-Food System 2017.* https://www.agr.gc.ca/eng/canadian-agri-food-sector/an-overview-of-the-canadian-agriculture-and-agri-food-system-2017/?id=1510326669269. Accessed 19 June 2020.

———. 2017e. *A portrait of a 21st century agricultural operation.* https://www.statcan.gc.ca/pub/95-640-x/2016001/article/14811-eng.htm. Accessed 19 June 2020.

———. 2019a. *Dictionary, census of population, 2016.* https://www12.statcan.gc.ca/census-recensement/2016/ref/dict/pop032-eng.cfm. Accessed 13 Feb 2022.

———. 2019b. *The changing face of the immigrant farm operator.* https://www150.statcan.gc.ca/n1/pub/96-325-x/2019001/article/00003-eng.htm. Accessed 15 Aug 2022.

———. 2019c. *Value per acre of farm land and buildings at July 1.* Table 32-10-0047-01. https://www150.statcan.gc.ca/t1/tbl1/en/tv.action?pid=321000470. Accessed 19 June 2020.

———. 2022. *Estimated areas, yield, production, average farm price and total farm value of principal field crops, in metric and imperial units.* Table: 32-10-0359-01. https://www150.statcan.gc.ca/t1/tbl1/en/tv.action?pid=3210035901. Accessed 13 Feb 2022.

Su, C. 2017. *Buy, sell or rent the farm: An agent based simulation of farm succession and land valuation*. MA thesis, University of Saskatchewan. https://ecommons.usask.ca/bitstream/handle/10388/7920/SU-THESIS-2017.pdf?sequence=1&isAllowed=y.

Taylor, J.E., J.E. Norris, and W.H. Howard. 1998. Succession patterns of farmer and successor in Canadian farm families. *Rural Sociology* 63 (4): 553.

Uchiyama, T., M. Lobley, A. Errington, and S. Yanagimura. 2008. Dimensions of intergenerational farm business transfers in Canada, England, the USA and Japan. *The Japanese Journal of Rural Economics* 10: 33–48.

Wasney, J.D. 1992. *Intergenerational transfers of farm property: The development and testing of an instrument to measure values and goals*. MA thesis, University of Manitoba. https://mspace.lib.umanitoba.ca/bitstream/handle/1993/12176/Wasney_Intergenerational transfers.pdf?sequence=1.

Wilson, A. 2015. *Sowing the seeds of a collective autonomy: an analysis of post-capitalist possibilities in food-based livelihoods*. PhD thesis, Carleton University. https://curve.carleton.ca/system/files/etd/26bb378d-fcfc-44b1-8972-ba947c55987c/etd_pdf/93f7ad151c16c02f181b208b4a218ee3/wilson-sowingtheseedsofacollectiveautonomyananalysis.pdf.

Xiong, D. 2017. Millennial farmer bridges cultures with organic Chinese veggies. *Richmond News*, October 26. https://www.richmond-news.com/local-business/millennial-farmer-bridges-cultures-with-organic-chinese-veggies-3060969. Accessed 13 Feb 2022.

3

"Regenerating" Agriculture: Becoming a Young Farmer in Manitoba, Canada

Hannah Jess Bihun and Annette Aurélie Desmarais

This chapter analyses the pathways, motivations, and challenges of young people who are bucking the trend of urban migration and instead are choosing to continue to farm or enter farming as new entrants in the Canadian province of Manitoba. We are particularly interested in addressing the following questions: Who are the young people entering and continuing in agriculture in Manitoba today? How did they enter agriculture? What motivates them to farm? How do they farm and market their products? What barriers are they facing as young farmers in the province? What kind of support do they need?

H. J. Bihun
Department of Environment and Geography, University of Manitoba, Winnipeg, MB, Canada

A. A. Desmarais (✉)
Department of Sociology and Criminology, University of Manitoba, Winnipeg, MB, Canada
e-mail: annette.desmarais@umanitoba.ca

© The Author(s) 2024   **65**
S. Srinivasan (ed.), *Becoming A Young Farmer*, Rethinking Rural,
https://doi.org/10.1007/978-3-031-15233-7_3

Our research highlights the experiences of two distinct groups of young farmers in Manitoba: those using direct marketing[1] and operating small-scale farms and those producing for conventional markets and operating medium to large-scale farms. While there are certainly some similarities and blurring of lines between these two groups of young farmers, there are important differences as well, particularly in their upbringing, pathways into agriculture, production models, ability to fit into the dominant industrial agriculture paradigm, applicability of regulations, and acceptance into rural communities. Although our study also revealed some important gender dimensions, due to space limitations, we cannot address these here.[2]

The chapter is structured as follows. We first situate the research by providing a brief description of the major agricultural trends in Manitoba, Canada. We then describe our research methods and research participants. This is followed by a discussion of the various pathways through which young people enter agriculture, their motivations, and the barriers they face. We conclude by summarizing some major differences and similarities between the two major types of young farmers and signalling a potential pathway to help ensure the "regeneration" of agriculture in the province.

Situating the Research

Agriculture in the province of Manitoba epitomizes the industrial, neo-liberal paradigm of agriculture in Canada described in Chap. 2 on the situational analysis of young farmers in Canada. Grains, oilseeds, and hog production dominate the province's agriculture industry and account for the largest portion of agri-food exports (Manitoba Agriculture 2021).

---

[1] Direct marketers are those who market their products directly to consumers through models such as farmgate, farmers' markets, or community-supported agriculture (CSA). In CSAs, consumers pay/invest in the farm at the beginning of the season and receive weekly shares of food throughout the growing season. For our purposes, a direct marketer may also be selling and marketing directly (in person) to local restaurants and retailers, which will sell to local consumers.

[2] The specific challenges that women face in the province are examined in a forthcoming book on young women farmers.

Manitoba has the largest pig farms and the highest percentage of dairy farms adopting robotic milking technology in the country (Manitoba Agriculture 2017). In 2013, Manitoba was among the top three contributors to the Canadian agriculture economy, contributing 10.3 per cent to the total Gross Domestic Product (GDP) in the sector (cited in Statistics Canada 2017). Farms grossing less than C$250,000[3] fell by 15.3 per cent between 2011 and 2016, and those grossing more than C$250,000 increased by 14.2 per cent over the same period (Manitoba Agriculture 2017). Furthermore, in 2015, only 6.1 per cent of farms reported selling products directly to consumers (Manitoba Agriculture 2017), which suggests that the majority of medium and large-scale farms in Manitoba are selling their products through conventional markets.

In line with national trends, Manitoba lost 70 per cent of farmers under 35 years old between 1991 and 2016 (Statistics Canada 2018), and rural communities are deeply concerned about the ongoing outmigration of rural youth (Bacon and Brewin 2008). The price of farmland has skyrocketed; while prices averaged C$720 per acre in 2006, a decade later, it was two-and-a-half times as high (Qualman et al. 2018). Land concentration too has increased significantly. For example, according to the Census of Agriculture in 1986, the percentage of the land that was farmed by smaller farms of up to 999 acres in size was 43 per cent and by the 2016 Census of Agriculture, this category of farms covered only 16 per cent of the land (Qualman et al. 2020). Meanwhile, larger-scale farms had increased the percentage of land they farmed considerably. For example, in 1986, farms of 5000 acres and 10,000 acres and above covered only 4 per cent and 1 per cent of the land, respectively; by 2016, the former category had jumped to 24 per cent and the latter to 8 per cent (Qualman et al. 2020). Meanwhile, farm debt in Manitoba has never been higher (Government of Manitoba n.d.) and it is expected to continue to climb in the foreseeable future.

While the agricultural situation we have just described may appear bleak, there are also glimmers of hope for the province. Between 2011 and 2016, the number of young farmers under 35 increased by 11 per cent (Qualman et al. 2018) and the average age of farmers in the province

---

[3] Using a US dollar to Canadian dollar exchange rate of 1.3269.

is the second lowest across Canada (Statistics Canada 2018). There is also a growing and vibrant local food movement that includes small-scale farmers, direct marketers, and their urban-based allies (Anderson et al. 2017; Small Scale Food Manitoba working group 2015; Laforge et al. 2017; Sivilay 2019). Finally, in prioritizing the voices of young farmers and allowing their stories and experiences to guide our analysis of young farmers in Manitoba today, we heard a lot about their desire and commitment to farm; we also heard about the hard work and the tenacity that it takes to be a young farmer.

Methods and Research Participants

Our findings are based on qualitative research involving 60 semi-structured interviews as well as quantitative data from a survey used to collect information to supplement interview data. Between April and July 2017, we interviewed 48 young farmers (18–40 years old),[4] 9 older farmers (over 40 years old), and 3 others involved in the agriculture industry in Manitoba. In the interests of capturing diverse experiences, we sought to interview farmers in various parts of the province (see Fig. 3.1). We also selected research participants using the snowball method.

Participants were selected using a combination of purposive, volunteer, and snowball methods in order to gain a diverse sample in terms of gender, types of production, and marketing strategies. Many of the young farmers volunteered to join our study by responding to the invitation that we had posted on Twitter and Facebook. Some participants contacted us as a result of obtaining information about the study through the Keystone Agricultural Producers, the National Farmers Union, and the Agriculture diploma programme at the University of Manitoba. A local radio

---

[4] The 18–40 age range is based on parameters set by farmer organizations and by the federal government of Canada in terms of who is eligible for "young farmer" services. One of the female farmers interviewed was 45 years old; however, she was interviewed with her husband who is 40 years old. She is a new entrant to farming through marriage to her husband and she provided insight about entering farming as a young woman. For these reasons, we included her in our results for young farmers.

Fig. 3.1 Locations of interviews conducted in Manitoba. (Note: The pins on this map represent all of the locations in Manitoba where we interviewed farmers and industry representatives. In some cases, we interviewed a number of farmers from the same location and sometimes only one. Source: Google Maps (2018))

programme in the province also interviewed Meghan Entz, the research assistant who conducted the interviews, and this media also attracted some farmers to the research.

In selecting the participants, we tried as much as possible to obtain a sample reflective of the overall agricultural landscape in Manitoba. For example, in line with the gender distribution in the landscape of Canadian agriculture, our sample of young farmers includes just over 30 per cent women. Additionally, Manitoba Agriculture (2017) reports that only 6.1 per cent of all farms in Manitoba reported using some form of direct marketing. Accordingly, in efforts to best represent the landscape of agriculture in Manitoba, we included a larger number of conventional producers using conventional marketing and a much smaller number of alternative and organic producers using direct marketing strategies.

Ultimately, the research involved the following 60 interviewees: 10 new entrants (first generation) and 38 continuing (from a farming family) young farmers offer their experiences and perceptions of entering into farming in Manitoba. Among these 48 young farmers interviewed, 13 are direct marketers and 35 sell through conventional markets. For the purpose of our study, we classified direct marketers as those who sell at least 51 per cent of their products directly to consumers or directly to local retailers who will sell to local consumers. In total, 32 men and 16 women participated as young farmers, 18 of whom were farming couples that were interviewed together. We also spoke with nine older farmers (between the ages of 41 and 63), and three additional people involved in the agriculture industry, including an employee of an agriculture diploma programme, an instructor involved in a university farm, and a representative involved with Manitoba Agriculture Services Corporation (MASC).

In analysing the data, we categorized young farmers based on their production methods and their marketing strategies (Figs. 3.2 and 3.3). It is worth noting that these categories do not capture all of the nuances and diversity that exist within each category and the lines do blur between them. However, the categories do help to draw distinctions between young farmers depending on what and how they produce and market their products.

We identified the following four categories of young farmers. The *conventional* grain, livestock, and mixed farm category includes those farming grains, legumes, forage seed, hay, cattle, hogs, sheep, and any mixture of those using conventional methods of production, on a larger scale.[5]

---

[5] Based on the Small Scale Food Manitoba working group's (2015) definition, small-scale farms are those producing a variety of products and marketing directly to consumers. Larger-scale producers grow commodity crops primarily for larger markets that are either provincial, regional, national, and/or export and they do not market their products directly to consumers.

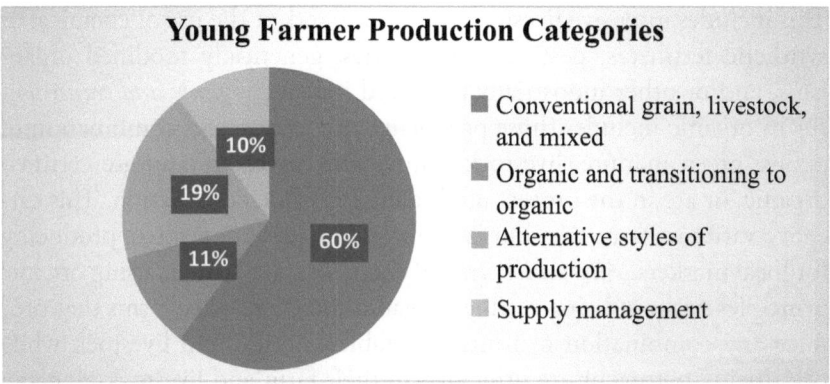

Fig. 3.2 The distribution of young farmers based on their style of production

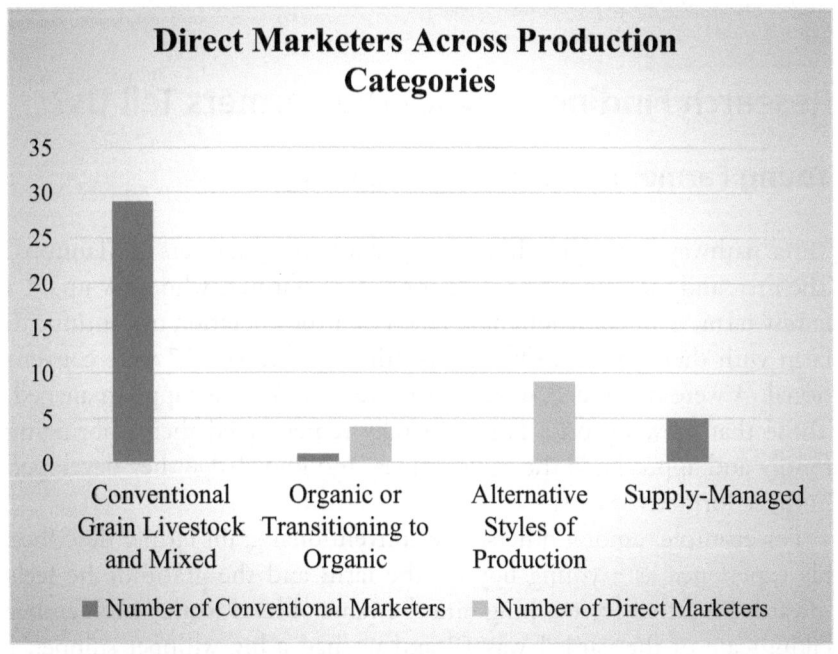

Fig. 3.3 The distribution of direct marketers within the four production categories. For the purpose of this study, direct marketers are classified as any farmer that markets 51 per cent or more of their products directly to the consumer. This can be through, for example, u-picks, community-supported agriculture (CSA) programmes, or farmers' markets

This includes intensive livestock operations and/or the use of chemical or synthetic fertilizers, pesticides, herbicides, genetically modified organisms, and/or other industrially produced inputs. *Organic and transitioning* to organic includes those producing any variety and combination of grains, green manure, livestock, fruits, and vegetables and are certified organic, or are in the process of obtaining organic certification. This category varies in farm size, but is primarily small-scale farmers producing for local markets. *Alternative styles of production* are farmers using organic principles without being certified organic and small-scale farms that produce any combination of fruits, vegetables, honey, and livestock while prioritizing permaculture principles in their farm and livestock management. The *supply management* category includes dairy, egg, and poultry producers whose operations are subject to quotas under supply management regulations.

Research Findings—What Did Farmers Tell Us?

Young Farmer Pathways into Farming

Three pathways into agriculture emerged for young farmers in Manitoba. The first, and most common, is that of young farmers who grew up on a family farm, many of whom have taken over the operation or continue to farm with their parents. Of these continuing farmers, 27 were conventional, 3 were organic, 3 were alternative, and 5 were supply-managed. Those that grew up on a farm generally remembered their upbringing fondly and appreciated the skills, values, and knowledge they developed in those formative years.

For example, among others, one conventional grain farmer described his experience as a young boy on the farm and the gratitude he feels towards his father for teaching him the value of hard work: "I remember taking care of the yard, I was 12 and we had a big whipper snipper. I remember coming back to the shop to take a break and trying to stay in the shop and my Dad said, 'No, get back out there.' It took another 12 years for me to be so happy that he made me do that. It just taught me perseverance and to keep pushing."

A dairy farmer explained that she and her siblings shared responsibilities on the farm as children and emphasized the need for teamwork to ensure everything ran smoothly: "My sister loved the grunt work, I loved the cows, she didn't mind doing the feeding and the grunt work, we each had gifts to disperse among [the farm]. My brother enjoyed the machinery part, so if we got a tractor stuck he'd haul out a chain. We work together, you have to work together."

Of the 38 continuing farmers, 84 per cent farm using conventional methods of production and marketing. This includes supply-managed producers who also grew up on a family farm.[6] Most of the continuing farmers receive significant help from their families and rely on them for access to resources such as machinery, land, knowledge, and financial support. The six continuing farmers using organic or alternative methods of production also receive some form of support from their parents. However, in most cases, these young farmers started a farm independent from their family farm after an extended period of time away, or their parents stopped farming before they returned to agriculture. None of the continuing young farmers who practise conventional or supply-managed production are considered direct marketers as they primarily sell their products through conventional markets. Of the organic and alternative continuing farmers, five are primarily direct marketers and one has a large-scale conventional farm where he recently converted over 4000 acres of farmland used for grain production to certified organic for sale in conventional markets.

In total, we spoke with 10 new entrants, that is, young farmers who did not grow up on a farm and who entered agriculture via two distinct pathways. Five new entrants entered agriculture through marriage to a spouse who comes from a farm family.[7] All five of these young farmers are women; two married men from large-scale conventional farms (the only

---

[6] Only five out of the 48 young farmers in this study are farming solely under supply management. Two other farmers (a couple) have a mixed dairy and grain operation, and therefore, they are classified under conventional mixed, livestock, and grain since they are a mixed farm. Nonetheless, they are both from family farms and produce using conventional methods of production.

[7] One of these women did not directly identify as a farmer; however, she agreed to an interview along with her husband. She lives and works on the farm, does not have any off-farm work, and spoke extensively about the challenges that they face as farmers. For these reasons, we included her in our sample of young farmers.

two new entrants in this farming category) and joined their spouses' family farms. The other three women married men from family farms that are no longer in operation; they are all farming using organic or alternative methods of production and built their farm together with their spouse. All of the new entrants who entered farming through marriage highlighted the value of their spouses' farming knowledge as a result of having grown up on a farm, and most indicated that they depend on their spouse and his family for advice and guidance. For example, although she has been on the farm with her spouse for over 10 years, one first-generation conventional livestock farmer explained that she looks to her husband to make the decisions on the farm due to his lifelong experience as a farmer: "In all honesty, I rely on [him] very much. Because he just has been doing it so long that when it comes down to a big decision, I think it just goes through [him] … Not that I think [he] wouldn't be willing to let me make some decisions, but I just feel that he's the more qualified decision maker."

All five of the new entrants through marriage received some form of support from their spouse's parents, whether it was financial support, knowledge-sharing, or access to machinery and affordable farmland to share, rent, or own.

The third pathway into agriculture are the new entrants who entered into agriculture without any recent family history in farming, without any family-owned farmland or machinery, without farm assets to use as collateral for bank loans, and without years of intergenerational knowledge transfer under their belt. In this group there are two couples and one single man, all from non-farming backgrounds. All of these farmers are practising alternative methods of production and are farming on a small-scale using direct marketing to consumers as their primary avenue for sales. Most farmers who entered agriculture in this way spent time interning or working on farms prior to starting their own and rely on their neighbours and farm community networks and resources to access knowledge from experienced farmers.

A mixed livestock new entrant who had followed this third pathway described the learning process that she and her husband, who is also a new entrant, went through to become farmers. Besides her husband's experience completing an agriculture diploma, they rely heavily on the knowledge and experience of farmers that they have met along the way, through grazing management school, internships, or just friendly neighbours who are willing to pass on knowledge: "Everyone we've ever worked for or with has taught us quite a bit. And it doesn't happen, things don't happen in isolation, you don't just learn about cattle when you're working on a cattle ranch. You learn how to run the bailer and how to fix a fence."

This same couple has a unique scenario in that they have managed to develop a relationship with an older farmer in the area who also has a livestock operation. The couple practises holistic management on their farm, while the older farmer holds many of these same values. Together they agreed on a succession plan whereby the young couple are in the process of taking over the older farmer's farm. While this type of scenario is not common, it does point to interesting possibilities if more retiring farmers passed on their knowledge, and their farms, to aspiring and new farmers.

What Motivates Young Farmers to Farm?

Young farmers identified a number of motivating factors in their decision to farm (Table 3.1). The most commonly mentioned motivation was family, that is, that they value being able to work near their parents, children, or other family members. Some also indicated that they were happy to be able to raise their children on the farm and instil in them the value of hard work. All but one of the young farmers indicated that they would like to see their children farm in the future. The second most common motivator is the quiet life that farming affords and living in close proximity to nature. They talked about valuing the privacy of rural living, working with their animals, and connecting with nature. The third most common motivating factor was the diverse set of skills and tasks associated with farming and a sense of pride. For example, they spoke about the satisfaction of watching things grow, feeding people, and being proud of the work that they do as farmers. They also indicated that they appreciate being able to do something different each day and that they are constantly learning new things.

Table 3.1 Young farmer motivations

Motivation (% of farmers who mentioned this at least once)	Young farmer quotes
Family (79%)	"Part of it is raising our kids on the farm, that's probably one of the reasons why do what we do"
Quiet life in nature (60%)	"Out in the country it's just quiet ... 99 per cent of the time if the truck drives by on the road I know who it is"
Pride and the variety in their work (58%)	"I enjoy the fact that every day is different, the challenges are different ... it's a multi-faceted job, you go from one thing to the next—you're a veterinarian, engineer, welder, accountant, business person, [you also do some] marketing. And then growing food for people"
Social responsibility (52%)	"I feel as though a lot of my peers and younger generations are going to look up to what I do. So, I feel that I have a great responsibility in doing it the right way"
Parents and family support (46%)	"My brother and I wouldn't be farming if it wasn't for my Dad, no question about that"
Rural culture (46%)	"But I think in some ways it's harder than we thought but easier in other ways. I think the relationship part is easier, like you find there's a lot more support in a rural community"
Viability or profitability (44%)	"Each year I sell more CSA shares and sell them faster and I don't even advertise anymore, they just keep coming so I think there's a lot of interest. I've got a customer base now"
Autonomy (40%)	"I own my time ... and [have] flexible hours. I like to get up at 3:30 a.m. and work, and have a three-hour nap in the middle of the afternoon, and go to bed at 10 p.m. at night. You can't do that at a normal job"

Table 3.1 also indicates a number of other important motivations that young farmers articulated. Some farmers conveyed a feeling of altruism or social responsibility when they talked about their role and the need to farm in a sustainable way for the continued health of the land and for the next generation of farmers. Interestingly, social responsibility was mentioned by 85 per cent of direct market farmers, while only 40 per cent of conventional market farmers mentioned this as a motivation. Many of

the continuing young farmers mentioned that their parents enabled or required them to farm while also stating that their parents instilled in them a love for farming at a young age or gave them financial or other necessary supports as they entered into farming.[8] Being their own boss and setting their own work schedule was important to almost half of the participants as they appreciate the autonomy that farming affords. Finally, some young farmers indicated that they are motivated to farm because it is viable or profitable for them. It is important to note that when farmers talked about viability, they were not necessarily pointing to the profitability of farming. That is, they were not speaking directly about making money; instead, they talked about farming as a sustainable livelihood for them and their families.

Barriers to Farming

Young Manitoba farmers face myriad barriers in entering and continuing to farm. Although the challenges they face are complex and, in some cases, specific to their unique pathways and situations as farmers, Table 3.2 lists 12 of the main barriers that farmers raised during our interviews and we discuss the 5 that emerged as being the most important.

Access to Farmland

Access to land was the barrier mentioned most frequently during the interviews with 41 out of 48 young farmers mentioning this as a barrier at least once (Table 3.2). Gaining access to land as a barrier was common across all farm sizes, marketing strategies, and production categories. Young farmers in Manitoba are accessing land in diverse ways and many farmers simultaneously own, rent, and share the land that they farm. Our study shows that 76 per cent of the young farmers interviewed own land, while 70 per cent rent at least some of the land that they farm. Only five young farmers inherited land, although 32 of the young farmers

---

[8] This could be parents co-signing for the young farmer's first loans or something as simple as letting them use their machinery and giving advice when needed.

Table 3.2 Key barriers identified by young farmers in Manitoba

Barrier	# of young farmers out of 48 who mentioned this barrier at least once
1. Access to land	41 (85%)
2. Access to credit/finances/financial management	37 (77%)
3. Policy and government regulations or programmes	36 (75%)
4. Risk/weather	24 (50%)
5. Social and/or physical isolation	24 (50%)
6. Succession planning	23 (48%)
7. Public perceptions of agriculture	22 (46%)
8. Finding labour/good labour	19 (40%)
9. Balancing work and home life	19 (40%)
10. Profitability or viability of farming	17 (35%)
11. Finding or balancing off-farm work	13 (27%)
12. Effective marketing	9 (19%)

indicated that they were likely to inherit some land in the future. All of the young farmers who said that they had already inherited or were likely to inherit land were from a farming family or married into a farming family. The 13 young farmers who indicated that they are not likely to inherit land include 6 new entrants to farming and seven farmers who come from farm families but either started their farm completely independent from their parents or their family farm shut down before they re-entered agriculture.

Among some of the specific barriers related to accessing farmland were the rising price of farmland, increased competition for land, large corporate farms exercising their buying power to secure land, private farmland sales, less availability of farmland overall, the insecurity associated with renting land, and the inability to access credit and/or build the capital required to buy land. In summarizing some of the key issues related to land, a conventional grain farmer put it like this:

It's getting more and more unobtainable, even to rent. It's very competitive, very cut throat. I've even been told by someone, who will remain anonymous, that I'm stepping on people's toes because I went and bid on a piece of land. But, you know, if land is at $6500 an acre, I don't see how a young farmer can afford to buy land, so what do you think we're going to do? We have to try to rent it.

An older organic farmer explained that the imperative to expand the farm continues to direct farmers' actions:

[There is] pressure from the megafarms, the really big farms, to buy land. The big farms are in a position where, if they don't expand, they are going to go bankrupt and so they are expanding, and that leaves no land, or very little land, for the young farmer and it also drives the price up, above the cost of production. So, you are going into debt to buy a piece of land that is not going to make enough to pay your payments. So it's pretty hard to find land.

A conventional grain and livestock farmer agreed with this and linked the price of farmland to the lack of new farmers in her area:

There are no newcomers that are coming to agriculture ... I consider myself extremely lucky to be able to farm with my dad because if there is somebody my age who does not come from a farming family, with the price of land and equipment, you don't really have a chance, unless you're really wealthy, to buy farmland and start a farm. I don't know if there'd be an opportunity for new young people to come [here] and farm.

The most common barrier to accessing land was accessing the necessary credit and capital to purchase land when the prices are so high. We found that 69 per cent of the time when a young farmer was speaking about accessing land as a barrier, they talked about land as it relates to finances.

Access to Credit and Financial Management

The second most important barrier for young farmers is accessing credit and managing finances. Again, this was a common barrier mentioned across all farming styles, marketing strategies, and sizes. Young farmers indicated that they are struggling with increasing debt loads, efforts to build capital and equity, accessing financing, difficulties in understanding financial management, and issues related to cash flow. A supply-managed dairy farmer, for example, explained that the key problem is acquiring "equity to buy quota. The knowledge you can get, there're lots of people around who can help you gain the knowledge you need to be a dairy farmer, and there's a lot of great herdsmen out there who would be great farmers. But they just don't have the equity to actually do it."

This same farmer went on to explain that although the Manitoba New Entrant programme for dairy farmers was a big help, one still needs a lot of money to get started in the province. A conventional grain farmer explained her struggle to build equity and access credit:

> When I first started, I kept wanting to buy land and I just didn't have enough equity. You rent land and you can't borrow enough money to put enough inputs in the ground to grow your crop. It was a constant battle to come up with the revenues to be able to plant the crop and get established, and then build equity so that you could buy land. I would say in my experience that's the biggest barrier of getting in; it is just getting established without [already] having someone in the industry.

While this particular young farmer did have help from her father in the industry, she is sympathetic to those trying to begin farming without that support.

A MASC representative and farmer, whose job it is to review applications for loans and farm support, believes that:

> In Manitoba … you're either getting to the corporate farm type structure or the small direct marketing …. I think that's kind of where we're gonna end up, is either big or small, and I don't know if the margins are better or worse on either side, but you can't buy a million dollar combine if you're not running over enough acres to sustain it.

Furthermore, he acknowledged that young farmers need to have some capital, education, and experience to access financial support programmes, further limiting the ability of new entrants to secure financing:

> They would need a down payment, for sure; there's just no way you can borrow money without some form of down payment. Now MASC programmes are a little bit more favourable ... because we do just require 20 per cent now. Some of our young farmer programmes can be eligible for as little as 10 per cent down, but you do have to have the financial strength and business plan, goal, and knowledge to be able to qualify for that .... [E]ducation would be another one you probably need, to have some form of experience or education to be able to be considered viable, right?

Although the same representative said MASC tries to "make it economical for people to borrow money" by having 25-year fixed interest rates, no pre-payment penalties, and low application fees, he stressed that young farmers still need to be aware of the risks of farming as they enter these financial agreements: "Sometimes I wonder if the really new farmers understand what they're getting into because the machinery costs are so high, the land prices are so high, the input costs are so high. I don't know if they really fully understand what happens if you don't get a crop."

Government Regulations and Policies

The third most common barrier that young farmers in Manitoba identified is government regulations or programmes. Many young farmers felt that there are no political parties that really understand the needs of young farmers. As one farmer put it:

> Some of them [politicians] use [farmer interests] for their benefit, to get the vote. But, in terms of actual action, I haven't really seen in recent history any party that's really jumped, that's really tried to understand the issues enough to really do something ... I don't believe ... just because you have less taxes on farmers, that all of a sudden everyone's just going to be hunky dory.

A significant number of those interviewed expressed the view that young farmers in Manitoba are not feeling heard or understood by their political representatives and policy makers. As one alternative livestock farmer stated: "No government that I know of is trying to address the looming succession crisis in most of the western world .... There are no programmes to help starting or new entrants in farming. Provincial, federal ... you have to be moneyed to a certain extent and established to a certain extent to be able to access those programmes."

Supply management and regulations were other important barriers that young farmers mentioned. Small-scale farmers, in particular, found the regulations restrictive primarily because they do not recognize different farm scales and stages. For example, a new entrant said that she and her partner "support supply management as a national protection ... but it's not set up appropriately for new entrants who are also having tenure issues and financing issues to get in."

It is not only dairy producers that are affected by these issues; a vegetable grower who sells his produce using a community-supported agriculture (CSA) model explained the difficulties that he faces:

> [The] current cap on potato acres and carrot acres is detrimental. We're only allowed to do five acres of potatoes and after that we have to get a quota from Peak of the Market. Once you do Peak of the Market, then you can only sell through Peak of the Market. And that changes your whole pricing. So, as soon as you get past five acres and you get to six, your selling price drops and then you have to compete with the bigger guys, so you either go from five acres and you'd have to do at least 60 or 70. As soon as you start doing that, you have to start thinking chemicals and it just changes everything. So, it's quite frustrating. That policy just protects the larger farms and that drives me nuts.

Even conventional farmers who perhaps fit more readily within government policies and regulations are sensitive to the struggles that new entrants and small-scale producers face. A young conventional grain farmer stated that:

Government programmes are usually geared towards larger farmers. So FCC [Farm Credit Canada] has an option that we've been able to access, non-interest loans … So if we have a good financial stance, they can offer interest-free loans. In my mind that should be offered to those passionate young new entrant farmers first. But it's a huge security risk. But the other security risk is not having farmers.

An older conventional grain farmer who has been involved with policy development through Keystone Agricultural Producers for the past 20 years lamented the lack of equal access to programmes and support for small-scale farmers versus large-scale farmers, stating that all young farmers matter: "And I really think that if you're a young farmer and you're willing to farm, then you should have equal access to all these programmes. Whether you farm 5 acres of sweet corn or raspberries, direct marketed, or you farm 35,000 acres, you know, it doesn't matter. You're still a young person farming."

Social Isolation

When I started farming, let's say in a five-mile radius, there might have been 40 farmers. Now there're four left. (Older farmer)

Manitoba farmers are feeling the impacts of bigger and fewer farms in the province as rural communities become more spread out, their neighbours are further away, and there are fewer opportunities to build community. An older alternative farmer talks about how social isolation occurring in rural communities prompted him to change production models: "There is no sense of community anymore. So that got me irritated that that part was taken away from farming. So that's why I went to micro-farming if you want to call it that, where you get to see your end consumer, you get to know the people you're delivering to."

Social isolation was mentioned by 77 per cent of the small-scale, direct marketers, while only 40 per cent of the conventional marketers mentioned isolation as a barrier. An alternative vegetable farmer who is a new entrant talked about how difficult it has been considering the lack of communication and interaction in his rural community: "Especially in the winter time, I find that I'm just living here alone … the neighbours are all really nice, but there's no interaction other than the odd hello kind of thing. And I feel like everyone just sticks to themselves. Yeah actually, I find it pretty tough."

An organic vegetable farmer who is a new entrant through marriage talks about the struggles of living rurally, especially since they are surrounded by large farms that have very different goals and strategies than they do on their own farm: "Mine would definitely be the lack of community that, you know, you see in other provinces. Farms like ours would be surrounded with other farms like ours, a different type of agriculture but still small scale, like-minded." Her husband agreed with her and explained that although there are many small-scale farms in Manitoba, they are too spread out to really form a cohesive community.

Succession

The barriers identified by the nine older farmers that we interviewed are in line with those identified by young farmers; older farmers pointed to access to finances and credit as the most common barrier with land at a close second. However, difficulties with succession planning took precedence over government regulations and policies, probably reflecting the fact that many older farmers are currently staring at the very real prospect of retiring in the near future. Many older farmers struggled with ensuring that their succession planning was fair for all their children, especially in cases where one child is farming and the others are working in the city: "That's the million-dollar question, figuring out how to make things fair. You know, fair doesn't mean equal, but how do you make things fair for everybody—those questions are not all answered yet."

A young conventional grain farmer echoed similar concerns as he and his family work out their own succession planning:

It's way more complicated than we ever thought it would be … My parents have a fairly good definition of equal versus fair. It's not necessarily going to be equal, it's going to be fair. Which means, theoretically, I'll get more in to the farm because I put in to build it and if I hadn't, it wouldn't have grown as quickly. But it also wouldn't be fair for my sisters to get nothing.

Meanwhile, an older alternative livestock farmer is creating a succession plan with a young couple to whom he is not related in order to ensure that the farm will continue into the future:

I don't know that I ever really thought about it that much. What's important to me, more than anything, is to see that the land continues to be farmed in the way you want it to be farmed. From that point of view, it's seeking somebody with a similar mindset and management philosophies. That's really what drives it, it's not about the money, it's not about being able to live beyond the grave or anything. It's just a desire to see the land protected in the sort of multi-generational mindset we've had.

Although this kind of arrangement is rare, it does offer solutions to some of the barriers that first-generation entrants are facing as well as the challenges faced by farmers who want to retire without seeing their farm disappear.

Discussion

Our research reveals that young farmers in Manitoba come from diverse upbringings and entered into agriculture via distinct pathways and that these are heavily influenced by whether one has historical connections to farming, comes from an urban or rural setting, and is a man or a woman. In addition, the pathways deeply influence the type and scale of agriculture that are available to young farmers. For example, a young person entering agriculture without any previous connection to a farm has to acquire land, machinery, and knowledge, while someone coming from a farm family may access these resources somewhat more readily. The financial requirements to get into conventional farming mean that this type of agriculture is essentially off limits to new entrants. There are a number of young people in Manitoba who have grown up on conventional farms and are poised to take over from their parents; however, as land prices continue to rise and farms continue to increase in size, there are fewer and fewer farmers.

Although this process may create "efficiencies" by increasing economies of scale (Pouliot 2011; Weis 2012), it ultimately benefits the dominant agri-business companies that control much of the agriculture sector in Canada, leaving farmers overworked, stressed, and in debt (Bacon and Brewin 2008; Qualman et al. 2018). With each generation that ceases to farm, there is a loss of generations of knowledge about the land and farming, and the vitality of rural communities is further jeopardized (Laforge et al. 2017). This has left some feeling pessimistic about a future in agriculture in the province (Bacon and Brewin 2008).

Many of the young farmers we interviewed said that their own parents had not encouraged them to do so. For example, a young organic vegetable farmer from a multi-generation farm who we interviewed said that parents in his family, for generations, have urged their children not to farm. Meanwhile, when asked if he believes that young people see a promising or attractive future in farming in Manitoba, the instructor involved with a university farm who we spoke with said: "I don't. I don't, and I lament that … I think I'm always amazed when young people find their way into it."

Yet, there is good reason to feel positive about and for young farmers in Manitoba. The latest Census of Agriculture demonstrates an 11 per cent increase in young farmers in the province. An older farmer who we spoke with who works as a succession planning consultant said: "When you said at the outset that this whole study is around the concern that there aren't going to be enough young farmers … that's totally not my reality. And when you look at the dynamic of the farmers who you'll be interviewing in this area, there's heaps of young farmers here. Heaps and heaps of them."

Additionally, the majority of the 48 young farmers we interviewed indicated they are strongly motivated to farm by the lifestyle that farming affords in terms of family time, opportunities for learning every day, managing their own time, carrying on traditions, and being advocates for the environment. These motivations suggest that they are in it for the long haul. Significantly, all but one of the young farmers in our study indicated that they would like their children to farm.

All young farmers in our study who are currently taking over the family farm come from conventional or supply-managed farms and thus are engaged in agricultural practices involving more intensive livestock raising, synthetic/chemical inputs, and distant markets.[9] However, the majority expressed worries about the public's perceptions of the environmental and health consequences of these practices. There are also some who are aware of the unsustainable aspects of conventional farming; for example, one conventional grain farmer working 2500 acres of grains and legumes said: "When you read the statistics about how much fossil fuels we use on every acre for our fertilizer and our chemical, and transporting those products, I found it unbelievable. I think we can reduce that, produce our own nitrogen, have healthy soil. I think that's something to strive for."

Meanwhile, first-generation farmers are engaging in small-scale and ecologically sustainable farming, perhaps due to their own morals and interests in creating food system change or because they have insufficient capital. While the lack of access to the resources needed for conventional farming does limit options for new and young farmers, it also provides opportunities to build more sustainable food systems at a smaller scale, as young people entering the sector are increasingly using organic or alternative methods to produce food while also engaging with and selling directly to consumers, thus building rural-urban connections and creating more opportunities for food literacy. If the new entrants represented in our study and others across Canada (Haalboom 2013; Laforge et al. 2017, 2018) are any indication of the future of agriculture in Manitoba, then the future holds some promise for more small-scale farms producing for local markets. However, for these to flourish, appropriate government policies are needed.

Although there are clear distinctions in Manitoba agriculture in terms of the size of farms, goals, and production models, government policies fail to distinguish between the two existing agricultural paradigms: small-scale for local markets and medium- to large-scale conventional

---

[9] The young farmers in our study who come from a farming background and are practising alternative styles of production came from family farms that are no longer in operation, or they started a farm independent from their family farm.

production for distant markets. Consequently, most policies do not address the specific needs of the different types of farmers and are most often biased towards larger-scale industrial farms producing for distant markets. For example, earlier research on the impact of the regulatory framework on small-scale farmers in Manitoba (Laforge et al. 2017; Sivilay 2019) and interviews with direct marketers in our study reveal that small-scale farmers struggle to afford, understand, and abide by regulations that are simply not applicable to their production methods and marketing strategies.

While these two agricultural paradigms are not in competition with each other in terms of the markets they target, they do, however, compete for land and other resources. The larger-scale conventional farmers are much more likely to have generations of farm experience and greater access to resources (land, credit, knowledge) that are critical to helping them become established, whereas the small-scale new entrant has little to fall back on. Policy makers need to recognize that by pitting the distinct types of producers against one another while prioritizing one type of farming over another, it is the small-scale farmers that are disadvantaged. Conventional and direct market farmers in our study both voiced concern for a lack of young farmers and called for equal access to programmes and financial assistance, even for new entrants who may not have the same experience and collateral as larger farmers.

Conclusion

Much of the recent research in Canada about young and new/beginner farmers focuses on a particular voice—that of the farmer growing on a small scale and producing for local markets. While this is a critically important part of the farming sector since it is central to a growing local food movement that emphasizes food sovereignty and environmental well-being, it is not the only voice of young farmers. The life stories of 48 young farmers in the province indicate that they come from diverse upbringings, have different levels of experiences and varying motivations, and they are engaging in different kinds of agricultural models. While there are important, powerful tensions and differences between small-scale producers engaged in local direct marketing and those who engage in medium to large-scale conventional farming for distant markets, they do share some similar challenges. As the price of farmland continues to rise, start-up and input costs become ever more burdensome, and current government policy is inadequate and works to create and/or perpetuate inequalities in the countryside, young farmers producing food and commodities often feel like they are left to fend for themselves.

This points to the need for a more wholistic approach that recognizes the diversity of farmers who are attempting to make a living from farming. As Ngo and Brklacich argue in their study of Ontario farmers:

> An "us" versus "them" mentality may offer the path of least resistance; however, the future of these various agricultural actors are intricately linked through the space they share. It would seem that there is an opportunity to engage different members of the rural farming community in the LFM [local food movement] conversations. This would not necessarily be about finding a united voice but more about moving across boundaries, finding commonalities, and pioneering collaborations that have the potential to strengthen the resiliency of rural communities as a whole. (2014, 65)

Considering that Manitoba has lost 70 per cent of its young farmers over the last three decades (Statistics Canada 2018), an approach in farm activism, research, and policy that focuses on recognizing differences, establishing commonalities, and fostering collaboration may well contribute to regenerating agriculture into the future and ensuring vibrant and thriving rural communities for the remaining young farmers in the province.

References

Anderson, C., J. Sivilay, and K. Lobe. 2017. Organizing for food systems change. In *Everyday experts: How people's knowledge can transform the food system*, ed. T. Wakeford, J. Sanchez-Rodriguez, M. Chang, C. Buchanan, and C. Anderson, 169–187. Coventry: Coventry University. https://www.coventry.ac.uk/globalassets/media/global/08-new-research-section/signposts/everyday-experts-ch11.pdf.

Bacon, B., and D. Brewin. 2008. *Rural community viability: Lessons from 4 communities*. Working paper, Faculty of Social Work, University of Manitoba.

Google Maps. 2018. *Manitoba*. https://www.google.ca/maps/place/Manitoba /@54.1666436,-104.4140414,5z/data=!3m1!4b1!4m5!3m4!1s0x526de06f4 5e7b513:0x1c0f55f2abc1c768!8m2!3d53.7608608!4d-98.8138762?dcr=0. Accessed 10 April 2020.

Government of Manitoba. n.d. *Manitoba agriculture and agrifood statistics*. https://www.gov.mb.ca/agriculture/markets-and-statistics/pubs/mb-agriculture-statistics-factsheet-2019.pdf.

Haalboom, S. 2013. *Young agrarian culture in nova scotia: The initial and ongoing motivations for young farmers from non-agricultural backgrounds*. BSc honours thesis, Dalhousie University. http://dalspace.library.dal.ca/bitstream/handle/10222/21740/Thesis%20S.%20Haalboom.pdf?sequence=1& isAllowed=y.

Laforge, J.M.L., C.R. Anderson, and S.M. McLachlan. 2017. Governments, grassroots, and the struggle for local food systems: Containing, coopting, contesting and collaborating. *Agriculture and Human Values* 34 (3): 663–681.

Laforge, J., A. Fenton, V. Lavalée-Picard, and S. Mclachlan. 2018. New farmers and food policies in Canada. *Canadian Food Studies* 5 (3): 128–152.

Manitoba Agriculture. 2017. *Manitoba: Agricultural profile, 2016 census*. https:// www.gov.mb.ca/agriculture/markets-and-statistics/statistics-tables/pubs/ census-of-agriculture-mb-profile.pdf. Accessed 16 Feb 2022.

———. 2021. *Manitoba agriculture and agri-food statistics*. https://www.gov. mb.ca/agriculture/markets-and-statistics/publications-and-proceedings/ pubs/mb-ag-stats-factsheet-november-2021.pdf. Accessed 16 Feb 2022.

Ngo, M., and M. Brklacich. 2014. New farmers' efforts to create a sense of place in rural communities: Insights from southern Ontario, Canada. *Agriculture and Human Values* 31 (1): 53–67.

Pouliot, S.. 2011. *The beginning farmers' problem in Canada*. Structure and Performance of Agriculture and Agri-products industry Network (SPAA) Network working paper #2011-9. Québec: SPAA Network. https://ageconsearch.umn.edu/bitstream/118019/2/Beginning%20farmers%20-%20 Pouliot%20Nov%202011.pdf.

Qualman, D., A. Akram-Lodhi, A.A. Desmarais, and S. Srinivasan. 2018. Forever young? The crisis of generational renewal on Canada's farms. *Canadian Food Studies* 5 (3): 100–127.

Qualman, D., A.A. Desmarais, A. Magnan, and M. Wendimu. 2020. *Concentration matters: Farmland inequality on the prairies*." Saskatoon and

Winnipeg: Canadian Centre for Policy Alternatives. https://policyalternatives.ca/publications/reports/concentration-matters.

Sivilay, J. 2019. *Organizing for food sovereignty in Manitoba: Small-scale farmers, communities of resistance, and the local food movement.* MA thesis, University of Manitoba.

Small Scale Food Manitoba working group, Government of Manitoba. 2015. *Advancing the small scale, local food sector in Manitoba.* https://www.gov.mb.ca/agriculture/food-and-ag-processing/pubs/small-scale-food-report.pdf.

Statistics Canada. 2017. *Canadian agriculture: Evolution and innovation. Contribution of the agricultural sector to the economy.* https://www150.statcan.gc.ca/n1/pub/11-631-x/11-631-x2017006-eng.htm#a1. Accessed 16 Feb 2022.

———. 2018. *Number of farm operators by sex, age and paid non-farm work, historical data.* Table 32-10-0169-01 (formerly CANSIM 004-0017). https://www150.statcan.gc.ca/t1/tbl1/en/tv.action?pid=3210016901. Accessed 11 May 2020.

Weis, T. 2012. A political ecology approach to industrial food production. In *Critical Perspectives in Food Studies,* ed. M. Koç, J. Sumner, and A. Winson, 104–121. Don Mills, ON: Oxford University Press.

4

Impervious Odds and Complicated Legacies: Young People's Pathways into Farming in Ontario, Canada

Travis Jansen, Sharada Srinivasan, and A. Haroon Akram-Lodhi

Introduction

Ontario is the most populated province in Canada and has some of its most productive agricultural soils. However, Ontario faces problems attracting youth into agriculture. Since 1991, the province has lost about two-thirds of its 18,440 young farmers, and less than 10 per cent of Ontario's current farmers are under the age of 35. Part of these losses can be attributed to the challenges that Ontario's youth face in becoming a

T. Jansen
University of Guelph, Guelph, ON, Canada
e-mail: traviswjansen@gmail.com

S. Srinivasan (✉)
Department of Sociology & Anthropology, University of Guelph,
Guelph, ON, Canada
e-mail: sharada@uoguelph.ca

A. H. Akram-Lodhi
Trent University, Peterborough, ON, Canada
e-mail: haroonakramlodhi@trentu.ca

© The Author(s) 2024   **93**
S. Srinivasan (ed.), *Becoming A Young Farmer*, Rethinking Rural,
https://doi.org/10.1007/978-3-031-15233-7_4

farmer. To understand these challenges, it is necessary to understand the different pathways that young people take to becoming a farmer. By understanding these pathways, it becomes possible to create more opportunities and reduce the challenges that young people must overcome. This chapter will differentiate the pathways of entry into farming for young people in Ontario, highlighting differences in how they access resources, their motivations for farming, and the type of farming that they carry out. Understanding the unique circumstances facing Ontario's young farmers can help identify ways to encourage young people to begin a career in farming.

This chapter is organized into three sections. The first section provides a description of the agrarian context within which Ontario's young farmers operate. This is followed by a profile of the group of young farmers that were interviewed for this chapter. The third section discusses how the respondents became (young) farmers, focusing on the different experiences and challenges of those who are new to farming, those who returned to work on family farms, and those who fall somewhere in between. The conclusion reflects upon the constraints that must be addressed if more young people are to be encouraged to enter farming.

The analysis in this chapter is based on data from Statistics Canada, insights from scholarly literature, and the information collected from interviews with 49 young farmers from across southern Ontario. Most of the interviews were conducted within 100 kilometres of the city of Guelph (Fig. 4.1). The interviews were undertaken by three research assistants who were themselves farmers; many of the farmers interviewed were found through their networks. Interviews ranged from one to three hours. The length of the interview depended on the amount of detail that the farmer was willing to share and the amount of time that the farmer devoted to the interview. While the sample cannot be deemed representative of young farmers in Ontario, the information from these interviews offers rich insights into the experiences and challenges of young farmers and when complemented by scholarly literature and data from Statistics Canada, provides a more well-rounded understanding of young farmers in southern Ontario.

Fig. 4.1 Geospatial map of young farmers interviewed in southern Ontario

Ontario's Agrarian Context

Food and agriculture play a large role in Ontario's economy. In the 2016 agricultural census, Ontario had 49,600 of the 193,492 farms in Canada and more than half of the highest-quality Class 1 farmland in the country. Ontario farmers accounted for almost one quarter of all farm revenue in Canada in 2016 (Statistics Canada 2016; Government of Ontario 2019) and Ontario farmers contribute C$7.6 billion[1] to Ontario's total Gross Domestic Product (GDP) (Ontario Ministry of Agriculture, Food and Rural Affairs 2018).

---

[1] Using a US dollar to Canadian dollar exchange rate of 1.3269.

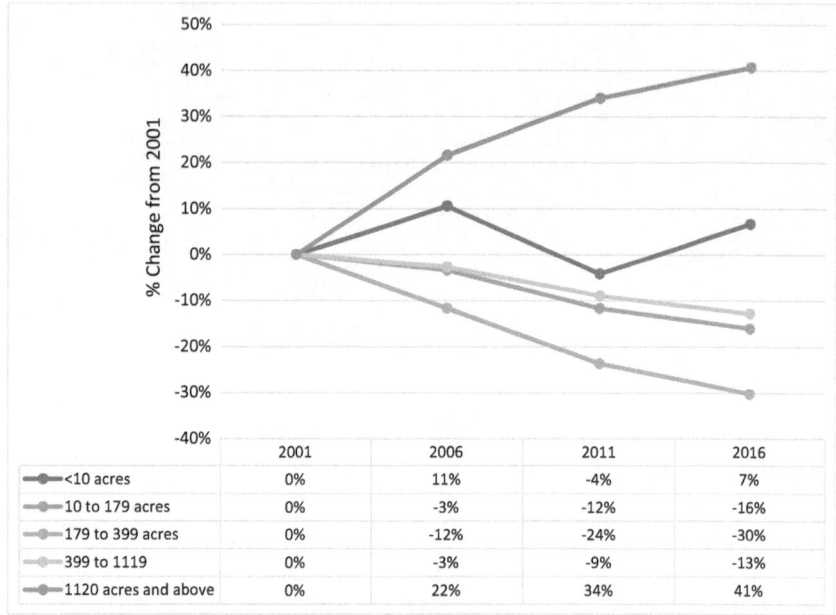

	2001	2006	2011	2016
<10 acres	0%	11%	-4%	7%
10 to 179 acres	0%	-3%	-12%	-16%
179 to 399 acres	0%	-12%	-24%	-30%
399 to 1119	0%	-3%	-9%	-13%
1120 acres and above	0%	22%	34%	41%

Fig. 4.2 Change in number of farms by acreage (2001–2016). (Source: Table: 32-10-0156-01 (formerly CANSIM 004-0005))

The average farm size in Ontario is close to 250 acres, while the average Canadian farm is 820 acres. Nonetheless, over the past 15 years, Ontario's farming communities have witnessed the trend towards fewer and larger farms. Figure 4.2 demonstrates this trend based on the number of acres a farm has, while Table 4.1 demonstrates this trend based on a farmer's livestock count. From 2001 to 2016, the number of farm operators fell by 17 per cent, from 85,020 to 70,470, and the number of farms also fell by 17 per cent, from 59,728 to 49,600. Figure 4.1 shows that while the number of small farms (<10 acres) has varied, the number of medium-sized farms (10–1119 acres) has fallen dramatically, and the number of large farms (>1120 acres) has grown steadily. The number of small farms (<10 acres) tends to be more variable as the smaller acreage and lower cost make it relatively easier for farmers to enter and exit the sector. By way of contrast, the growth of larger farms has required a significant number of other farms to exit the sector; the correlation with the decline of

Table 4.1 Number and size of Ontario livestock farms (2001–2016)

	2001	2006	2011	2016	Δ 2001–2016
Farms reporting					
Cattle and calves	28,209	25,040	20,349	17,452	-38%
Pigs	4972	4070	2556	2760	-44%
Sheep and lambs	3978	3408	3569	3119	-22%
Hens and chickens	8306	7397	7263	8246	-1%
Turkeys	1159	983	926	851	-27%
Average animals					
Cattle and calves	76	79	86	20	22%
Pigs	695	971	1208	1280	84%
Sheep and lambs	85	91	99	103	21%
Hens and chickens	5252	5962	6458	6156	17%
Turkeys	2936	3618	3762	4433	51%

Source: Table: 32-10-0155-01 (formerly CANSIM 004-0004)

medium-sized farms is suggestive. Larger farms sell homogeneous goods into commercial markets and benefit greatly from economies of scale.[2]

While farms get bigger, the distribution of owned and rented land has remained relatively consistent; about 70 per cent of farmland is owned and the other 30 per cent is rented. The distribution of rented land, however, has changed. Between 2011 and 2016, land under sharecropping increased from 305,202 acres to 361,575 acres, while the number of acres leased from the government decreased from 97,779 to 89,140 acres.

For farmers with access to land, both the earnings from farming and value of Ontario farmland have increased. Table 4.2 provides a summary of how Ontario farmers measure up with respect to their values for key financial indicators. The current ratio can be defined as a measure of a farm's ability to pay off its short-term debts with its liquid assets. Current assets are cash or other assets that can be converted into cash within a

---

[2] Commercial agriculture refers to producers growing a commodity that is essentially non-differentiable from other similar crops in the international market (maize, soybeans, wheat, pig, etc.). This commodity is sold to an intermediary and then mixed with other farmers' products before being sold to another intermediary or end user. Direct marketing refers to a farmer's product being identified as from a specific farm and then sold directly to a customer like a restaurant or consumer. Farms that undertake commercial agriculture tend to be larger and focused on efficiency and volume. By contrast, smaller farms tend to "direct market" their products to end users for a premium. While these markets aren't mutually exclusive, they will be juxtaposed throughout the chapter as returning farmers tend to be a part of commercial agriculture while new farmers often direct market their products.

Table 4.2 Average financial values for farms in Ontario

Value	Ontario				
	2009	2011	2013	2015	2017
Current assets	131,211	162,606	188,011	229,498	214,970
Current liabilities	53,830	66,991	70,069	98,965	106,074
Current ratio	**2.44**	**2.43**	**2.68**	**2.32**	**2.03**
Long-term assets	1,832,517	2,266,803	2,787,424	3,451,874	3,524,481
Long-term liabilities	342,495	361,085	480,396	554,804	666,949
Long-term debt-to-asset ratio	**0.19**	**0.16**	**0.17**	**0.16**	**0.19**
Total assets	1,963,728	2,429,409	2,975,435	3,681,372	3,739,452
Total liabilities	396,325	428,076	550,465	653,769	773,023
Net worth	**1,567,404**	**2,001,334**	**2,424,970**	**3,027,603**	**2,966,429**
Total revenue	363,002	422,326	511,131	614,970	538,822
Total expenses	321,894	361,740	433,738	541,752	449,151
Net cash farm income	**41,107**	**60,586**	**77,393**	**73,217**	**89,670**

Source: Table: 32-10-0102-01 (formerly CANSIM 002-0072)

year, while current liabilities are debts that must be paid within a year. As demonstrated in Table 4.2, on average Ontario farms have a current ratio of >2, meaning that Ontario farms have at least C$2 available to them in the form of liquid assets for every dollar of expected debt that they will have for that year. Moreover, Table 4.2 demonstrates that Ontario farms have a long-term asset to long-term liability ratio of around <0.2, meaning that less than 20 per cent of long-term assets would need to be sold in order to pay off all outstanding long-term debt. Both of these ratios suggest that Ontario farmers face a low level of risk and have a correspondingly strong ability to be able to manage medium-term financial challenges that they may face.[3] Subtracting short and long-term liabilities from short and long-term assets provides an average net worth of farms. In Ontario, this value almost doubled between 2009 and 2017, as estimated farm net worth climbed from $1,567,404 to $2,966,429. Furthermore, net farm income more than doubled from $41,107 in 2009 to $89,670 in 2017. Given that the average annual income in Ontario for those over 16 years old was $46,700, Ontario farmers appear to be in a good position relative to the rest of the population (Statistics Canada

---

[3] The long-term debt-to-asset ratio is generally an indication of how established a farm is as older farms tend to have more of their mortgages paid off and as a result are more financially stable.

2017). There appear to be two drivers of this higher net farm income. The first is that commodity prices have significantly increased from 2009 onwards. Figures 4.3, 4.4, and 4.5 show that commodity prices for Ontario's common agricultural products have been higher than usual, but whether this trend continues remains unknown. Furthermore, as discussed earlier, Ontario's farms are becoming larger, which means that they will tend to have a larger net farm income.

The strong farm real estate market is also an indicator of the financial strength of the sector. In 2018, the average value of land and buildings was more than C$11,000 per acre in Ontario, which is close to double the value of land in any other province; the next closest is Québec, where average land values are around $6000 per acre (Statistics Canada 2018). Moreover, it is more than double the average value of Ontario farmland in 2009, which was approximately C$5000 per acre. Like commodity prices, Fig. 4.6 demonstrates that Ontario farmland values have followed a similar, increasing trend from 2009 onwards. This increase in the value of land can make it harder for new farmers to access the land that they need to get their start in farming.

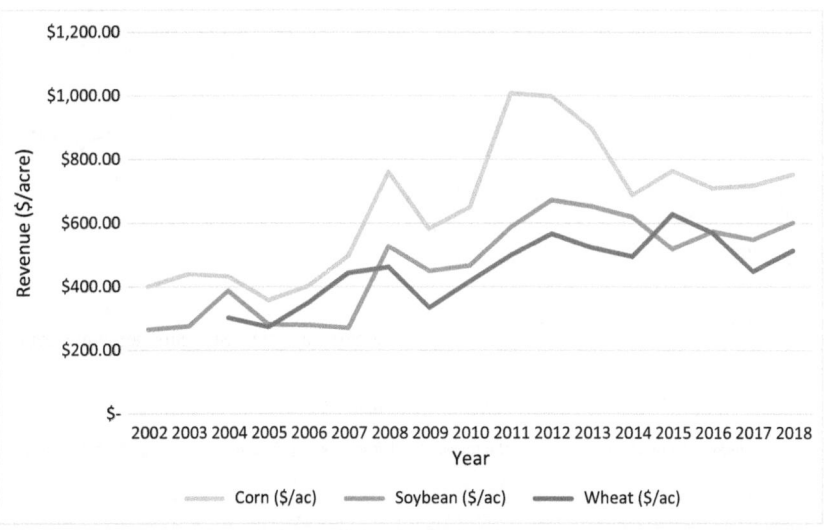

Fig. 4.3 Ontario corn, soybean, and wheat revenues. (Source: Table: 32-10-0359-01 (formerly CANSIM 001-0017))

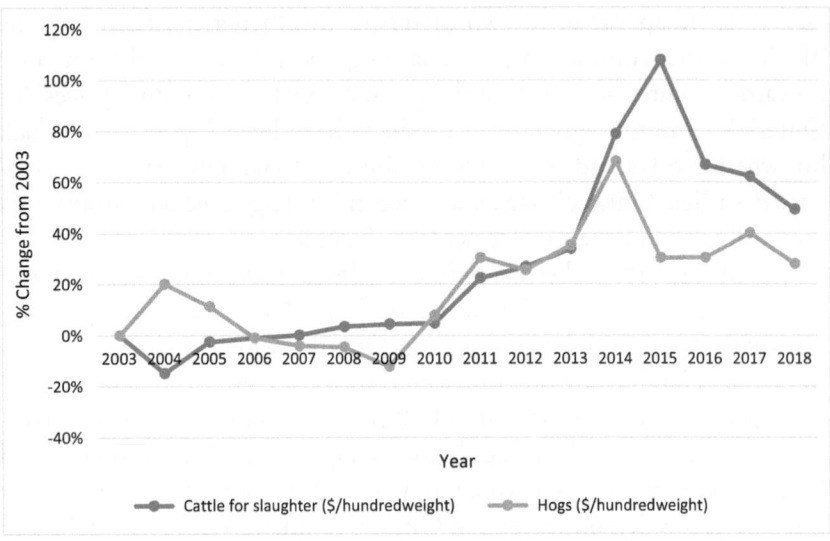

Fig. 4.4 Relative change in red meat prices (2003–2018). (Source: Table: 32-10-0077-01 (formerly CANSIM 002-0043))

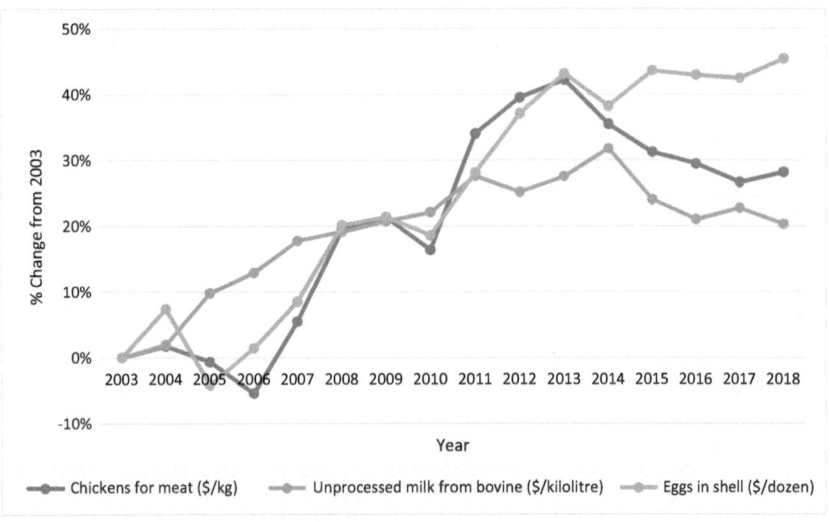

Fig. 4.5 Relative change in prices for supply-managed goods (2003–2018). (Source: Table: 32-10-0077-01 (formerly CANSIM 002-0043))

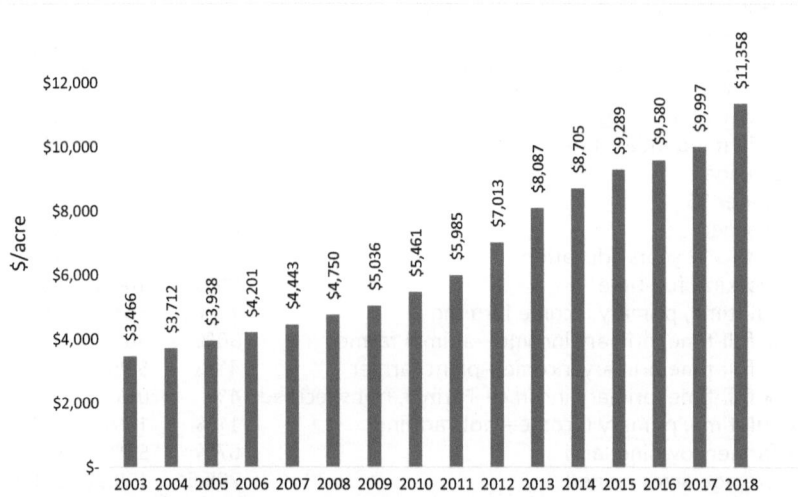

Fig. 4.6 Value of Ontario land and buildings ($/Acre). (Source: Table: 32-10-0047-01 (formerly CANSIM 002-0003))

Despite challenges in accessing land, changing consumer preferences in Canada have opened up a number of opportunities for smaller, less intensive models of agriculture where farmers can directly market their products to end users. These opportunities are better suited for new farmers as they can function on a smaller land base and don't require the expensive barns and equipment needed for commercial agriculture. While we see increasing farm sizes and consolidation among those in commercial agriculture, there are an increasing number of farms that run smaller operations and sell specialty products, directly marketing to local customers. These different approaches to farming bring different challenges for new farmers looking to enter the sector.

Who Are the Young Farmers?

A descriptive summary of the young farmers interviewed for this study can be found in Table 4.3. We define a "returning" or "continuing" young farmer as someone who had grown up on a farm and their farming career was in some way connected to their family's farm. A "new" young farmer

Table 4.3 Summary of Ontario young farmers in our study

	Female	Male	Total
Number of farmers surveyed	27	22	49
Age started farming	19	15	17
Age farming independently	26	25	25
Mean age	34	34	34
% under 35	56%	64%	59%
% married	63%	59%	61%
% with >12 years education	96%	91%	94%
% working full-time	85%	100%	92%
% full-time, primary income farming	74%	86%	80%
% full-time, primary income—animal farmer	30%	36%	33%
% full-time primary income—plant farmer	41%	50%	45%
% full-time, primary income—Farmer, not specified	4%	0%	2%
% full-time, primary income—not farming	11%	14%	12%
% farmers owning land	67%	50%	59%
Average acres owned	285.64	129.70	215.63
% farmers that have inherited land	0%	9%	4%
Average acres inherited	0.00	81.25	36.48
% farmers likely to inherit land	41%	50%	45%
Average acres likely to be inherited	97.00	798.57	411.99
% farmers renting in land	48%	55%	51%
Average acres rented in	125.46	989.08	513.21
% farmers sharing land	19%	36%	27%
Average acres shared	69.20	121.91	92.86
% with access to community land	0%	0%	0%
Average of community land			

is someone who was farming but who did not grow up on a farm. In total, 49 young farmers were interviewed: 28 were considered returning farmers (16 female, 12 male), while 21 were new farmers (11 female, 10 male). Returning farmers were an average age of 33. They had often spent more than a decade learning to farm before beginning to farm independently at around age 24. On average, new farmers had a later start. After completing some form of post-secondary education, they typically began to learn to farm at age 22 and were farming independently within five years, when they were 27. As a result, returning farmers had the advantage of taking more time to learn how to farm and starting to farm independently at an earlier age. These findings are similar to those of Monllor

(2012), who found that continuing farmers in Southern Ontario (and Catalonia) tended to start farming more than five years earlier than newcomers.[4]

Monllor (2012) also describes how continuers (returning in our study) and newcomers (new in our study) follow two distinct models of farming. Monllor highlights how continuers tend to follow commercial agriculture while newcomers tend to practise small-scale farming. Statistics Canada does not explicitly differentiate between continuing and new farmers when collecting census data. However, the finding that continuers and newcomers tended to follow distinct farming models was largely replicated among the present study's young farmers.

Based on the interviews, being a returning or new farmer also influences the type of agricultural products that the young farmer produces. Fourteen of the 16 livestock farmers were returning farmers, while 16 of the 22 plant farmers were new farmers. It makes sense that returning farmers would be more likely to have livestock because of the significant costs of building barns. For new farmers, the start-up costs for a small plant or vegetable farm are much lower than those faced when entering livestock production.

Of the 49 farmers interviewed, 45 said that they worked full-time on the farm and 39 of them derived most of their income from working on the farm; six respondents said that most of their income came from non-farm activities. Given that average net off-farm income was more than three times the amount of net farm income for Ontario farmers in 2013, the number of farmers in this research reliant on farm income is very high compared to the national data (Statistics Canada 2013).

Surprisingly, the returning and new young farmers in our study were just as likely to own land, with about 60 per cent of each group being landowners.[5] Excluding an outlier who owned 2000 acres through marriage, the average number of acres owned for returning farmers was 232 and for new farmers was 56. New farmers were more likely to rent in or

---

[4] Monllor's (2012) "Continuers" and "Newcomers" are comparable to the "Returning" and "New" farmers discussed in this chapter.

[5] To the best of our knowledge, there are no other studies that evaluate land use patterns of young farmers in Ontario or Canada.

sharecrop land than returning farmers. Part of the reason may be that new farmers face greater difficulty finding good pieces of land to buy that are the appropriate size for their business. For new farmers, this can be a challenge because municipalities favour larger farm sizes as a means of retaining farmland. Sharecropping is an alternative way for new farmers to access land as they share management and profits with the landowner but do not have to pay a mortgage or land rental fees. While contradictory to our findings, knowledge of sharecropping opportunities may be more limited for new farmers, given their lack of community networks. In terms of land rental, knowledge of government-owned land for rent is relatively more accessible, which means that the reduction in government-owned acres that are available for rent eliminates some of these opportunities for new farmers. A few young farmers had inherited land and the only ones who had were returning farmers. Returning farmers were, however, more optimistic about the likelihood of inheriting land compared to new farmers, and the size of their expected inheritance was much larger than the expected inheritance of new farmers.

Becoming a Young Farmer

The data collected from interviews paired with the characteristics of Ontario's agrarian landscape suggest a wide diversity in the size and types of farms that exist in Ontario. While commercial agriculture continues to expand and consolidate, there is an increasing number of farmers who run smaller operations and sell specialty products by directly marketing to local customers. This range not only produces different types of farmers but also different pathways into becoming a farmer. It is by understanding these pathways that it becomes possible to create more opportunities for young people to enter the sector. Therefore, this section will draw upon the interviews to differentiate the pathways of new and returning young farmers, highlighting the differences in how they access resources, their motivations for farming, and the type of farming that they carry out.

Becoming a Returning Farmer

Returning young farmers grew up on a family farm and learned to farm by helping their parents with the farm work. Farming became a strong part of their identity at an early age and many returned to farm within a few years of completing post-secondary education. While some pursued diplomas and degrees outside of agriculture, it was common for returning farmers to complete their schooling in a university or college agricultural programme. Typical of this pathway is Rachel,[6] a 27-year-old egg farmer from London, Ontario. Growing up, Rachel worked on her family's egg farm before attending the University of Guelph to study Animal Nutrition. During that time, she participated in the school's poultry club and worked for the university's poultry research facility. After finishing her undergraduate degree, the family rule was that the children were not allowed to come home and farm until they had worked off-farm for a couple of years. This is a common rule among farm families and most returning farmers see the value in taking the time to work somewhere else, "to get experience to make sure that farming is what you want to do." After working at a local feed company for a couple of years, Rachel returned home to work on the family farm. In Ontario, many family farms are incorporated businesses and one way to transition the farm is to have the younger generation gradually buy shares in the business. For Rachel, these shares were a part of the compensation that she received for her work in addition to an hourly wage. Now, Rachel manages the newest egg barn, helps with other parts of the family operation, and spends a lot of time learning at sector meetings, conferences, and through courses.

Not all returning farmers follow a preplanned, organized transition. Mike is a young dairy goat farmer who lives on his father's farm. Growing up, Mike helped his father with taking care of the goats. Despite his father owning most of the farm assets, Mike knew that he would have to take charge of the use of those assets if he wanted a chance at full-time farming. Thus, shortly after finishing high school, he took matters into his own hands:

> It was all me for the decision to switch it over. I said like, I don't want to be 50 years old and have you own it and be just you pay me. I don't like that.

---

[6] Actual names are not used.

I don't see the point in it, to be honest. So, I'm like I want to switch it all over to my name, and he's like, okay well yeah we can talk about that and blah blah blah … Then I phoned Gay Lea (the milk purchaser) and got them to switch the milk cheque to my bank account. And then I went out and got all of the bills that were getting billed to dad, switched it all over to my name and then the next month the milk cheque went into my bank account and then all the bills came to me. He was like, oh that was pretty easy.

Although this process worked for Mike and his family, one of the reasons that it was possible was that Mike's father had become less interested in farming and had begun to establish a career in landscaping. His father also gave him a discount when he purchased the dairy goat herd in lieu of being paid for work that he did on the farm growing up.

Not all transitions are as smooth and quick as they were for Rachel and Mike. For many returning farmers, the transition is often the most challenging part of their career. This was Jessica's experience as she looked to come home and work on her parent's dairy farm. As with Rachel, Jessica grew up working on her parent's farm before attending university. During university, she worked part-time at a local car dealership and once she graduated, she continued to work there for two years. Jessica quit the dealership when she became pregnant and began farming with her husband on her family's sheep, dairy, and cash crop farm. At the time, her father and her uncle owned the farm. Although money was tight, things were going relatively well until one morning, without warning, her father abandoned the farm and left the area. With her uncle on vacation, this left Jessica, her husband, and her cousin to look after the farm. As a result, Jessica and her husband ended up buying a large portion of her father's shares in the business. As Jessica took over the accounting for the business, she knew that she and her husband needed to radically change its focus:

We were doing about 1500 acres of cash cropping. Which at that time … it was not cash cropping. It was cropping and losing money …. I knew what was happening financially and I was the one who had the rapport with the bank and I knew what was coming down the pipe and I said to my partner if we are going to make a go of this then we have to do something now. So, I looked at him and I said, the cows are what's making us money. We need to buy that, we need that.

In the end, Jessica and her husband ended up splitting the business with her uncle and moving the cows to eastern Ontario where they bought a new farm. Despite taking on a significant amount of debt and moving away from their homestead, this was the only way that Jessica could continue her family's legacy as dairy farmers. Although a returning farmer, Jessica's story shows that not all pathways from farm child to farmer are smooth transitions.

One thing that can help to ease the pathways of returning farmers is the collection of risk management programmes that governments offer to farmers. Delivered by an organization called Agricorp,[7] there is a unique risk management programme for cattle, edible horticulture, grains and oilseeds, hogs, sheep, and veal farmers. These programmes are designed to help mitigate against the production and price risks that farmers face. Payments are made when prices fall below the annual support level, defined by average sector cost of production, and producers can choose the amount of coverage that they would like to purchase. As Chad, who is a young returning grain farmer working a large amount of land, put it: "One of the biggest things that is a success story that our government has done is our crop insurance programme for grains and oil seeds. It has been great. It has been a saviour for us. It gives you the confidence year in year out to continue to put a crop in."

While this programme is useful for returning/commercial farmers, it does not mitigate the risks that new, smaller-scale farmers face. This is because these farmers do not reflect the sector average in terms of cost of production. New farmers tend to have a higher cost of production as they do not have the big equipment or barns found in commercial agriculture. As a result, the coverage offered through these programmes is inadequate, given the risk profile that new farmers face. This is just one of their many unique challenges when compared to returning farmers.

---

[7] Agricorp is a crown corporation that facilitates the delivery of a variety of programmes and payments on behalf of the federal and provincial governments to Ontario farmers.

Becoming a "New" Farmer

New farmers do not have a family farm that they were raised on or to which they can return. As a result, they typically learn about and develop an interest in farming later in life compared to returning farmers. Their reasons for entering farming vary, but none of the new farmers interviewed said that they entered the sector because they expected to earn a high financial return for their work. Rather, it was commonly a combination of lifestyle choice and the desire for work that felt meaningful that motivated them to enter farming:

> I got into farming for a lot of different reasons. One of the primary ones is enjoying working outside and another was studying environmentalism and ecological farm systems and having an affinity for what was happening on ecological farms. I was at the University of Guelph studying rural agricultural development and so I was exposed to things that were happening rurally in other countries and didn't know a lot about Canada so I started volunteering on farms here to get a sense of what was happening here and then it kind of evolved from there. I also had some health problems, so I was shifting my diet to more whole foods, more vegetables, and felt like everything kind of coming together in what I enjoyed … Every time I worked on a farm, I felt very content and felt like at the end of the day, I got so many different tasks done that you could just see, and it was more satisfying than other work I had done.—Katrina, young woman farmer

At 28, Katrina (in the above quote) and her husband own and operate a small vegetable farm just outside of Guelph. Together they took out a loan to buy the farm in 2016 and directly market their products to local restaurants and to the members of their community-supported agriculture programme. Prior to buying their farm, they rented land and buildings, keeping costs down while building their customer base. Indeed, it was their customer base that partly enabled them to buy a farm: "Because we had a whole bunch of clients, we actually did some crowdfunding because they already knew that we wanted to farm and so we did that. And that was part of our down payment." However, securing the loan would never have been possible without family support, which provided them with the additional money that they needed for the down payment. Katrina highlighted this as one of the biggest challenges for new farmers: "Buying land, I think that's the number one. Yes, we could secure a loan, but the size of the down payment that is required these days for farms is unrealistic for the average person."

For new farmers, the price of land makes ownership next to impossible. For those who do not inherit land or who do not have a financial benefactor, renting is the only option. However, whether buying or renting, fields in Ontario are much typically larger than that which any new farmer would need, further increasing the cost of entering farming. While landowners that rent out could sub-divide their fields into different sections, most landowners avoid renting small sections to more than one tenant because renting the entire farm to one commercial agricultural producer is easier. Municipalities also prefer land not to be subdivided.

Denise is a 32-year-old new farmer who was able to find a work around for this issue of accessing appropriately sized land parcels. She was able to find a publicly owned piece of land that was used as an incubator facility for new farmers. This land was divided into smaller pieces and was made available to rent through an organization called FarmStart. Although this organization has since dissolved, the property was purchased by another organization, which continues to rent part of the original farmland to small producers such as Denise: "The reason I'm there is that it's a relic property of FarmStart. So, FarmStart established this relationship with them when they (the hospitality organization) purchased the farm property. Basically, I rent directly from them, I have access to land, communal barn space, and then there is a staff member that will do tractor work that I pay for by the hour."

Other new farmers find land to rent from family members, friends, or landowners who have properties in regions with more variable land types (soil, rock, foliage, waterways), with smaller sections of arable land. This fragmentation makes the land less convenient for commercial agriculture. However, finding these rental farms can still be challenging. Stacey, a 33-year-old farmer from Peterborough, describes her experience searching for land:

We were still looking for land that winter after we moved, and nothing was coming up. There is this website here called "Farms at Work" and they try to connect landowners with land-seekers and so we had contacted a couple of people. There were not that many people offering land. But nothing really seemed like quite the right fit and there was no affordable land close enough to Peterborough, although we did look. So, I sort of came pleadingly to my family, my aunt and uncle and cousin. My aunt and uncle farm this land, which belongs to my cousin, and they were like we could probably like find an acre of land somewhere in some field that would work for you.

New farmers must also find ways to learn and develop the skills needed to farm successfully. Many new farmers acquire formal agriculture education through various college and university programmes that tend to focus specifically on non-conventional farming practices. Many respondents found these programmes essential to becoming a farmer as they learn how to farm and also gained access to a network of farmers who operated in this way. Many programmes required internships where students could work on a farm and apply what they had learned in school. In addition to this formal education and work experience, many new farmers continued their training through internships and work placements after graduation. Everdale Organic Farm is one farm in particular that new farmers repeatedly mentioned. For many, Everdale provided them with the training and work experience that they needed to develop the skills to run a successful farm. Stacey enjoyed the benefits that working at Everdale provided:

> I would say the main ways that I acquired my farming skills were (through) my internship. That was like my introduction. I learned the growing and harvesting but not any of the planning at Ignatius (another farm), and then so my real opportunity to learn the planning was at Everdale working as farm staff … that was perfect! Because there I was a farm manager and that gave me responsibilities, but people were there to guide me and help me if I didn't know what to do. I participated in some of the education days that the interns got to do (at Everdale).

While these internships provided training, respondents repeatedly pointed to the fact that they were either unpaid or paid very little. For new farmers looking to start their own operation, this made working as an intern somewhat incompatible with building their own business. Not only was the pay insufficient to supplement the start-up costs of their own farm, but the peak periods of work when they would be busy on their own farms would also be the peak periods when they would be required for their internship. As a result, many new farmers had to work two jobs, on their own farm and as an intern.

Between Returning and New

For some returning young farmers, simply buying farm assets is not an option as the farm (for whatever reason) is not able to bring or keep them on as farm labourers. For these young farmers, it was up to them to find a way to create a full-time farming opportunity for themselves. Consider Natalie, a 38-year-old sheep farmer from eastern Ontario. Natalie grew up on dairy farm in southwestern Ontario, and like many returning farmers, she worked for two years after completing her university degree. After that, she and her husband Ben started to work full time on her parent's dairy farm with the rest of her family. Ben describes the difficulties that this created over the five years that it lasted:

> I started working with her parents and her brother and we moved into a farmhouse as part of our salary. That kind of proceeded for about a year-and-a-half and then we got pregnant and were expecting our first child. Natalie went on maternity leave and made the decision, after more discussion with her family, that there would be opportunity for her there as well. So, she never went back to corporate John Deere. So then, there were five adults full time on the farm, we were milking about 140 Holsteins. And the barn wasn't old by any means; it was only like 10 years old. But we decided that we should milk more cows to have more cash flow to support more families. So, we designed and built a new 220-head barn to milk 240 cows, which we did for a short while. And then, it was family complications, differences of opinions, succession planning was progressing but failing sort of. Communication was not good and there were issues and troubles and people were bottling it up and just working with each other all ticked off. Day in day out, and it just came to a breaking point. And it was our marriage that was suffering immensely, we basically had to …

Well, I had to pick between my wife and her farm. So, that was a no-brainer. And, at the same time, we were told by her parents that there was never any conflict or issues before we came around. Therefore, in their opinion, they saw us as the stem of the troubles. So, we said okay, fine, then we will leave. So, as much as it was devastating and heartbreaking, we walked away from her family farm and there were hard feelings for a couple of years. We saw very little of each other.

For a few years Ben and Natalie worked in the agricultural sector while they looked for a farm that they could purchase. When an affordable farm became available close to Ben's parent's place, they attended the auction, and despite not getting the farm, they caught the attention of an elderly sheep farming couple who were looking to sell their business. Although Natalie and Ben were not new farmers, they were new to sheep. Here is how Natalie described the process:

So, that was April 21st (auction day) and on Mother's Day, the middle of May, we came back here and met with the people who owned the farm and we spent eight hours here you know, going over their books. They opened up everything to us. We walked all 300 acres that were on this farm, they pretty much walked us through the whole business and gave us a pretty good idea of what we would be buying, and you know, what it would be like. And yeah, so, we left here and kind of shook hands and said yes. So, we bought the farm and moved in September of 2012 and the couple that we bought the farm from, kept 70 acres at the back of the farm on the next road and built a house and part of the deal was that they would mentor us. You know, on the business so that we would know because we knew nothing about sheep and when we bought it, it came with 650 sheep. So, they came over here every single day for the first several months ... Into the second year is when they slowed down in coming and we had kind of said to them, you know, we are ready to take our approach. And we made some little changes the first year and then the third year we made a few more changes and we don't really see them at all now.

Natalie and Ben had the opportunity to work on the family farm for five years, but some young farmers never get this opportunity. Maggie grew up on a beef farm in eastern Ontario. After deciding that veterinary school wasn't for her, she graduated from university and began working for a feed company. Throughout her childhood, Maggie had been slowly building up her beef herd, which her parents allowed her to keep on their farm in lieu of paying her for her labour. A few years after graduating, she and her husband bought a farm close to her parents and moved her beef herd over from her parent's place. Unfortunately, operating in commercial agriculture didn't generate enough money for Maggie or her husband to quit their jobs and farm full time. So, after Maggie had her first child, they created a plan to allow at least one of them to become a full-time farmer:

> It was too much to both try and work full time off the farm plus manage the farm and the cow herd and try and raise a child so we decided to put together a plan that would allow the farm to at least be sustainable for itself and then with plans of hopefully bringing one person home and then eventually the second person. So, we put together a plan to start direct marketing our beef so that's how we got started. We started attending farmers' markets, one in Toronto and one in Kingston.

Creating this plan was the first step to creating a farm business servicing a very specific market niche. As Maggie describes it:

> It was interesting because you were obviously able to get a premium for your beef, but it was a really low volume and it was really unpredictable. If it happened to rain and people didn't come ... So, there were challenges, but it was a really good way to get ourselves introduced into the different cuts and what consumers were interested in and really just trying to get our product to match what the consumers were asking for. We did that for a year or two. We were in the one (farmer's market) at Kingston when we had a guy come to our booth and he wanted to know if he could get a sirloin steak and I was like yeah for sure. And then he was telling me how he was going to use this for his appetizer and serve it over the salad. And I thought, oh that sounds amazing, I want to eat at your house. So, he left with his steak and then the vendor beside me was like, do you know who that is? And I said no, I have no idea. It was just a normal guy dressed in normal clothes. And she says: that's this chef at this really high-end restaurant, Le Chateau Noir in Kingston, and I was like oh that's so cool. So, I went home

and told my husband and mom and dad that our beef was going to be sold at this restaurant and that was kind of cool. So, the next week, lo and behold he comes walking back to the booth again and said that beef was incredible, can I get more, can I get this consistently in a larger volume? And I was like sure, this is what I need in a larger volume. That basically started the next evolution of our business and we started direct marketing to mainly restaurants—it really fit our lifestyle a lot better.

Maggie's opportunity to farm was a consequence of her upbringing, but it was her individual creativity and drive that allowed her to create a farm that sells its beef to restaurants all over Ontario and farm full time.

Another unique returning farmer was Jared, a 33-year-old pig farmer from west of Guelph. Unlike most returning farmers, Jared disliked farming: "Growing up on the farm, I wanted nothing to do with it, I hated everything about it, and I couldn't get out of there soon enough." He did, however, enjoy agriculture and decided to pursue a business degree that focused on the sector at university. However, after graduating, he came to realize that this wasn't the life he wanted either: "It took me like eight months of sitting in an office to be like, well, this isn't going to work. Because I hate sitting still and I couldn't handle an office mentality where it was like I'm done work at three, why can't I go home. I'm just going to sit here staring at my screen and I thought to myself, you know, this is asinine."

After a year of work, Jared returned to graduate school where he studied agricultural economics. It was then that he decided that maybe he would be interested in taking over his parent's pig farm. After graduating, he returned home, but it did not go as planned:

I started working for Mom and Dad in 2010 and that was kind of in the midst of a brief pause of an otherwise very terrible five years in pig farming. So, we started expanding my parents' operation in 2012 and that did not go well. It was still all owned by them, but I was driving expansion. This was my first real life experience where an Excel model didn't work in a real-life barn … So, when the expansion didn't work out, I kind of pulled right back and I did some travelling and I ended up getting approached by an Ontario political party who asked to run for them in the upcoming election.

At this point Jared had gone from hating farming to driving the expansion on his family's farm to running to be a member of Ontario's parliament. Although Jared didn't win, he did become a political advisor for the province's Minister of Agriculture for a couple of years while also beginning to buy some of his parent's farm assets from them. Then a shift happened:

> Over time we decided that I wasn't going to run in the upcoming election and that I would exit from politics because it was not sustainable to be commuting to Toronto. We basically started looking at what our goals were for household income and in the meantime, I was able to find a market where we could sell niche pigs. This was to sell certified humane pork to a packer in the US. Traditionally, you would have to go through a feed company to find those contracts, but I border on being too stubborn and I hate being told where I have to buy my feed. So, I was able to get a direct contract with the processor and through this I was able to find a full-time role for myself on the farm. I had to do this because if I was going to come back to farm full time, we had two full-time employees and I didn't want to let either of them go. Beyond the fact that I didn't want to let them go, working in a barn full time isn't my cup of tea.

In addition to buying his parent's commercial pig herd, Jared now successfully manages an organization of about 20 other farmers that he contracts to grow certified humane pigs for him. Most of these farmers were already following the standards for certified humane pork but were selling their pigs into the commercial agricultural market. By working with Jared, these farmers were able to earn a premium for their product and Jared was able to earn a living by creating a business based on his ability to facilitate this connection. Although a returning farmer, Jared took a unique route to becoming a full-time farmer himself.

Conclusion

Young people face significant challenges in becoming and being a successful farmer. While many returning farmers already have a farm in place to help springboard their career, these young people need to navigate family relationships, inheritances, and fairness while transitioning the farm from one generation to the next. Farms are also changing, and returning young farmers need to create new opportunities to ensure that the business stays viable, competitive, and relevant to what consumers are seeking.

New farmers face different challenges. For them, it is difficult to acquire both the physical resources and technical training needed. Most begin as interns on other farms where they gain valuable experience but do not make enough to save up and start their own farm. As a result, many new farmers rely on a benefactor to help them get started. When buying is not an option, they must seek out rent or sharecropping opportunities, which can be difficult to find.

These differences highlight the significant diversity in the way that young people farm across the province. Supporting these different farmers will require targeted policies that suit their specific needs. Young farmers are looking for opportunities: an opportunity to learn, an opportunity to find land and equipment, and an opportunity to try farming for themselves. Policies should ease the capacity of young farmers to take advantage of potential opportunities, which in turn means understanding their needs and the structural challenges that different young farmers face.

List of Primary Sources (not actual names)

Chad, male, 33 years, married, returning, grain farmer, runs 4500 acres.

Denise, female, 32 years, married, new, flower farmer, sells directly at markets.

Jared, male, 33 years, married, returning, pig farmer, owns pigs and runs certified humane pig business.

Jessica, female, 37 years, married, returning, dairy farmer, 100 cows.

Katrina, female, 28 years, married, new, vegetable farmer, sells to restaurants and through her Community-Supported Agriculture (CSA)

Maggie, female, 37, married, returning, direct markets beef to Ontario restaurants.

Mike, male, 22 years, single, returning, dairy goat farmer, 350 milking goats.

Natalie, female, 38 years, married, returning, sheep farmer, 1000 ewes.

Rachel, female, 27 years, single, returning, egg farmer, manages part of operations on family farm.

Stacey, female, 33 years, married, new, vegetable farmer, sells at market.

References

Government of Ontario. 2019. *About Ontario.* https://www.ontario.ca/page/about-ontario. Accessed 30 Sept 2019.

Monllor, N. 2012. *Farm entry: A comparative analysis of young farmers, their pathways, attitudes, and practices in Ontario (Canada) and Catalunya (Spain).* http://www.accesstoland.eu/IMG/pdf/monllor_farm_entry_report.

Ontario Ministry of Agriculture, Food, and Rural Affairs. 2018. *Agri-food industries: GDP, sales, employment summary, Ontario.* http://www.omafra.gov.on.ca/english/stats/economy/gdp_agrifood.xls. Accessed 1 Oct 2019.

Statistics Canada. 2013. *Total and average off-farm income by source.* Table 32-10-0057-01. https://www150.statcan.gc.ca/t1/tbl1/en/tv.action?pid=3210005701&pickMembers%5B0%5D= 1.7. Accessed 5 Oct 2019.

―――. 2016. *Total number of farms and farm operators.* Table 32-10-0440-01. https://www150.statcan.gc.ca/t1/tbl1/en/cv.action?pid=3210044001#timeframe. Accessed 1 Oct 2019.

―――. 2017. *Income of individuals by age group, sex and income source, Canada, provinces and selected census metropolitan areas.* Table 11-10-0239-01. https://www150.statcan.gc.ca/t1/tbl1/en/tv.action?pid=1110023901. Accessed 31 Jan 2020.

―――. 2018. *Value per acre of farmland and buildings at July 1.* Table 32-10-0047-01. https://www150.statcan.gc.ca/t1/tbl1/en/tv.action?pid=3210004701. Accessed 30 Sept 2019.

5

Young Farmers and the Dynamics of Agrarian Transition in China

Lu Pan

Introduction

For a transformative country like China that still has a large rural population and significant agricultural sector, the role of young farmers in the
future of the country's agrarian transition has extraordinary importance.
The social awareness of young farmers and their important role varies in
China in the last decade depending on the dynamics of agrarian change
and socio-economic transformations. This issue is particularly relevant in
the second decade of the twentieth century when the government is
actively encouraging young people to commit to agriculture and rural
revitalization in China; farming families are in a period of increased
opportunity but not without related challenges. This chapter begins with
a brief review of agrarian transition in China before exploring the general

L. Pan (✉)
College of Humanities and Development Studies (COHD),
China Agricultural University, Beijing, China
e-mail: panlu@cau.edu.cn

© The Author(s) 2024
S. Srinivasan (ed.), *Becoming A Young Farmer*, Rethinking Rural,
https://doi.org/10.1007/978-3-031-15233-7_5

situation of rural youth in terms of education, vocation, and social mobility in order to provide the reader with a comprehensive background of the situation for young farmers in China. Before concluding, the final section will address the demographic challenges of Chinese agriculture.

Development of Chinese Agriculture

In contrast to the large farm sizes in North America and Australia, and the comparatively middle-size farms in Europe and central Asia, farming in Southeast and East Asian countries can be characterized as smallholding agriculture. Small-scale family farming in China came into being as an adaptation to the high population density and relatively scarce agricultural land resources (Zhang 2011). It has remained the dominant organizational form of agricultural production for over hundred years and has been specifically strengthened since the 1980s. In 2016, family farming accounted for nearly 97 per cent of all farming units, overwhelmingly surpassing large-scale farms that employ hired labour (Ministry of Agriculture and Rural Affairs of China 2017). On an average holding of 5 mu (about one-third of a hectare) of land, Chinese agriculture has exceptional performance and produces about one-third of the world's grain, one-sixth of its wheat, and one-fifth of its maize (Wang 2013). Chinese farmers have not only secured domestic food supply and contributed to economic growth but also significantly contributed to the food export market. Agro-products exports in 2020 were over US$76 billion and mainly went to Japan, Korea, Vietnam, the USA, and some other east Asian countries (Ministry of Agriculture and Rural Affairs of China 2022).

Contemporary Achievements of Chinese Agriculture

In the history of modern China over the last 100 years, the basic and primary operative unit of farming has always been the rural household. Small-scale family farming has demonstrated the centrality of the family in agricultural production and its embeddedness in local society. For agricultural households, the ultimate objective was "health and wellness of whole family, income is only an intermediate objective" (Soda 2003). In a word, subsistence was a family's priority. In realizing this goal, a series of features and merits arise that are associated with family farming, including biodiversity, resource efficiency, co-production with nature, multiple jobholding, and a resilient community. The dynamics and agency of peasant agriculture and rural society have been especially reactivated after 1978 when agriculture collectivization was abolished and economic reform in rural China was piloted. The result was great transformation and changes in agriculture and in the countryside. Grain production, for example, increased from 3.05 million tons in 1978 to over 669.49 million tons in 2020 (National Bureau of Statistics of China 2021a). The net income of rural residents has risen from CNY[1] 171 per capita in 1978 to CNY 32,189 in 2020, which contributed to poverty alleviation in many parts of the country (Central Government of China 2021c).

The state has played an important role in delivering public goods and services for agriculture development. In 2011, fiscal expenditure on agriculture and rural development by central government was CNY 2972.72 billion (Ministry of Finance of China 2012) and the annual amount reached over CNY 3000 billion in recent years (*People's Daily* 2021). State support to agriculture is also reflected in taxation and subsidy improvements. Since 2004, the government has annually increased direct subsidies that support grain production. Each farming household can be subsidized according to their land area of cultivation. In 2006, the government took the historic step to abolish the agricultural tax—farmers do not have to pay tax for agricultural products.

The increases in income from farming and agricultural activities and the aforementioned economic growth in the countryside are closely associated with the occupational differentiation of the rural population. Before the 1980s, the Chinese countryside was a peasant society, with characteristics as defined by Shanin (1990). They were the small agricultural producers who, with the help of simple equipment and family labour, produced mainly for their own consumption and for the fulfilment of obligations to the holders of political and economic power. Rural

---

[1] 1 CNY equals about US$0.15 at the time of writing.

society was quite homogenous, especially at the end of agriculture collectivization in the 1970s in terms of wealth and occupation. It was only after 1978 and during a period of increasing rural-urban economic interaction that these peasants were able to find employment beyond the agricultural sector and outside their villages or counties. The consequences were the social mobility of the individual and the socio-economic differentiation of rural society. In China, the term "peasant" (*nongmin*) has combined meanings beyond the academic debate, and simply, it ascribes identity to people who were registered as rural inhabitants in the *hukou* system.[2] The majority of those who live in rural areas are still engaged in agricultural production. They are farmers, though if one examines their specific mode of farming, some can still be classed as peasants, while others are commercial farmers. Depeasantization in rural China witnessed on one hand those who used to work on land for subsistence constantly flowing into urban areas and on the other hand the mode of farming that relied on family-controlled land and labour for subsistence-oriented production disappearing.

Some scholars attributed the agricultural growth between 1980 and 2010 to fertility rate decline, large-scale non-farm employment, urbanization, and changing patterns of food consumption related to rising incomes. Philip Huang (2016) concludes that an enormous mass of small family farms, rather than capitalist farming enterprises, have led a "hidden agricultural revolution." It is "hidden" because its growing production value is largely driven by the switch from grain production to increasingly higher-value agricultural products like meat-poultry-fish, milk-eggs, and quality fruits and vegetables. It is so-called new agriculture because it is both capital and labour intensive. Changes can be seen in the increased area dedicated to cultivation and higher-valued agricultural products' production value, for example, from vegetables and fruits. The cultivation area for the latter has increased by 606 per cent and 680 per cent, respectively, between 1980 and 2010, accounting for 18.9 per cent of 2 billion mu[3] of cultivated land nationally in 2010. In the same

---

[2] The hukou system has been implemented since the 1950s in China as a tool for social control that artificially differentiates people into rural and residential registration. Rural residential registration, or rural hukou, denies farmers the same advantages and rights as those in urban areas.

[3] Mu is a Chinese unit to measure land area and 15 mu equals 1 hectare.

year, the production value of vegetables, fruits, meat-poultry-fish, and milk-eggs accounted for 66 per cent of total agricultural production value in China, compared to 15.9 per cent for grain production (Huang 2016).

Despite the remarkable achievements in agricultural production in China since 1978, there remain many challenges for farmers and for agriculture more generally. Rapid urbanization over the last decades has not only attracted millions of capable rural labourers to cities but also created competition for arable land. Less than 30 per cent of the total population lived in counties and cities in the 1990s (Jian and Huang 2010), while in 2020 it reached 64.72 per cent (Central government of China 2022). Urban expansion is predicated on the availability of arable land, which has shrunk by over 7.53 million hectares between 2007 and 2018 (Central Government of China 2021b). According to the *Communiqué on Land and Resources of China 2015*, there were 0.135 billion hectares of arable land in that year, a figure that is expected to drop to 0.12 billion hectare by 2030 (Central Government of China 2016). This ongoing decrease in the availability of arable land is a key constraint for agriculture nationwide. Climate change and global warming are also challenges for farmers in China and around the world. Crops and regions are impacted differently, but overall, the result is production losses and potential instability of food production. Experts predict that food production could be reduced by 10 per cent and the chance of instability increased by 15 per cent by 2050 (Pan et al. 2011). The loss of arable land and the unstable climate foreshadow a severe threat to food security and agricultural sustainability in China. Such challenges require the institutional protection of farmland as well as innovative technologies and farming methods to cope with these realities.

Transition Towards Modern Agriculture

Alongside the consolidation and growth of peasant agriculture, there are other tendencies in China's agriculture. For decades, Chinese authorities have been promoting a "modernization of agriculture" that has created tensions as well as opportunities for small-scale family farming. Modernization is generally assumed to entail a double process: a large part of the agricultural labour force moves from the countryside to the cities and there is a simultaneous restructuring of agriculture since the work is now done with fewer people. This modernized agriculture is far more integrated with the wider processes of capital accumulation than peasant agriculture. According to van der Ploeg and Ye (2016), the integration of modern agriculture with capital accumulation usually occurs through: (1) increased indebtedness; (2) greater use of external inputs and new, more sophisticated technologies; (3) delivery to, and increased dependency upon, food industries and large retail organizations; and (4) state taxation. All of these factors result in a profound repatterning of farming practices. It is widely assumed that a new model of "entrepreneurial farming" or a model of "capitalist farming" in China is likely to become dominant. These processes clearly reconstitute farming on a new basis: land and labour are converted into commodities, farming is increasingly grounded upon multiple commodity flows, and the units of production become part of overarching and complex financial operations (van der Ploeg and Ye 2016).

In the last two decades, China's agrarian transition has occurred in two stages: first, the commoditization of agricultural production and the reproduction of farming households and, second, the organization of agricultural production beyond the household boundary (variously known as vertical integration, industrialization, and scaling-up) (Zhang and Donaldson 2010). Since 2000, the fixed capital stock in agriculture has risen from CNY 484.013 billion to CNY 1448.49 billion in 2011. Before 2007, the growth rate of fixed capital stock in agriculture was below 10 per cent and then increased to 12 per cent after 2007, reaching 15.37 per cent in 2009 (Luo 2013). The capital-labour ratio in agriculture has increased from CNY 480 per capita in 1990 to CNY 670 per capita in 2000 and CNY 1670 yuan per capita a decade later (Luo 2013). This progression implies that capital is becoming increasingly important in China's agriculture. Capitalization in agriculture is also reflected in the acceleration of land transfers. Transfers of farmland increased from 67 million mu in 2007 to 471 million mu in 2016, accounting for 35 per cent of total area of rural households that are a part of the contract farming system (Han 2016). The involvement of industrial capital in agriculture is viewed by some as a significant outcome of policy incentives. In

2015, the *No. 1 Document* of the Chinese government[4] clearly states that holders of commercial capital should be encouraged to participate in entrepreneurial agriculture, processing and circulation of agro-products, and socialized service in rural areas.

It is clear that all of these changes are the consequence of the political steering of modern agriculture since the country's founding in 1949. In 2007, the government's *No. 1 Document* comprehensively illustrated agriculture modernization, that is to "equip agriculture with modern materials, to reconstruct agriculture with modern science and technology, to upgrade agriculture with modern industry, to promote agriculture with modern forms of management, to lead agriculture with modern ideology of development, to cultivate new farmers to develop agriculture." The 18th National Congress of the Communist Party of China (NCCPC) in 2012 also put forward the integration of agricultural modernization with urbanization, industrialization, and informatization. In a word, the way to modern agriculture is the process of reconstructing "traditional" agriculture and changing its growth pattern. This process implies reduced space for the vast number of family-based smallholders, especially for rural youth who have the potential and willingness to become involved in agriculture. As Li (2017) shows in her research on oil palm expansion in Indonesia, there are intergenerational effects of large-scale agriculture—young people will face constricted access to land and deteriorating rewards for their labour.

Accelerated Farmland Transfer

Land reform in China was completed after 1949; all arable land was distributed among peasants who obtained use rights according to the number of people in one's household. Village collectives periodically redistribute land use rights to guarantee the rights of those who have not transferred their residence away from the village. Such a multi-functional

---

[4] The *No. 1 Document* is the Chinese government's first official document to illustrate the overall political goals and key development issues. Between 2004 and 2017, the *No. 1 Document* was published annually with the aim of closely examining agrarian issues important for agriculture, peasants, and the countryside.

right naturally created a rationality that helped to absorb the cost of external risks through mechanisms within the villages (Houtart and Wen 2013). This was soon followed by government efforts to develop large, collective operations, and by 1956, most of China's agricultural production was done on a collective basis. However, the collective farms drew on many organizational features of the family farms that they brought together. Some two decades later, the family farms, recreated by the division of communal land, have also reacquired an independence in decision-making that has become more robust in subsequent years (Brookfield 2008). In 1978, the government began to decentralize agricultural production from the commune system to individuals and farm households. By 1984, more than 99 per cent of production units had adopted the Household Production Responsibility System (HRS). Under the HRS, rural households do not have ownership of land, instead they have land-use rights and the freedom of decision-making on major production and marketing activities (Fan and Chan-Kang 2003).

The first round of rural household contracting under HRS in most regions of China started in 1983 and ended in 1997. In that year, the government issued related policies to prolong the previous contract period for rural land use by a further 30 years. The second round of contracting period extended from 1998 to the close of 2018 and now it is in the third round of 30 years since 2018. The Rural Land Contract Law issued in 2002 affirms this contract period of 30 years for farmland; the period for grassland and forestry land is even longer.

Regardless of the contract period, there is always conflict between shifting family size and comparatively fixed land arrangements. In order to avoid fragmented farmland adjustments to counter population change and stabilize investment in farmland, Meitan County in Guizhou province initiated an institutional arrangement that farmland contracts for each rural household did not change with an increase or decrease of family member,[5] that is, farmland size would not change when a family member marries and leaves the farm, is born, dies, or there are residential changes. Since 1993, the government has gradually institutionalized and legalized this arrangement in order to stabilize farmland contracts and

---

[5] This refers to the so-called *zeng ren bu zeng di, jian ren bu jian di.*

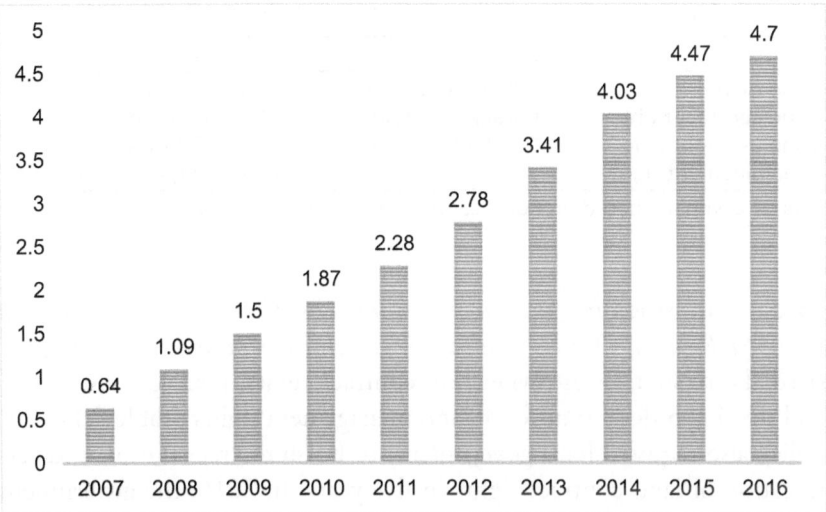

Fig. 5.1 Area of farmland transfer by year (unit: hundred million mu). (Source: Tuliu Net 2021)

agricultural production (Shao 2015). This means that young people who were born after 1998 do not receive a farmland allocation through the village collective. For these young people, the only way to access farmland is to sub-contract land from others or to work on their family land (Fig. 5.1).

Informal land transfer between rural households can be traced back to the 1990s when rural labour began to migrate and agricultural tax remained a heavy burden on rural people. Such land transfer was spontaneous, informal, and low cost. Households that rented out land could retrieve their land whenever necessary. Households who rented land paid very low rent, sometimes nothing. Rural households rarely have their rented land adjacent to their own land to make a larger plot, and therefore their operational scale is usually not very large (Tan and Sun 2014). Peasants' spontaneous land transfers gradually accelerated an institutional update on the right to a land contract. The Rural Land Contract Law, issued in 2003, regulates land contracts for a variety of transfers, including sub-contract, rent, exchange, and transfer. The village collective gradually lost its rights in the arena of land readjustment. In 2014, the

Table 5.1 Separation of three rights on farmland in China

	Before 2014	After 2014
Land ownership	Village collective	Village collective
Land contract right	Rural household	Rural household
Land use right/land management right	Rural household	Individual, household, enterprise, corporate, etc.

Source: Compiled based on Rural Land Contract Law of China

government issued *Instructions on Leading Orderly Transfer of Rural Land Contract Right to Develop Scaled Agricultural Operation*, which emphasized the separation of ownership, contract rights, management rights, and steering orderly transfer of land management rights (Table 5.1). This policy also required local governments to finish the registration and certification of land contract rights in five years. In 2016, the government again issued a policy to improve the separation of these three land rights. Theoretically, rural households are free to dispose of their land contract rights according to legal regulations. Some scholars argue that registration and certification of land contract rights would strengthen peasants' capacity in disposing of their rights and not definitely accelerate transfer of land contract rights (Luo and Li 2014). However, many are concerned that registration is just a disguise for land privatization and its aim is not pro-peasants but to promote the concentration of farmland for scaled operation (He 2015).

From official discourse, the promotion of land transfer highly conforms to the state's pursuit of modern agriculture. In the *Instructions on Leading the Orderly Transfer of Rural Land Contract Right to Develop Scaled Agricultural Operation* (Central Government of China 2014), it was clearly stated that:

> … land transfer and moderate scaled operation are the inevitable path to develop modern agriculture and are in favour of optimizing allocation of land resources and raising labour productivity, in favour of guaranteeing food security and major products supply, in favour of promoting technological extension and increasing peasants' income.

In general, rural land transfer won wide support from different levels of governments and scholars who believe that land transfers will help to create agricultural management efficiencies, guarantee food security, and promote labour migration and local employment of rural labour (Ye et al. 2016).

In recent years, the rate of land transfers has been accelerating. In the 1980s, household-based land contracting accounted for 95 per cent of the total farmland and only 8.2 per cent of rural households had sub-contracted land from others. In 1999, only 2.53 per cent of total farmland was transferred and the ratio was 4.57 per cent in 2006, 8.6 per cent in 2008, and 17.80 per cent in 2011. As stated earlier in the chapter, transfers of farmland increased from 67 million mu in 2007 to 471 million mu in 2016, accounting for 35 per cent of total area of rural households that are a part of the contract farming system (Han 2016).

Most land transfers are short term for a period of less than 5 years. In Heilongjiang province, for example, among the 22.54 million mu of transferred land, 71 per cent was granted on a one-year lease and only 6 per cent was leased for a period of five years or more. Usually transfers of land for recreational agriculture and perennial cash crops have a longer period of transfer, some for up to 20 years (Lu and Chen 2015). Most of the land transfers are negotiated without duress, but there are problems with the system, including forced transfers, changing the purpose of land use (e.g., convert land for construction), altered cultivation structure after the transfer (from grain crops to non-grain crops), and unequal distribution of land revenue for rural households (Han 2012).

Emergence of New Entities in Agricultural Operation

In the governmental pursuit of modern agriculture, one important strategy is cultivating new actors/bodies suitable for scaled farming and market-oriented agriculture. The government's *No. 1 Document* (Central Government of China 2013) explicitly indicates that "specialized large-holders and family farms should be supported and promoted through favourable policy and legal environment and subsidies and grants." The *No.1 Document* in 2014 pushed further in that direction, encouraging

the establishment of farmers' cooperatives through specialized cooperation, joint share cooperation, and so on. In so doing, government financial programme funding can be invested in qualified cooperatives and
local government and private capital are encouraged to setup financing
guarantee companies to provide loans to new agro-operators.
Governmental subsidies have favoured family farms and large holders of
land in recent years. For example, in 2016, a farm over 200 mu for vegetable and fruits production qualified for a CNY 5000 per mu subsidy
from the Ministry of Agriculture. Agri-businesses, agricultural demonstration bases, and the other processing businesses could receive various
subsidies from the Departments of Agriculture, Finance, Poverty
Alleviation, and others (Tuliu Net 2016).

Among the promotion of new agricultural operating entities, the "family farm" is the most controversial. With thousands of years of the family
farming tradition, the central government officially proposed the term
"family farm" in 2008 during the third plenary session of the Seventeenth
Central Committee and encouraged to develop such entity of agriculture. As a consequence of this policy incentive, the definition, identification, and registration process of a family farm became a focal point in
academia and practice. The Ministry of Agriculture defines in 2013 the
family farm as "a new type of operative entity in agriculture that is mainly
based on family labour to pursue large-scale, intensified and commodified agricultural production and operation." Through this definition,
politicians have been promoting the "family farm" as a new entity that is
meant to be largely separated from existing family farms. Extensive academic discussion developed in the country and almost all scholars who
are pro "family farm" agree that family farms should be a legal entity like
any other business or enterprise and subject to marketization and modernization. They argue that the "family farm" should be characterized as
a family operation of moderate scale, operating as part of the market and
under entrepreneurial management. In terms of production factors,
labour, and product, such "family farms" are very close to corporate
farms. The fundamental difference between them is that the former
depends more on family labour in production and operation. The essential difference between "family farms" and empirically existing family
farming households is that the former completely participate in market

exchange with their specialized commodity production, that is, marketized operation.

Although the government has emphasized the term "family farm" in various policies and documents, it is not uniformly defined. Standards and criteria for identification and registration of a "family farm" are inconsistent in different provinces, municipalities, and even counties. In March 2013, the Ministry of Agriculture in China carried out its inaugural national survey on the development of family farms. Farming units satisfying the criteria summarized in Box 5.1 qualified as a "family farm."

The survey results show that 877,000 "family farms" were working 13.4 per cent of contracted farmland in China in 2013. The average working labour on a family farm was 6.01 persons, among which long-term hired labour accounts for 1.68 persons. Most family farms specialize their production and operation in either crop cultivation or husbandry; only 6 per cent engage in a diversified operation. The average production scale of the surveyed family farms is 13.3 hectares, nearly 27 times the national farming land scale per household (see Table 5.2).

Box 5.1 Criteria for "Family Farm" from a Ministry of Agriculture Survey

- Operators of the family farm should be registered as rural residents (hukou).
- Family labour dominates, no long-term hired labour; the amount of hired labour does not exceed family labour.
- Income from agriculture comprises the major source of income for the family and net income from agriculture accounts for over 80 per cent of the farm's total income.
- Scale of production reaches a certain standard and remains stable. Size of farmland (with a contract period longer than five years) in grain production should be over 3.33 ha (double cropping) or over 6.66 ha (single cropping). For farms with cash crops, husbandry, or both modes of production, operational size should reach the standards that local departments of agriculture set.
- Operators of the family farm should have received technical training in agriculture.
- Family farm should have complete financial records.
- Family farm should have a demonstration effect to other farmers and agricultural households.

Table 5.2 Scale of officially identified "family farms" in China

Scale (unit: ha)	Amounts (unit: 10,000)	Percentage (%)
<3.33	48.42	55.2
3.33–6.66	18.98	21.6
6.66–33.3	17.07	19.5
33.3–66.6	1.58	1.8
>66.6	1.65	1.9
Total	87.7	100

Source: Ministry of Agriculture and Rural Affairs (2013)

Based on government policies and the criteria used in the 2013 national survey, it is obvious that the politically promoted "family farm" is, in essence, a capitalized family farm. Its capitalist features imply a predatory impact on the livelihoods of many small-scale farms. The promotion of large-scale "family farms" is based on the premise of land transfer. In the demonstration area for "family farms" in the Songjiang district of Shanghai, for example, 99.4 per cent of farmland had been transferred to "family farms" by 2011 (Chen 2013a). Such politically directed transfers to a minority who hold large amounts of land disrupt the land transfer that ordinary farming households need to operate profitably. The unequal power relationship between large holders of land (who are usually social elites in rural communities) and other family farmers makes the latter vulnerable in contract negotiation. Additionally, the normalization and long-term (usually 5–10 years) nature of land transfer contracts that the "family farm" registration requires imply that rural households who contract out their land use right cannot abort the contract even if their basic livelihoods are urgently endangered. Land concentration to "family farms" and those farms' preference for capital-intensive farming is exacerbating the employment issue for rural labour in the contemporary period of intensive urbanization in China. The large-scale nature of capitalist farming should not be the future of Chinese agriculture.

Rural Youth: Hovering Between Rural and Urban

Young people are the key actors and an important variable in the ongoing development of both the countryside and agriculture. The negative impact of mass labour migration on the countryside is obvious, especially in terms of social cohesion and collective action in community development. The absence of young successors in agriculture is leading to an increasing concentration of land by large holders as well as to a different trajectory for Chinese agriculture. Most young people with a rural background do not want to remain in the countryside or work in agriculture, but they still live in marginalized conditions after they move to cities. As rural-urban interaction has become the most important feature of China's societal transformation, rural youth are floating between rural and urban societies throughout their education and as they begin their working lives. Rural-urban integration offers prospects to rural youth, but also problems and challenges.

Educational Constraints for Social Mobility

For young people in China, education is the most important path for social mobility. It is particularly true for children in rural families who lack the social and financial capital required to secure a decent job in the labour market. Since the 1980s, the government's development of the education system, especially post-secondary education, has offered increased opportunities for young people to study and realize their social mobility. Since the Ministry of Education issued its college enrolment expansion plan in 1999, the enrolment rate has leapt from 12.5 per cent in 2000 to 48.1 per cent in 2018. This expansion has very important and positive implications for rural youth. According to 2018 Ministry of Education statistics, over 60 per cent of the nearly 40 million college students are from rural areas (Ministry of Education of China 2019). This investment in human capital is just one step in remedying the rural-urban difference in education. Considering the larger rural population than the urban and the government's urban-biased educational resource allocation, rural youth face more difficulties in order to access a good education. Research reveals that in 2018 the rural workforce received, on average, 9 years of schooling compared to 11.3 years for the urban workforce (*Workers' Daily* 2020). It is widely recognized that rural youth face increased challenges to achieving social mobility through education.

Rural youth who cannot enrol in post-secondary studies at the college level find it similarly difficult to receive high quality vocational education in agriculture or non-agricultural disciplines. The development of agricultural vocational education has evolved through three stages since 1949. The core policy objective in the first stage was to promote the development of rural vocational education to restore and develop agriculture. The second stage—1978 to the end of the 1990s—was a period when the state attached great importance to economic growth and modernization. Vocational education in this period focused on cultivating professional talents to accelerate overall economic growth of rural areas, not only to develop agriculture but to cultivate farmers with new skills for rural development (Xie 2010). Since the mid-1990s, agricultural vocational education has experienced a serious decline. In 2007, there were 141 agroforestry technical secondary schools, a drop from 365 in 1987 (Tian 2010). The enrolment of students in the agroforestry discipline accounted for only 3.41 per cent of all students enrolled in 2007 in a secondary vocational school. Many of these schools faced a myriad of problems including the loss of teachers and students and poor teaching conditions (Xia and Peng 2004). Outside of vocational education, technological training for the rural population is limited. In 2008, only 20 per cent of farmers received short-term training and even less, only 3.4 per cent, received basic vocational training in agriculture (Xu and Wang 2009).

Failure within the education system is an important force driving rural youth migration. When these young people finish (or drop out of) senior (or even junior) high school, they often follow their parents or other family members to the city to find employment. It is illegal for an employer to hire children under the age of 16, and as such, many rural children need to wait in the village for this birthday to secure the needed ID card that will facilitate their job search. For rural youth who have the chance to attend college, many have negative perceptions of agriculture and are reluctant to return to the countryside to work on the land. Throughout their education, the mantra of their teachers, and especially their parents, has been the goal of securing a non-farm job with a stable income in the city. The government has issued some national policies in order to encourage college students to return to and serve in the countryside, including the Opinions on Guiding and Encouraging College Graduates to Work in Grass Roots (2005). Local governments followed suit but the outcome was disappointing. Even for graduates of agriculture-related disciplines or universities, few students have the intention to work in a rural area, even though it is a larger percentage than those who attend non-agricultural universities or study-related disciplines—55.7 per cent versus 23.54 per cent (Fei and Wang 2013). In a survey done by Chen Shensheng (2013b) with 398 college students who worked in agri-business, 67.6 per cent did not understand agricultural production or national policies on agriculture; 38.1 per cent were unfamiliar with the social economic context about agriculture and countryside; and only 49.6 per cent had participated in production activities in a rural area. College students who work in agri-business are usually paid low salaries—85.37 per cent earn an annual salary that is less than CNY 50,000. Only 58.79 per cent of the students work in a position that is related to their discipline. Aside from the subjective enthusiasm of college students and youth, material guarantees and room for career advancement are equally important in order to attract young people into agriculture-related fields (Chen 2013b).

Rural Labour Migration and New Generation of Migrants

China's economic restructuring, a key phase in its development, initiated the outflow of rural labour in the late 1980s. Before the reform and opening up, the government strictly limited any spontaneous flow of people under urban-rural dualism which implies two different systems of social policies in rural and urban areas. This was gradually cancelled after the reform to facilitate labour provision for urban industries. In 2020, there were 285.6 million rural labourers working in non-agriculture sectors, among which 169.6 million were migrant workers (Central Government of China 2021a). The mid-west provinces such as Hunan, Sichuan, Jiangsu, Henan, and Guangxi are the main sending areas. The destination cities are mostly large and medium-sized cities. Due to their low level of education and lack of skills, most of the rural labour migrants are engaged in the labour-intensive secondary and tertiary industries (Zhang et al. 2004).

Incentives for rural labour migrants are complicated and diverse. Income gap between urban and rural, the adjustment of industrial structure, public policy, the urbanization process, and other factors all have important impacts on the flow of rural labour. Migration has evolved to be almost a "rite of passage" that most rural youth, male or female, encounter on their path to adulthood (Ye et al. 2014). Many migrants and rural households gradually form their life course around migration—leaving their home when they are young in search of employment and returning to the countryside when they are old as a farmer, wage worker, or business operator (Li 2012).

As rural migration continues without abatement, this population is also experiencing intergenerational transition. Scholars suggest that this population could be divided into three generations at an interval of 15 years. The older generation (over 46 years old), middle generation (31–45 years old), and the new generation (16–30 years old) are coexisting (Duan and Ma 2011). According to *Report on Monitoring and Investigation of Rural Migrant Workers in 2020* (Central Government of China 2021a), rural migrants below 40 years old accounted for 49.4 per cent of rural migrants, indicating that young people have unsurprisingly become the majority within this population. A similar result was found for rural youth's social mobility.

Young migrants then quickly became a key focus for research in youth studies and urban studies. Studies have explored all aspects of this new generation of rural migrants: work and employment, social welfare, social integration in the city, lifestyle and consumption, marriage, and crime, among others. Some researchers summarized the characteristics of this group as low occupational reputation, low income, low social security, low standard labour time, low identity recognition, low level of employment, and poor living conditions (Yang 2010a). Researchers also report several differences between this group and the older generation of rural migrants. For example, the new generation are more familiar with cities and urban areas and when compared to their parents, have greater ambitions to relocate permanently (Pun et al. 2009). They are not migrating for subsistence as their parents did but migrating for their future (Yang 2010b). Work performance also differs. For example, the younger generation is unable to adapt to intensive labour and harsh working conditions or frequent job changes, which increase tensions in the work-place, especially in factories and other blue-collar positions (Wang and Huang 2014). Despite difficulties in the city, the new generation is reluc-tant to return to the countryside. A survey shows that only 8 per cent of young rural migrants considered themselves to be peasants (as defined earlier) and most of those surveyed felt no emotional attachment to land or to agriculture (Yang 2010a). Another study found that about 80 per cent of young rural migrants did not plan to return to their home county, let alone their home village (Duan and Ma 2011). This situation can be defined as "dual disembededness" as some scholars argued—disembed-ded from rural society on one hand and disembedded from the labour regime in the city on the other. Such dual disembeddedness is unstable and the future of rural youth remains a critical issue in the development of China's cities and its countryside.

Split Family Reproduction for the Young Generation

The reproduction of the rural family and the population left-behind, as they are called, are two casualties of the outflow of rural labour to cities. Under the hukou system, government departments that are responsible for labour, social security, public education, and urban administrative management exclude rural labour migrants from the entitlements of "cit-izen" when they are resident outside of their home village. This marginal-ized situation exacerbates "split labour reproduction," which makes it nearly impossible to migrate as a household and some family members—especially women, children, and the elderly—have to remain in the coun-tryside. This well-known phenomenon has resulted in a form of split family. Researchers estimated that by 2006 there were 87 million left-behind population in China, including 47 million women (Ye and Pan 2008). This family strategy is affected by the institutional restrictions on the one hand—for example, women's disadvantages in education, the labour market, underdeveloped social services, and supportive net-works—and shaped by fundamental cultural values and social norms such as familism and differentiated gender roles and motherhood obligations (Ye et al. 2014). This changing demographic structure also contributes to grey agriculture and agriculture feminization.

For many rural households, the split family is not only a passive response to migration constraints under rural-urban dualism, but also a livelihood strategy for family reproduction, a "half work half till" livelihood strategy. Wages from migrant work are for cash income, while farm work in the countryside is mainly for household food security. Rural households need family members to work in the city and on the farm in order to support the family and realize family reproduction; the absence of either will lead rural households into poverty or other difficulties. Historically, Chinese families have relied on agriculture and non-agricultural activities for their livelihood, but it is only in the past 40 years that this type of multiple livelihood pattern has shifted from "agriculture + handicraft + sideline activities" to "hoe + salary." The "hoe + salary" strategy usually relates to intergenerational and gender labour divisions within a family (Xia 2014). In split families, young and middle-aged family members usually relocate to work in the city go to the city while the elderly remain in the village to farm.

When compared to the traditional family relationship, split families and three-generation families are different in many aspects. In traditional society, when sons married, their parents would distribute land, animals, housing, and other properties among their children. Family division is an important landmark for the beginning of a new family. Household division is not only about property redistribution but also about sharing responsibility for elder care. In contemporary rural society, this family division is still practised, but it gradually loses its meaning in the context of rural labour migration. After marriage and separation from parents, sons who migrate cannot fulfil their duty to elderly parents. Even though the sons have separated from parental households and received their share of family land, in many cases, it is still the old parents who farm the land and take care of their sons' children. Family reproduction in the countryside relies considerably on the older generation. However, compared to the older generation of rural migrants, young migrants (around 30 years old) can often not provide sufficient support to their rural families—some even rely completely on their parents for marriage preparation, child rearing, and house construction. The older generation of rural migrants could tolerate harsh working condition in the city in order to improve the quality of life of their rural family, while the younger generation pays more attention to individual enjoyment. From the author's observations in the countryside, many young migrants do not have savings, do not send remittances home, and sometimes are even in need of their parents' money to support their life in the city. Given young migrants' unstable work and life in urban areas and their limited support to their rural families, family reproduction of the young generation has serious problems. Alongside the heavy burden that the left-behind elderly endure, left-behind children face the risks of anomie behaviours and some engage in crime as a result of their parents' physical absence and lack of guidance. These are the social costs of rural youth's migration. They reflect the intergenerational differences between old and young in the countryside but at the same time reveal the dilemma of reproduction for rural households and society at large.

Demographic Challenge of Agricultural Labour: Positioning the Youth in Farming

Land, capital, and labour are the three important ingredients for agricultural production and development. Except for the changes in land transfer and capitalization mentioned above, the changing structure of the farming population is the most serious concern for the country and is having a profound impact on both agriculture and countryside development. Since the institution of its "one child policy" in the late 1970s, the government has strictly controlled the birth rate, and as a result, population increases have gradually slowed. The birth rate within the rural population has significantly declined from over 30 per cent in 1970 to 12 per cent in 2012. The juvenile dependency ratio in the countryside is 30.65 per cent, much lower than the global average of 46 per cent. Meanwhile, the old-age dependency ratio is 12.04 per cent, almost as high as the global average of 13 per cent (Li and Qiu 2012). According to the latest demographic census in 2020, people over 60 years old accounted for 32.2 per cent of the total population. Compared with the census outcome in 2010, the proportion of population over 60 years old increased by 10 per cent (National Bureau of Statistics of China 2021b). These demographic changes imply an increasing shortage of young people in rural areas, continually being widened by the lasting migration of rural youth. Labour remains a critical issue for agriculture sustainability in China.

Debate on Grey Agriculture and Feminization of Agriculture

According to nationwide agricultural censuses that the government conducted in 1996 and 2006, older labour over 60 years old comprised a larger portion of the rural labour force than before, which is proof of the ageing of the country's rural labour force. If we look at the structure of the agricultural labour force horizontally, there is also a high tendency of ageing, according to 2014 Chinese Academy of Social Science research. It found that people over the age of 40 accounted for 61 per cent of the agricultural labour force[6] and those over 50 years old accounted for 34.6 per cent (Chinese Academy of Social Science 2013). According to the findings of the sixth nationwide census in 2010, the population over 60 years of age in the countryside was 99.28 million or 15 per cent of the total rural population. The number of older rural people was 1.3 times the urban contingent. Dang (2014) and other scholars predict that this trend will continue to surpass the age 60 plus population in urban areas until 2050 (Table 5.3).

Table 5.3 Comparison of rural labour composition

	Rural labour aged 51–60 years of age		Rural labour above age 61		Total rural labour (million)
	No. (million)	% of total	No. (million)	% of total	
1996	58.73	10.15	39.16	6.97	561.47
2006	–	20.7	–	25	531.00

Source: National Bureau of Statistics of China (2001, 2008)

---

[6] In China's agricultural censuses to date, the age bracket (earlier than 15 years old to older than 60 years old) for the agricultural labour force is open as people can work with their family on the farm until a later age than a worker employed in industry.

With the shifting demographic structure of the rural population leading to the changing composition of the farming population, there is a persistent debate on grey agriculture and the feminization of agriculture. The increasing participation of rural women and the elderly in agriculture is a direct result of the composition of rural migrants. Research shows that female migrants accounted for 33.6 per cent of rural migrants in 2015, with more women living in the countryside than men. The percentage of migrants in the rural population rises before peaking at age 29 and only 10 per cent of migrants are above age 60. This verifies the general practice of young rural people migrating at a young age, usually after senior middle school at age 16 and then returning to their home area when they are older or have lost their working capacity for migration (Gai et al. 2014). Another study reveals that 57 was the average age of a farmer, which corroborates the aforementioned ageing trend in the farming population (Zhu 2013).

Debates around grey agriculture and the feminization of agriculture have produced controversial viewpoints. Those with pessimistic viewpoints focus on the negative impacts of male labour migration that some females cannot cope with migration-induced labour loss and that they leave the land fallow or extensively cultivated (Zhu and Yang 2011; Gai et al. 2014). Others focus on the older generation's lower levels of education, which can become an obstacle for agricultural extension and the development of modern agriculture because they face more difficulties in applying new techniques (Ning 2013). All of these negative impacts potentially endanger the country's food security (Zhu and Yang 2011).

On the contrary, other scholars assert the positive impacts of rural labour migration for agriculture. Their arguments include: (1) the outflow of rural labour can adjust for the highly unbalanced land-person ratio and increase production per unit of labour; (2) rural migrants' remittances are fed back into agricultural production and accelerate its mechanization; (3) an outflow of rural labour can expedite land transfers, allowing for scaled farming and specialized agriculture. They argue that rural labour migration creates opportunities for "modern agriculture" as well as entrepreneurial agriculture and therefore should be encouraged.

Yet other scholars do not see the feminization of agriculture or the ageing of the farming population as disastrous for Chinese agriculture. For the current average scale of family farming, neither has much influence on the sector due to the increased use of machinery and accessibility to agricultural social service. One study notes how female participation in agriculture can have positive effects in terms of increasing cultivation area (Wen 2014). Even with this changing composition of the farming population, grain production in China has been increasing in the 2010s. However, the outflow of rural labour, especially male labour, has unquestionably had a profound influence on family and social relationships, bringing to the fore the issue of absent husbands/men in care and family reproduction.

Children's Involvement in Agriculture

In contrast to the ongoing academic focus on female and the elderly's participation in agriculture, children's involvement has been a relatively marginal issue despite the profound implications. The issue of child labour was very prominent in China in the early 1900s as the manufacturing industrialization was developing alongside the entrenched poverty of ordinary people. According to a China Industrial Survey report based on a survey of 1206 enterprises in China in the 1930s, there were 115,000 child labourers in China, accounting for 14.8 per cent of the labour force (Li 2018). Labour research in this field has been concentrated on youth employed in manufacturing. If children's participation in handicrafts and agriculture was taken into account, the ratio of child labour to adult would be significantly higher. Although systematic research on child labour in agriculture in the early 1900s is not available, researchers in the period did examine the relationship between agricultural production performance and child labour involvement. In northeast China, for example, when the soybean harvest was poor, rural households had to adjust their family labour division and increase children's labour input in farming to cope with the lost revenues from this major cash crop. This research shows that when an area of farmland that could not be harvested increased by 1 per cent, the use of child labour (generally aged 11–14) also increased by 0.04 persons (Li 2018).

The situation of child labour has evolved due to rapid urbanization, improved socio-economic conditions, and the country's education system. As rural children spend more time in schools and as school locations are increasingly being shifted from villages to townships and counties, many scholars, especially educators, bemoan children's segregation from rural communities and detachment from agriculture and labour work. Close to nature and agriculture and the children's daily interactions in their neighbourhood are seen as important components of socialization and education for rural children. Underneath this general trend of gain in formal education but detachment from agriculture and the community, however, children's labour participation in some less developed areas is still common. Rural children are involved in various kinds of labour work through interwoven macro-level labour regimes and family gender labour division. In the split labour regime for migrant workers, the burden of labour reproduction (such as childrearing and elder care) is externalized and transferred to rural households. The left-behind people have to take up the domestic and economic work in order to sustain labour regeneration. The children who remain at home may help their grandparents in farm work and domestic work. Ye and Pan (2008) show a significant increase in the labour burden for left-behind children after their parents' migration. About 45.6 per cent of these youth regularly participate in farm work. The increase in labour burden is most dramatic for children between age 6 and 14 since their older siblings (aged 15–18) usually live away from home at boarding school. Children's participation in farm work not only limits their time for play and study, but also induces complaints about the workload, increases pressure on them, and can result in conflicts with grandparents (Ye and Pan 2008).

Affected by traditional gender norms and family gender labour division, left-behind children's labour work in the household is highly gendered. Research shows that young girls need to take up care work, domestic work, and farm work with very little time for recreation, while boys could keep their distance from labour work and spend more time at play. The burden of labour reproduction induced by the split labour regime of rural migrants was first transferred to the rural elderly and then strengthened the gender labour divisions among children (Wang 2019). Other studies illustrate the extent of children's involvement, not only in domestic and farm work but also in the household economy and in wage labour. Flexible labour employment worldwide has reconstructed social relations and in the Chinese countryside, this has meant the wide-ranging involvement of rural children in the labour process (Ren and Zhang 2015). Regardless of whether the work is waged or domestic labour, the increased responsibilities owing to their parents' or other family members' migration is negatively influencing children's physical and mental well-being, especially that of girls who studies show carry a larger burden.

Cultivating Young Farmers

Youth is a concept that is defined differently in terms of chronological age and has various definitions depending on the region, culture, or life world of the person using the term. China is no exception. The Chinese Communist Youth League is one of the national organizations working for young people. In its articles of organization, young people who are between the ages of 14 and 28 can join the League. The upper age limit for committee members who belong to the All-China Youth Federation is 40 years old. For many awards or other committee across the country, the upper age limit for applications is age 39. For example, since 1996, the Ministry of Agriculture has selected and rewarded outstanding young farmers from across the country. Qualified young farmers are those between 18 and 39 years old who have been working in agriculture for more than three years and have outstanding economic performance and demonstration effect to other farmers on application of new techniques or new way of farming. The Ministry's definition of a "young farmer" is just one of many—there is no common definition among government departments. This is evident in the national agriculture censuses in 2006 and 2016. As Table 5.4 shows, agricultural labourers in family farming were divided by the age difference of 10 years in the 2006 census and there was no clear demarcation or illustration of a "young farmer." In the 2016 census (see Table 5.5), as the presentation and significance of young farmers was better appreciated, the age classification of farmers was simplified to three groups: young farmers (below age 35), middle-aged farmers (those aged 36–54), and old farmers (aged 55 and above). In the

Table 5.4 Age composition of agricultural labour in family farming (2006)

Age (years)	Amount (10,000 persons)	Percentage (%)
20 and below	1820.96	5.3
21–30	5111.02	14.9
31–40	8266.11	24.2
41–50	7892.21	23.1
51–60	7279.93	21.3
60 and above	3846.77	11.2
Total	34217.02	100

Source: National Bureau of Statistics of China (2008)
Note: *Data on agricultural labour by age working on commercial farms are not available*

Table 5.5 Age composition of agricultural labour (2016)

Age (years)	Percentage (%)
35 and below	21.1
36–54	58.3
55 and above	20.7

Source: National Bureau of Statistics of China (2017)

Medium and Long-term Youth Development Plan, issued by the government and the State Council of China in 2017, they define youth as individuals between age 14 and 35 (Central Government of China 2017). This definition has since become the standard in policymaking and service delivery for young people, especially in policies relating to agricultural and rural development.

Tables 5.4 and 5.5 also show the significant positioning of young people in the agricultural labour force. Although the 2016 census data did not differentiate young people in family farming and young people as farm workers, it did confirm the existence and contribution of young farmers in agriculture. However, the internal dynamics of this group have not been fully recognized in social discourse or scholarly research. Attention on young farmers was merely revived in recent years in the context of rapid agrarian transition. In 2012, official government documents advocated "cultivating new vocational farmers" as a countermeasure for rural people's outflow of agriculture and the potential challenge of food security. There is no uniform definition of "new vocational farmer," but many scholars have explored the ideal characteristics of this group. Some highlight that the new vocational farmer should differ from the traditional peasant with their access to and knowledge of modern technology and modern ideology. Others argue that new farmers are an outstanding part of the larger farmer/peasant population that can organize and mobilize the other as a model (Lu and Zhu 2006). Some anticipate that the cultivation of new farmers can halt the drain of agricultural successors, but also attract more people to the countryside to establish businesses (Shen et al. 2014). The cultivation of this group involves systematic training, instruction, and financial support. Although standards for new vocational farmers and their cultivation vary between provinces, candidates and applicants should have farming experiences and meet criteria related to scales and sources of income (i.e., 80 per cent of household income from agriculture) as indicators of professional farmer. In reality, these requirements are untenable for young people who are starting out in agriculture and are often unqualified in terms of certification and lack institutional support.

In 2015, in accordance with the government's national efforts to promote modern agriculture, the Ministry of Agriculture and Rural Affairs initiated a "Training Program for Modern Young Farmers" with a plan to involve 10,000 young farmers. In 2016, another 10,000 young farmers were included in the programme. Target groups included large holders in agriculture, operators of officially registered family farms, major organizers in agricultural co-ops, college students who returned to their rural hometown for business, high school graduates, and veterans. Participants needed to be under the age of 45. The programme provided training in business operations, family farm management, e-business, quality control of agricultural products, and rural construction, among others. Each cohort received three years of part-time training.

In addition to the aforementioned programme for young farmers, the government has implemented several supporting programmes for young farmers in recent years, including the Green Certificate Training Program, which offers practical skills training in agricultural production and operation.

Along with the training programmes by central government on young farmers, the phenomenon of new farming people (*xin nong ren*) is emerging. There is no single definition or uniformed characteristics of members of this group, but there are many expectations for those who have recently become involved in agriculture. Those who argue it is a new phenomenon state that individuals who fit in this group: (1) are making agriculture their career, (2) are pursuing agriculture in new modes/ways of production when compared to traditional agriculture, and (3) constitute new actors in the countryside that did not previously exist (Wang 2014). When compared to young people who returned to their rural hometowns, *xin nong ren* comprise individuals who spontaneously left their urban existence and transitioned from urban to rural, white collar to farmer. Second, they have a higher level of education than traditional small-scale farmers and become involved in agricultural operation through ecological plantation. The constitution process of *xin nong ren* is the process through which urban youth built a new way of life and new sense of belonging and identity (Li 2016). Some scholars from the Chinese Academy of Social Science summarize their composition as four sub-groups: (1) rural migrants who have a certain level of capital accumulation, have emotional attachment to land, and are determined to return to a rural setting; (2) young people who think agriculture could provide space for their talents and wish to start a career; (3) middle-aged individuals who have had a successful career and decide to engage in organic farming for their family; (4) social organizations or non-governmental organizations that are concerned with rural issues. A sample survey of *xin nong ren* shows similar demographic characteristics. Among the 155 respondents, more than 90 per cent were under age 40; 51.6 per cent were formerly white-collar workers before engaging in agriculture; about 22.6 per cent used to be researchers, teachers, media practitioners, or freelancers. *Xin nong ren* do not necessarily return to their rural hometown, and some relocate to a suburban area or another rural location. Although the concept of *xin nong ren* is still a topic of discussion among scholars, the phenomenon is now widely recognized. In February 2015, Aliresearch, the Alibaba group's research centre, issued the *Xin Nong Ren Research Report 2014*, which indicated that there were more than one million *xin nong ren* in China by 2015. This group is continuously growing and becoming vibrant actors in agriculture.

Conclusion

The issue of young farmers is gaining increasing political recognition, especially since the release of *Opinions of the Chinese Communist Party's Central Committee and the State Council on the Implementation of the Strategy for Rural Revitalization* in late 2017. As the most important comprehensive development strategy on agrarian issues in China, it aims to realize overall vitalization of the countryside by 2050 when the agriculture will be stronger and more promising, the countryside will be more pleasant for living, farmers will be affluent, and farming will be an attractive occupation. The strategy's release immediately invoked extensive discussion among scholars, policymakers, and practitioners who began to research the pathways to realize rural revitalization and the diversified practical innovations and experiences it includes.

It is in this context that young farmers, as the key players in this rural revitalization, have gained increased attention. The key to rural revitalization lies in people, people who love the countryside and are committed to agricultural and rural development. On the other hand, a developed countryside and agriculture system should provide enough space for young farmers to realize their ambitions and support household reproduction. Unfortunately, due to the long-standing rural-urban imbalance and its impact on the country's development in addition to significant rural labour migration, young farmers have been under researched and await the social recognition that their role in China's economy deserves. Many studies have focused on their marriages, employment, and social adaption in cities as migrant labour while overlooking the dynamics and internal complexities that shape farming and agrarian transition. There remains a knowledge gap in terms of the differences between young farmers in different regions, of different genders, their varying styles of farming, among others. The diversification of young farmers and their role in agrarian transition is yet to be explored.

For a country like China with vast territory and prominent regional differences, it is very difficult to provide a general picture of all young farmers in China. Rather, the following chapters are a way to present different stories of young farmers across the territorial scope. The first site is located in Sichuan province in China's southwest and the second is in the north of China, in Hebei province. Farmers at both sites specialize in vegetable production and smallholding family farming is the major unit of production. Our research shows that the young farmers we interviewed have different pathways into farming. Small plots, bad traffic conditions, and unstable agricultural revenues at the first site, which is located in a mountainous area, have pushed many young people to migrate to urban areas in search of employment. When they returned to care duties in the village, they were to some extent trapped in a marginal situation due to the markets, bad agricultural infrastructure, and all of the other challenges faced by small-scale family farmers engaged in commodification. Their story is repeated in many areas in the west of China. In contrast, young farmers at the second site live and work in an organized community, and on their return from the city, these young people are able to enter into farming more smoothly and consolidate their farming and livelihood. Through these case studies and comparing the different pathways of young farmers into farming in different regions of China, the aim is to enrich and diversify our limited understanding of young farmers worldwide.

References

Brookfield, Harold. 2008. Family farms are still around: Time to invert the old agrarian question. *Geography Compass* 10: 108–126.

Central Government of China. 2013. *The CPC Central Committee and The State Council on accelerating the development of modern agriculture further enhance rural development vitality of a number of views [Zhonggongzhongyang guowuyuan guanyu jiakuai fazhan xiandai nongye jinyibu zengqiang nongcun fazhan huoli de ruogan yijian].* http://www.gov.cn/jrzg/2013-01/31/content_2324293.htm.

———.2014. *Instructions on leading the orderly transfer of rural land contract right to develop scaled agricultural operation [Guanyu yindao nongcun tudi jingyingquan youxu liuzhuan fazhan nongye shidu guimo jingying de yijian].* http://www.gov.cn/xinwen/2014-11/20/content_2781544.htm.

———. 2016. *The 2015 national land change survey results were released [2015niandu quanguo tudi biangeng diaocha jieguo fabu].* http://www.gov.cn/xinwen/2016-08/11/content_5098857.htm.

————. 2017. *Medium and long-term youth development plan (2016–2025)* *[Zhongchangqi qingnian fazhan guihua (2016–2015)]*. http://www.gov.cn/xinwen/2017-04/13/content_5185555.htm#1.

————. 2021a. *Report on monitoring and Investigation of Rural Migrant Workers in 2020 [2020nian nongmingong jiance diaocha baogao]*. http://www.gov.cn/xinwen/2021-04/30/content_5604232.htm.

————. 2021b. The *Ministry of Natural Resources and the National Bureau of Statistics jointly held a press conference on the main data results of the third National land survey [Ziran ziyuanbu, guojia tongjiju lianhe zhaokai disanci quanguo guotu diaocha zhuyao shuju chengguo xinwen fabuhui]*. http://www.gov.cn/xinwen/2021-08/27/content_5633643.htm.

————. 2021c. *China's all-round well-off society [Zhongguo de quanmian xiaokang]*. http://www.gov.cn/zhengce/2021-09/28/content_5639778.htm.

————. 2022. *The total population increased and the urbanization rate continued to rise [Renkou zongliang yousuo zengjia, chengzhenhualv jixu tigao]*. http://www.gov.cn/xinwen/2022-01/17/content_5668914.htm.

Chen, Minghe. 2013a. Relations between land transfer and family farms: Case of Liaoning Province [Tudi liuzhuan yu jiating nongchang guanxi tantao]. *Journal for Party and Administrative Cadres [Dangzheng ganbu xuekan]* 8: 45–50.

Chen, Shensheng. 2013b. *Study on Agricultural Entrepreneurship of College Students in Zhejiang Province [Zhejiangsheng daxuesheng nongye chuangye yanjiu]*. Master thesis. Zhejiang A & F University.

Chinese Academy of Social Science. 2013. *Quantity of agricultural labor was greatly overestimated [Nongye laodongli shuliang bei yanzhong gaogu]*. http://www.yicai.com/news/3203529.html.

Dang, Junwu. 2014. The shift of ageing problem studies: From Gerontology to ageing science [Laoling wenti yanjiu de zhuanxiang: cong laonianxue dao laoling kexue]. *Scientific Research on Ageing [Laoling kexue yanjiu]* 2: 3–9.

Duan, Chenrong, and Xueyang Ma. 2011. The 'new' situation of current Chinese new generation of migrant workers [Dangqian woguo xinshengdai nongmingong de xinzhuangkuang]. *Population & Economics [Renkou yu jingji]* 4: 16–22.

Fan, Shenggen and Connie Chan-Kang. 2003. *Is small beautiful? Farm size, productivity and poverty in asian agriculture*. Plenary paper presented at the International Association of Agricultural Economists conference, Durban, South Africa, August.

Gai, Qingen, Xi Zhu, and Qinghua Shi. 2014. The impact of labor migration on Chinese agricultural production [Laogongli zhuanyi dui zhongguo nongye shengchan de yingxiang]. *China Economic Quarterly [Jingjixue]* 3: 1147–1170.

Han, Song. 2012. The practical problems and countermeasures of land transfer in the process of new rural construction [Xinnongcun jianshe zhong tudi liuzhuan de xianshi wenti jiqi duice]. *China Legal Science [Zhongguo faxue]* 1: 19–32.

Han, Changfu. 2016. *Opinions on Improving the Measures for the Separation of Rural Land Ownership Contract Rights and Management Rights [Guanyu wanshan nongcun tudi suoyouquan chengbaoquan jingyingquan fenzhi banfa de yijian da jizhewen].* http://www.moa.gov.cn/xw/tpxw/201611/t20161104_5350207.htm.

He, Xuefeng. 2015. For whose agricultural modernization? [Weile shui de nongye xiandaihua]. *Open Times [Kaifang shidai]* 5: 36–48.

Houtart, Francois, and Tiejun Wen, eds. 2013. *Peasant's Agriculture in Asia.* Panama City: Ruth Casa Editorial.

———. 2016. The hidden agricultural revolution in China (1980-2010): A historical and comparative perspective [Zhongguo de yinxing nongye geming (1980-2010)—yige lishi he bijiao de shiye]. *Open Times [Kaifang shidai]* 2: 11–35.

Jian, Xinhua, and Kun Huang. 2010. The empirical analysis and prospect prediction of the level and speed of Chinese urbanization [Zhongguo chengzhenhua shuiping he sudu de shizhengfenxi yu qianjing yuce]. *Economic Research Journal [Jingji yanjiu]* 3: 28–39.

Li, Qiang. 2012. Rural labor migration and rural-urban income gap analysis [Nongcun laodongli liudong yu chengxiang shouru chaju fenxi]. *Statistics & Decision [Tongji yu juece]* 6: 111–115.

Li, Luping. 2016. *Research on the identity construction of current 'new farmers' [Dangdai xinnongren de shenfen jiangou yanjiu].* Master Thesis. East China Normal University.

Li, Tania. 2017. Intergenerational displacement in Indonesia's oil palm plantation zone. *The Journal of Peasant Studies* 44 (6): 1158–1176.

Li, Nan. 2018. Agricultural shock, credit constrain and child labor: Evidence from the villages of Northeast China in the 1930s [Nongye chongji, xiangcun jiedai yu tonggong shiyong: laizi 20 shiji 30 niandai dongbei beibu xiangcun shehui de kaocha]. *Research in Chinese Economic History [Zhongguo jingjishi yanjiu]* 1: 30–42.

Li, Chengzheng, and Junjie Qiu. 2012. The research on the Chinese rural population structure and household consumption [Zhongguo nongcun renkou jiegou yu jumin xiaofei yanjiu]. *Population & Economics [Renkou yu jingji]* 1: 49–56.

Lu, Zeyu, and Xiaoping Chen. 2015. The status quo, problems and counter-measures of Chinese rural land transfer [Zhongguo nongcun tudi liuzhuan xianzhuang]. *Journal of Xinjiang Normal University (Edition of Philosophy and Social Sciences) [Xinjiang shifan daxue xuebao]* 4: 114–119.

Lu, Kerong, and Qizhen Zhu. 2006. Socialist new rural construction and culti-vation of new farmers [Shehuizhuyi xinnongcun jianshe yu xinxing nongmin peiyang]. *Future and Development [Weilai yu fazhan]* 9: 27–29.

Luo, Haoxuan. 2013. The empirical study of the impact of the deepening of Chinese agricultural capital on the agricultural economy [Zhongguo nongye ziben shenhua dui nongye jingji yingxiang de shizheng yanjiu]. *Issues in Agricultural Economy [Nongye jingji wenti]*. 9: 4–14.

Luo, Biliang, and Yuqin Li. 2014. Agricultural management system: The bot-tom line of institution, identification of its nature and innovation space—Some thoughts based on the 'Seminar on the Rural Household Management System' [Nongye jingying zhidu: zhidu dixian, xingzhi bianshi yu chuangxin kongjian –jiyu nongcun jiating jingying zhidu yantaohui de sikao]. *Issues in Agricultural Economy [Nongye jingji wenti]* 1: 8–18.

Ministry of Agriculture and Rural Affairs of China. 2013. *Survey result on devel-opment of 'family farms' in China by Ministry of Agriculture [Nongyebu shouci dui quanguo jiating nongchang fazhan qingkuang tongji diaocha jieguo].* http:// sd.ifeng.com/zt/sdjtnc/jdjtnc/detail_2013_06/21/919015_0.shtml.

———. 2017. *State Information Office held a policy briefing, Vice Minister of Agriculture Ye Zhenqin introduced the cultivation of new agricultural operating entities—accelerate the cultivation of new agricultural operating entities to pro-mote the common development of small farmers [Guoxinban juxing zhengce chuifenghui, nongyebu fubuzhang yezhenqin jieshao xinxing nongye jingying zhuti peiyu youguan qingkuang—jiakuai peiyu xinxing nongye jingying zhuti, daidong xiaononghu gongtong fazhan].* http://www.moa.gov.cn/xw/ zwdt/201712/t20171219_6123309.htm.

———. 2022. *Agricultural import and export data.* http://zdscxx.moa.gov. cn:8080/nyb/pc/index.jsp.

Ministry of Education of China. 2019. *2663 general institutions of higher education in China with gross enrollment rate of 48.1 percent [Quanguo putong gaoxiao 2663suo, gaodeng jiaoyu mao ruxuelv 48.1%].* http://edu.sina.com.cn/ gaokao/2019-07-24/doc-ihytcerm5876983.shtml?cre=tianyi&mod=pcpager_ news&loc=21&r=9&rfunc=76&tj=none&tr=9.

Ministry of Finance of China. 2012. *Fiscal support for agriculture, rural areas and farmers [Caizheng zhichi sannong qingkuang].* http://www.mof.gov.cn/zhuan-tihuigu/czjbqk2011/czzc2011/201208/t20120831_679920.html.

National Bureau of Statistics of China. 2001. *National Agricultural Census Office Communique on the rapid summary results of the first National Agricultural census: rural workers and agricultural machinery [Quanguo nongye pucha bangongshi guanyu diyici quanguo nongye pucha kuaisu huizong jieguo de gongbao—nongcun congye renyuan he nongye jixie]*. http://www.stats.gov.cn/tjsj/tjgb/nypcgb/qgnypcgb/200203/t20020331_30458.html.

———. 2008. *Main date bulletin of the second national agricultural census [Dierci quanguo nongye pucha zhuyao shuju gongbao]*. http://www.stats.gov.cn/tjsj/tjgb/nypcgb/.

———. 2017. *Main date bulletin of the third national agricultural census [Disanci quanguo nongye pucha zhuyao shuju gongbao]*. http://www.stats.gov.cn/tjsj/tjgb/nypcgb/.

———. 2021a. *Statistical bulletin of the People's Republic of China on national economic and social development 2020 [Zhonghua renmin gongheguo 2020nian guomin jingji he shehui fazhan tongji gongbao]*. http://www.stats.gov.cn/tjsj/zxfb/202102/t20210227_1814154.html.

———. 2021b. *Bulletin of the seventh National Census [Di qici quanguo renkou pucha gongbao]*. http://www.stats.gov.cn/tjsj/tjgb/rkpcgb/qgrkpcgb/202106/t20210628_1818827.html.

Ning, Haonan. 2013. Analysis on the issue of the large outmigration of youth from Chinese rural areas [Woguo nongcun nianqing laodongli daliang liushi de wenti fenxi]. *Productivity Research [Shengchanli yanjiu]* 7: 31–32.

Pan, Genxing, Min Gao, Guohua Hu, and Qingping Wei. 2011. The impact of climate change on Chinese agricultural production [Qihou bianhua dui zhongguo nongye shengchan de yingxiang]. *Journal of Agro-Environment Science [Nongye huanjing kexue xuebao]* 30 (9): 1698–1706.

People's Daily. 2021. *Government spending on people's wellbeing increased year by year [Caizheng dui minsheng touru zhunian zengjia]*, July 31.

van der Ploeg, Jan Douwe, and Jingzhong Ye. 2016. *China's peasant agriculture and rural society: Changing paradigms of farming*. Abingdon and New York: Routledge.

Pun, Ngai, Lu Huilin, Hairong Yan, et al. 2009. Rural migrants: Unfinished proletarianization [Nongmingong: weiwancheng de wuchanjiejihua]. *Open Times [Kaifang shidai]* 6: 5–35.

Ren, Yan, and Shasha Zhang. 2015. Child labor and family reproduction: An experiential study in a village in Western Guangdong [Ertong laodong yu jiating zaishengchan]. *Open Times [Kaifang shidai]* 6: 159–177.

Shanin, Teodor. 1990. *Defining peasants*. Oxford: Blackwell Publisher.

Shao, Xiazhen. 2015. *Not changing land with population change and permanancy of land contract [Zengren bu zengdi, jianren bu jiandi yu changjiu bu bian]*. http://theory.people.com.cn/n1/2015/1216/c49150-27934294.html.

Shen, Hongmei, Youguang Huo, and Guoxian Zhang. 2014. Research on the cultivation mechanism of new type professional farmers—Based on the vision of agricultural modernization [Xinxing zhiye nongmin peiyu jizhi yanjiu—jiyu nongye xiandaihua shiyu]. *Modern Economic Research [Xiandai Jingji Tantao]* 1: 65–69.

Soda, Osamu. 2003. *Philosophy of agriculture [Nongxue yuanlun]*. Beijing: Press of Renmin University. in China.

Tan, Linli, and Xinhua Sun. 2014. Three current paths to scaled-up agriculture. [Dangqian nongye guimo jingying de sanzhong lujing]. *Journal of Southwest University (Social Sciences Edition) [Xinan daxue xuebao (shehui kexue ban)]* 6: 50–56.

Tian, Meihua. 2010. Development path of secondary agricultural vocational education under the background of new rural construction [Xinnongcun jianshe beijingxia zhongxia zhongdeng nongye zhiye jiaoyu de fazhan lujing]. *Journal of Anhui Agricultural Science [Anhui nongye kexue]* 31: 17975–17977.

Tuliu Net. 2016. *Summary of planting subsidy policies for cooperatives and family farms in 2016 [2016nian hezuoshe he jiating nongchang zhongzhi butie zhengce huizong]*. https://www.tuliu.com/read-30158.html.

———. 2021. *Area of land transfer [Tudi liuzhuan mianji]*. https://www.tuliu.com/data/nationalProgress.html.

Wang, Li. 2013. *The Role of Agriculture in Economic Growth of China [Nongye zai zhongguo jingji zengzhang zhong de zuoyong]*. PhD thesis. Liaoning University.

Wang, Xiangdong. 2014. 'Xin Nong Ren' and the Xinnongren Phenomenon [Xinnongren yu xinnongren xianxiang]. *New Agriculture [Xin nongye]* 2: 18–20.

Wang, Ou. 2019. Left-behind experiences and the formation of gender division of labor: An empirical study based on fieldwork at sites of origin and destination for migrants. [Liushou jingli yu xingbie laodong fenhua—jiyu nongmingong shuchudi he dagongdi de shizheng yanjiu]. *Sociological Study [Shehuixue yanjiu]* 2: 123–146.

Wang, Jianhua, and Binhuan Huang. 2014. Left-behind experience and job change of new workers: How does rural migrants' labor regime put itself into dilemma [Liushou jingli yu xingongren de gongzuo liudong: nongmingong shengchan tizhi ruhe shi zishen mianlin kunjing]. *The Society [Shehui]* 5: 88–104.

Wen, Huacheng. 2014. Feminization of Chinese agricultural labor: Extent, causes and impacts—Based on historical macro cross-section data [Zhongguo nongye laodongli nvxinghua: chengdu, chengyin yu yingxiang—jiyu lishi hongguan jiemian shuju de yanzheng]. *Population Journal [Renkou xuekan]* 4: 64–73.

Workers' Daily. 2020. *Report: The average educational attainment of the national workforce rose by 4.2 years in 33 years [Baogao: 33nianjian quanguo laodongli renkou pingjun shou jiaoyu chengdu shangsheng 4.2nian].* https://baijiahao.baidu.com/s?id=1685578198943494068&wfr=spider&for=pc.

Xia, Zhuzhi. 2014. The sociological implications of 'half industrial and half agricultural working structure' [Lun bangong bangeng de shehuixue yihan]. *The Journal of Humanities [Renwen zazhi]* 7: 112–116.

Xia, Jinjing, and Ganzi Peng. 2004. Comparative studies of the development of agricultural vocational education [Nongye zhiye jiaoyu fazhan de bijiao fenxi]. *Vocational and Technical Education [Zhiye jishu jiaoyu]* 22: 52–55.

Xie, Longjian. 2010. *"Research on Rural Vocational Education Policy Changes and Implementing Countermeasures" [Nongcun zhiyejiaoyu zhengce bianqian ji shishi duice yanjiu].* Master thesis. East China Normal University.

Xu, Jieling, and Wenxiang Wang. 2009. The status quo of farmer quality under the vision of new rural construction and the cultivation of new farmers [Xinnongcun shiye xia woguo nongmin suzhi xianzhuang yu xinxing nongmin de peiyu]. *Journal of Anhui Agricultural Science [Anhui nongye kexue]* 13: 6176–6178.

Yang, Chunhua. 2010a. Some concerns about the new generation of rural migrants [Guanyu xinshengdai nongmingong wenti de sikao]. *Issues in Agricultural Economy [Nongye jingji wenti]* 4: 80–85.

Yang, Juhua. 2010b. Some misunderstandings on the new generation of rural migrants [Dui xinshengdai liudong renkou de renshi wuqu]. *Population Research [Renkou yanjiu]* 2: 44–56.

Ye, Jingzhong, and Lu Pan. 2008. *Differentiated childhoods: Left-behind children in rural China [Bieyang tongnian: zhongguo nongcun liushou ertong].* Beijing: Social Sciences Academic Press.

Ye, Jingzhong, Lu Pan, and Congzhi He. 2014. *Double coercion: Gender exclusion and inequality in rural left-behind [Xiangcun liushou zhong de xingbie paichi yu bupingdeng].* Beijing: Social Sciences Academy Press.

Ye, Jingzhong, Wu Huifang, Xu Huijiao, and Yan Jiang. 2016. The myth and realities of land transfer [Tudi liuzhuan de misi yu xianshi]. *Open Times [Kaifang shidai]* 5: 76–91.

Zhang, Yulin. 2011. Agriculture and rural society after 'modernization' in East Asia: Cases of Japan, Korea and Taiwan and their historical implications [Xiandaihua zhihou de dongya nongye he nongcun shehui]. *Academic Journal of Nanjing Agricultural University [Nanjing nongye daxue xuebao]* 11 (3): 1–8.

Zhang, Qian Forrest, and John A. Donaldson. 2010. From peasants to farmers: Peasant differentiation, labor regimes and land-rights institutions in China's agrarian transition. *Politics & Society* 38: 458–489.

Zhang, Zhiwei, Jingdong Luan, and Jie Shen. 2004. The status quo of Chinese rural labor migration and impact factors analysis [Woguo nongcun laodongli liudong de xianzhuang ji yingxiang yinsu fenxi]. *Journal of Anhui Administration Institute [Anhui xingzheng xueyuan xuebao]* 5: 5–6.

Zhu, Qizhen. 2013. New type professional farmers and family farm [Xinxing zhiye nongmin yu jiating nongchang]. *Academic Journal of China Agricultural University (Social Sciences) [Zhongguo nongye daxue xuebao (shehui kexueban)]* 2: 157–159.

Zhu, Qizhen, and Huiquan Yang. 2011. Who is farming the land? The survey and thoughts on agricultural labor [Shuizai zhongdi—dui nongye laodongli de diaocha yu sikao]. *Academic Journal of China Agricultural University (Social Sciences) [Zhongguo nongye daxue xuebao]* 1: 162–169.

6

Young Farmers' Difficulties and Adaptations in Agriculture: A Case Study from a Mountainous Town in Sichuan Province, Southwest China

Dong Liang and Lu Pan

Introduction

The unprecedented rural labour migration in China has resulted in the transformation of rural areas and the entire rural society. Migrant workers returning to their hometowns are a new force that is promoting social transformation and changing the urban-rural relationship in the context of integrated development. Since the early 2000s, the modernization of agriculture and the integration of urban and rural areas has meant significant advancements in China's modernization. The government has introduced policies to enforce agricultural supply structural reforms and

D. Liang
School of Philosophy and Social Development, Shandong University, Jinan, China
e-mail: Liangd@sdu.edu.cn

L. Pan (✉)
College of Humanities and Development Studies (COHD), China Agricultural University, Beijing, China
e-mail: panlu@cau.edu.cn

© The Author(s) 2024
S. Srinivasan (ed.), *Becoming A Young Farmer*, Rethinking Rural,
https://doi.org/10.1007/978-3-031-15233-7_6

157

support rural areas and the agricultural sector. Therefore, returning farmers, especially those who are young, have become the "generation of the entrepreneurs" in the countryside. Having facilitated urban-rural integration, these young farmers are innovating agricultural production and management and utilizing diverse practices to realize agriculture's integration with secondary and tertiary industries as the government has been advocating. Rural youth have been growing up as the backbone of rural revitalization and agricultural development. Thus, it is of far-reaching practical significance to identify young farmers' difficulties in market integration and organizing their production and explore young farmers' adaptations as they enter or return to farming.

This chapter is based on a survey and field research in the summer of 2017 in Lin town, which is located in Sichuan province in Southwest China. The town is located in the Zengjia Mountain as the map illustrates. The town's agricultural specialization is cabbage, but the low profit earned from farming resulted in many young people migrating for employment. Young farmers are unusual in the village and much of the research team's time in the town was spent identifying interviewees. With the help of town officials and village committee cadres, the research team was able to interview 28 young farmers[1] under the age of 45 in five villages of Lin town. Residences are scattered across this mountainous area, and coupled with the rareness of young farmers, the research team had to rely on introductions from village leaders in locating interviewees. It engendered a potential risk of a biased sample in that young farmers who lived in remote areas or did not have a close relationship with village leaders may not have been contacted. Despite this reality of fieldwork, the interviewees' stories reflect the common situation and challenges of young farmers in the area (Map 6.1).

Among the 28 interviewees, the youngest was 19 years old while the oldest was 45 years old. Most of the farmers were over 30 years old. The overall education level of the research subjects was not high: 19 of the interviewees finished primary school (6 years of schooling), 6 finished junior high school (9 years), only 1 finished senior high school (12 years), and 2 didn't finish primary school. In terms of gender ratio, 15

---

[1] All farmers' names and village names are pseudonyms in this chapter.

中国地图

Lin Town

Map 6.1 Study area. (Source: Ministry of Natural Resources Map Technical Review Center)

Table 6.1 Basic information of interviewees

Age			Education level				Experience of migration		Sex	
≤30	31–40	≥40	Below primary school	Primary school	Middle school	High school	Yes	No	Male	Female
5	6	17	2	19	6	1	24	4	15	13

interviewees were male and 13 were female. Twenty-four were former labour migrants, many of whom had returned home in the past two years. Their occupational experiences were quite varied: 21 interviewees worked in market-based vegetable cultivation; one served as a village director while tending his farmland in his spare time; two grew Chinese herbs that were used for traditional medicines; six of the returners were engaged in farm tourism; and one farmer was dedicated to livestock rearing—he had 60 horses and more than 30 sheep (Table 6.1).

This chapter divides young farmers into two categories: passive settlers and the young returnees, defined by their migration experience. Living in the countryside, these young farmers have different logics and actions when they confront common market risks and systematization difficulties. Each adapts to the changing social environment and the relationship between urban and rural areas, acquiring resources and drawing on knowledge earned during their migration experiences. They draw on their close relations and social networks as they embark on a road of embedded innovation and entrepreneurship in the agricultural industry. Young farmers' adaptations have increased their incomes while helping them to cope with the issues of family separation and labour shortages in the sector. Their involvement in agriculture and rural development has made important impacts on rural-urban integration.

Community Profile

Lin is located in a mountainous area of northern Sichuan, lying at an altitude of over 1000 metres. The town is 50 kilometres from the county seat and over 300 kilometres from the provincial capital of Chengdu. Lin is a typical mountainous agricultural town. Farmers make their living through the crop cultivation. Due to the cold climate and scattered farmland allocations in mountains that limit crop size, yield of grain plantation was low. Local farmers gradually found vegetables to be most suitable for cultivation, especially cabbage and hot peppers. In Lin town, there are also farmers who grow herbs and other plants used in Chinese herbal medicine, including *Gastrodia Elata*, horseradish, and schisandra, among others. These operations, though, are on a very small scale. Vegetable cultivation began in the town in the early 1990s at the dawn of China's market economic reform. Farmers started small, growing vegetables on land that otherwise offered low grain yields. The quality of local vegetables is very good due to the bigger temperature difference in mountainous areas. In the 1990s, there was also considerable market demand for vegetables in the plains area of the county, and local farmers formulated a "grain-vegetable" plantation system as a stable livelihood activity. By the end of 2008, Lin town had developed a specialization in vegetable cultivation with the promotional support of its local government. Three decades after the first vegetable crops, farmers in the town are still focused on vegetable cultivation. The planting area has expanded significantly, and it has become the backbone of the town's agricultural economy. The mountainous terrain, however, means that the per capita land area is so small that farming alone cannot support rural families. The dawn of market reform was also, for Lin town, the dawn of rural labour migration. Like villages and towns across rural China, a large number of rural labourers have left home over the years to work as wage labourers in urban areas. There are few labourers younger than age 45 who remain in the village. The left-behind elderly have no choice but to devote themselves to vegetable cultivation as younger and more able family members are working in the city.

With the sluggish labour market in recent years, especially after the global fiscal crisis in 2008, some migrants began to return home. At present, officials and farmers in Lin town are responding to the government's appeal for agricultural supply-side structural reform and are continuing to invest in the town's vegetable industry. In 2012, the local government proposed a new strategy of industrial integration—agrotourism. Local policies have a considerable impact on young migrants returning home. Some return to work the land and engage in market-oriented vegetable production, which comes with its own problems related to marketing and organizing production. At the same time, other young returnees have dedicated their time, energy, and experience to various agriculture-related industries, such as e-commerce and agro-products processing, and taken steps to alter the traditional peasant agriculture.

Young Farmers' Pathways into Farming: Why Would They Stay or Return?

In Lin town, but also across China's vast western countryside, the rural labour migration trend began in the 1990s. The social mobility of the rural population, which was discussed in Chap. 5, must be taken into account as we try to better understand the relationship between agriculture, farmers, and rural development in the town. We identified two key characteristics of young farmers in Lin town: their social mobility between the rural and the urban, keeping in mind that most of the young farmers are former labour migrants, and that this group is highly diversified and differentiated in their rural and farming activities. These two characteristics are related to the local natural environment, to commodified vegetable production, and to the local government's current rural development policies. It is necessary to distinguish between farmers with experience as migrant workers and farmers who have remained in the village in order to understand better contemporary young farmers and the differences in their rural lives and farming work. Although these two groups are living together in the villages, the mindset of returnees and those who have never migrated is not identical.

Reasons for Staying

In rural area, it is common that those who engage in agriculture, especially in traditional grain cultivation, live a hard life, due in part to scarce and fragmented farmland. As rural life has been gradually commodified and monetized in the market economy, farmers face increasing pressure to meet their basic needs and earning money thus becomes the primary goal for them. In Lin town, there are very few farmers under the age of 45 engaging in agricultural production. The town's cadres told researchers that they believe the fundamental reason for this lack of interest is that "agriculture is not attractive." They offered two key reasons: First, it is difficult for young people to gain a sense of self-realization from working on land as farmers, and second, it is difficult to live a well-off life with earnings from agriculture. Those who remain in the village often have no other option. They may not have enough physical strength to seek employment in the city or they may have to take care of the elders, children, or sick family members. Patriarchy and gender norms in some local areas prevent young women from leaving home. These young people who have never experienced migration are referred to as passive stayers or "left-behind" labourers.

Wang Yong from YL Village is 42 years old and has never migrated for work. In his production team,[2] all of the young people under the age of 45 are former migrants. In the late 1990s when migration was in fashion in his village, he had to remain in the village to care for his parents. After their passing, he remained in the village to care for his own children. In taking care of these commitments, Wang missed the best time to migrate. As vegetable cultivation has steadily developed, there was also less incentive for Wang to leave. He could make a living by planting his 4 mu[3] with vegetables and through his part-time job in a local park. In 2017 when we interviewed him, the vegetable market was sluggish. Wang wanted to work in the city, but it was difficult for him to get a job because he was no longer young and did not have any special skill.

In the countryside, strength is the main, if not the only, advantage that rural men have in labour market. The period between age 20 and age 45 is the "golden age" for migrant workers. If they miss their chance to migrate for work when they are young, like in Wang's case, it becomes very difficult for them to find a job in urban area or they are limited to jobs with low pay and harsh working conditions. Remaining in rural areas like Lin town may have been an active choice for young people in the late 1990s or early 2000s when the price of vegetables was relatively stable and migrant workers' wages were relatively low. During this period, there was no great difference between income from agriculture and from wage work in the cities. Since 2013, along with the government's promotion of the vegetable industry and the increasing fluctuations in vegetable prices, more and more farmers are forced to seek alternative sources of income. Their age, however, becomes a major obstacle for their migration since they are no longer "young" in terms of migrant labour.

For young women farmers, their options can be even more limited. In many cases, they are compelled to stay in countryside, weighed down by

---

[2] Production team is the basic organizational unit in administrative village for agricultural production and public affair management. It's an institutional legacy from agricultural collectivization period in the 1950s when residents of one village were divided into several production teams based on geographical adjacency to organize agricultural production collectively. Now production team is still the basic unit for villagers to organize many public affairs, such as irrigation, infrastructure construction and maintenance.

[3] mu is a Chinese measuring unit for farmland and 15 mu equals 1 hectare.

the gender labour division in family reproduction, family rationalism that seeks maximum material benefits, and cultural disciplines of villages and rural families (Liang and Wu 2017). This is especially true in western rural areas where society is relatively more conservative and traditional compared to the eastern part of China.

Guo Li from YL Village is 40 years old. Her family and her village have strong views about a woman's role in society. As a result, Guo was not afforded the opportunity to study, and since she finished primary school, she has been working on the farm. When she was younger, women were not encouraged to leave the village for work. Rather, a young women's duty was to care for her parents and help them with vegetable cultivation. Guo is married with two sons, aged 19 and 12. After her marriage, any opportunity for migration was further thwarted as she needed to remain in the village to care for her children.

Gradually, the taboo that prevented young women from leaving home to find employment was broken and the number of young female migrants from Lin town grew. Here, as in villages and towns across China, the burden of care for children and the elderly remains a huge barrier for would-be young women migrants.

Reasons for Returning

Amidst urban-rural integrated modernization and with the benefits of supportive policies for public entrepreneurship and innovation, returning to the countryside for business has becomes a new phenomenon among rural youth, especially for migrants. Xia's (2017) research on returning youth entrepreneurs suggests that migrant workers' returning to their hometowns accelerates urban-rural integration as both physical resources and human resources have been syphoned from the countryside into urban areas for decades which led to enlarging gap between rural and urban. Liu et al.'s (2015) study indicates that young farmer elites are important forces in new forms of agricultural production and operation and play a central role in promoting agricultural production and leading innovation in the countryside. In contrast, some case studies also reveal that the return of young migrants is not an example of counter-urbanization, but rather a manifestation of the failure of urbanization or integration into urban life for rural migrants (Liu and Li 2017). In Lin town, reasons for farmers to return to their home village and those of passive residents who stay in the village are similar in many ways. As explained earlier, child and elder care are the most common reasons that would-be migrants remain in the countryside. For returnees, though, there is a strong push-pull effect. These young people have been pushed out of the cities due to harsh labour work and stressful workplace expectations. At the same time, they find themselves being pulled back to their rural homes by the entrepreneurial climate in the countryside. Among the 24 interviewees who had experience of migration, 10 returned to care for family members, four because of the difficulties of migrant life in the city, and six were attracted to return by rural development opportunities and entrepreneurship policies. We will focus on the latter two groups as their situations better reflect the realities of this young generation.

Meaningless Working Life and Lack of Dignity

Young farmers who take up a variety of occupations in cities have little freedom to manage their own labour and sometimes even their own bodies. Unlike their parents' generation of migrants who endured difficult working conditions to earn the wages to sustain their families at home in the village, young migrants today are more aware of their social class position and more conscious of their autonomy beyond wage income. It is a sense of meaningless and deprivation as a commodified labourer while working in cities that outweighs economic benefits of staying.

Zhao Jun is 26 and lives in MZ Village. He found employment in a factory in Hubei province, a province over 1000 kilometres away from his hometown after he finished junior high school. He told us:

> I was not accustomed to the urban life. The fast pace of urban life gave me tremendous pressure. Although the salary in the city was high, working like a robot made me extremely depressed. After travelling to many different places as a migrant, I was increasingly bored with the work in the city and convinced that planting vegetables at home would also make money if the

market was prosperous. I felt that farming freely and earning some income in my own hometown could also provide a sense of accomplishment for us young people.

In the spring of 2017, he returned to his hometown. He started cultivating vegetables on his 12-mu contracted land and made a trial of agroecology.

Wang Shaoyong from YL Village is the only interviewee with a high school education. After graduation, he worked in Chengdu, Beijing, and in other big cities, including Nanjing. It was during his time here that he had an experience that made him feel quite insecure about working stability and reflect on the meaning of work.

One day it was raining and all workers could not work. My fellow workers were all taking a nap in the hallway when our boss suddenly came and saw them. He scolded us workers for our laziness and even sacked a man who had been working for him for six years. I was really shocked by that incident. It was not usual that a migrant worker works with one boss for such long time. If it was me, I wanted to be treated like a family or friend. However, we're still working labour for the boss no matter how hard we work.

Shaoyong's attitude changed after this incident, and he asserts that migrant workers have no dignity nor guarantees when they are in the city. He moved back to his home village shortly after this incident.

Encouragement from Entrepreneurship

With the improvement to and expansion of rural roadways and means of transport, rural people's ease of travel has increased and the urban-rural dual structure has loosened. Lin town's cool climate in the summer attracts many residents from neighbouring cities who spend their summer vacations in the area. The combination of agriculture and tourism in Lin town has flourished since 2013. This boom was enhanced by governmental promotions and politics, and made even more attractive by agrotourism pioneers' healthy profits. It encouraged young farmers to return to set up their own small business in this sector.

Jing Taicheng of QJ Village is 44 years old. After he married, he was working on a building site in Beijing. "Standing on the particularly high scaffold and looking down, it was the hustle and bustle of Beijing. And all these do not belong to me." The high risk and low salary of his job gave rise to an increased willingness to return home. The county had been expanding its place in the vegetable industry since 2008, and he decided to return to take up vegetable cultivation. As mentioned earlier, agrotourism began to take hold in 2013. In response, Jing renovated his wooden house and cleaned up eight rooms in which guests could stay. In 2014, thanks to the rural development programme by local government, the outer walls of villagers' houses have been painted and the road has been re-laid, which made the environment more attractive for tourists. "I will never go out to work anymore," Jing said.

The demands of family reproduction, the stresses of urban life, and the appeal of entrepreneurship are the main reasons that young farmers in Lin town offered for their return to the countryside. Regardless of the primary reason, there are always multiple factors at play and most often they involve family. One interviewee told us: "I have elderly and children in the family, and I can come back to grow vegetables." Another said: "I have made money by working outside. Now I'm satisfied to go home to open a guesthouse and stay with my wife and children." The family-oriented ideology of rural society is a key factor in the comprehensive motivations of young farmers to return home.

Young Farmers' Difficulties in Farming

After the 2008 global fiscal crisis, a large number of migrant workers in Lin town returned to their hometowns. At that time, the government and the county were promoting the industrialization of vegetable cultivation. Farmers who abandoned grain production to plant vegetable could receive a subsidy of 100 CNY[4] per mu for vegetable cultivation in the first year of cultivation in addition to the general agricultural subsidy of 90 CNY per mu. The vegetable industry in Lin town also engaged in

---

[4] 1 CNY equals about 0.15 USD at the time of writing.

sustainable development activities, and the planting area continued to expand. Most farmers in Lin town dedicated the vast majority of their production to vegetables by 2014. At the same time, the county's agricultural bureau and the Lin town government proposed to construct a large-scale vegetable plantation as an enterprise operation. The government also established a cooperative called Shuguang as part of a modern agriculture project. Between 2008 and 2014, farmers who grew cabbages, hot peppers, and other high-yield vegetables could earn an equivalent income to migrant workers in urban areas. Moreover, the agricultural tax had just been abolished, and the nation was paying it back to farmers in the form of various agricultural subsidies. The agricultural industrial development in Lin town has become one of the driving forces to attract young people to return, which came at a time when the migrant worker market was performing sluggishly. Twenty-one of the 31 young farmers that we interviewed are engaged in vegetable cultivation, while 6 operate rural tourism enterprises and grow vegetables for family consumption. As vegetable production grew after 2014, the town's old and small market space could not accommodate the local farmers' large-scale and centralized vegetable supply. These young farmers were victims of their own success, trapped by old infrastructure and thwarted by poor market integration.

Difficulties in Market

Structural Risks of the Market: Planting Is Like Gambling

Some scholars suggest that agricultural developments and improvement of farmers' income can only be realized through intimate involvement in the market (Gao 2003; Zhao 2005). The market mechanism seems to provide farmers with the opportunity to live a prosperous life, but in fact it is impairing farmers' control over their labour. The connection between farmers' labour input and their gain is weakened (Ye 2012). The structural risk for young farmers who engage in highly commoditized and market-oriented agriculture is the increasingly fragile relationship between labour and the harvest. Peasants have their own words for this predicament: "the result of vegetable cultivation depends on your luck," "growing vegetables is a kind of gambling," and "the lucky make money while the unlucky lose money." Of the 31 young farmers that we interviewed, an overwhelming majority (28) agreed that the main issue in agricultural production was the fluctuating market prices. Farmers are unable to enhance profits by raising yield or reducing costs, and the market structure restricts farmers' independence and initiatives. They can no longer obtain the corresponding income by controlling their own labour input or their contribution to means of production. Their income, instead, depends on others' production situation in other locales. Nevertheless, there are not enough non-agricultural channels to support one's family in the underdeveloped western countryside, so farmers have little choice but to stay in farming.

Peng Xing from MZ Village is 42 years old. In around 2013, he planted 10 mu of hot peppers and cabbage. In the past, only a few farmers cultivated these vegetables and Peng's profits reflected this scarcity. The price of vegetables has declined fiercely since 2016. He has heard that farmers in other locales, including Liangshan in Sichuan province and Dingxi in Gansu province, have begun large-scale alpine vegetable production, which means he has to compete with farmers thousands of kilometres away for a better price:

> Now more and more farmers are growing vegetables and the market becomes very competitive. Only when they (farmers elsewhere) are affected by natural disasters, our vegetables can be sold at a good price. Farming is too toilsome. We lose money in the sluggish market and break even in the prosperous market. We could only earn money for our labour expense. To be a farmer is not easy and interesting.

Peng's two brothers and sister have converted their farmland into woodland. "We planted for food and clothing in the past. We know nothing about the market. Thus, we were satisfied with grain plantation even if it only had very few yields. Now it's not the same. We need more cash income to live and feel that our input in farming should bring us more return."

Yan Wenguang from MZ Village planted 20 mu of cabbage, of which 10 mu was from neighbours who were working in the city. Yan was asked by his neighbours to take care of the land for them for free, and he would need to return the land to his neighbours' use once they returned.

> Growing vegetables is a kind of gambling. The lucky ones make money while the unlucky ones lose money. That's it. In the year before the last, the cabbage price was only 16 cents per kilo. I planted 10 mu of cabbages with the yield about 100,000 catty.[5] I sold nothing. All the cabbages were rotten in the ground. I even could not cover the cost of hiring labour to harvest the vegetables. Some people came to buy but the price was extremely low. At that price, the more vegetables I sold, the more money I lost. I could not understand why the price was so low. However, my children are studying at school and we need a large amount of money. Whether I'm the lucky one or not, I have to farm.

Lack of Risk Avoidance: The Marginalization of Farmers

The market is essential for highly industrialized vegetable production and structural risks are inevitable. If there are no mechanisms for market service or risk mitigation in production and marketing, the risks will be prominent in agricultural production and in young farmers' daily lives. In Lin town, crop insurance does not cover the loss when farmers encounter marketing problems as it applies only for losses caused by natural disaster. Almost every household in Lin town purchased crop insurance annually as required by local government. When the price that farmers receive for their vegetables declined in 2015, the insurance company refused to pay out any of the farmers' claims because of its terms.

Zhao Hongju from YL Village is 40 years old and planted 7 mu of cabbages in 2017. She also purchased crop insurance. Last year, she lost money due to the bad market price but received no compensation from the insurance company. She visited the company's office to follow up on her request, but the staff just made excuses and she left empty-handed. "When I bought the insurance, the (county) government gave us

---

[5] Catty is a Chinese measuring unit for weight and 1 catty equals 0.5 kilo.

guarantees, promising that the company would pay if we had a bad season. With the (county) government as endorser, we could not refuse to buy. Otherwise it will offend the government."

The Last Obstacle to the Market: Lack of Bargaining Power

Despite the great infrastructure improvements in Lin town in recent years, there are still some villages that remain relatively inaccessible. The unfavourable traffic conditions in mountainous area greatly weaken farmers' bargaining power because they don't have vehicles to deliver their products to the market. These dual disadvantages of geographic location and traffic conditions are further obstacles in farmers' last mile to the market. Thus, these farmers can only passively accept the low prices that external brokers and middlemen offer without feeling that they have the ability to bargain. The brokers that control the market have formed a decentralized power that controls vegetable prices.

Li from QL Village is 31 years old and planted nearly 9 mu of vegetables. He perpetually worries about how his vegetables will sell. The village has not built roads to the locations where his production team and the other two production teams are located—there are only dirt paths. "On rainy days, the paths are muddy and cars can't move. Take my production team for instance. We have fertile land but no good roads. We can only sell vegetables when external dealers drive in. These dealers purchase at the price of 40 cents per kilo and resell at the price of 1 yuan (100 cents). We lost much profits in this circulation process."

Lack of Organization

Difficulties with Cooperative Organizations

Joining or forming a cooperative is a common way in which farmers can reduce their market risks. Cooperatives are generally considered to reduce costs and strengthen farmers' bargaining power in the processes of production and marketing. Cooperatives are also facing the organizational dilemma of having alienated the people they were organized to serve. They remain an effective tool for urban and commercial capital and a small number of rural elites to appropriate profits from smallholders (Feng 2014). The risks of the infinite market have also forced some large holdings into failure, bankruptcy, or even run away from the countryside. In Lin town, the Shuguang cooperative does not function as expected and it has many practical difficulties. First, for small-scale farmers, some cooperatives have become a market subject independent to or overriding the member farmers from who it extracts profits. The large holders of land and rural elites are often the beneficiaries of co-ops (Feng 2014). Ironically, the local government continues to support such agricultural organizations since supporting agricultural businesses and cooperatives is the policy orientation by the central government.

Peng in MZ Village told us that "the Shuguang cooperative has no entry barriers. You can buy production inputs such as seeds and fertilizers from the co-op with a preferential price as long as you plant and pay the annual fee to the co-op. It is 100 yuan per year." For Peng, it is like applying for a membership card. He said that the cooperative might raise the price of materials when it resells to farmers. "The more products you buy, the more discounts you receive. It sounds like promotion. Joining in co-op doesn't really benefit farmer."

Consequently, cooperative organizations do not reduce farmers' costs significantly in the production process. In terms of selling one's product, the co-op did not function as it should. It could do nothing when the market for vegetables was depressed. One farmer told us: "In the first half of this year, the price of lettuce was only two cents per kilo. The co-op could only accept the price offered by external middlemen. It got a high fee from the dealers. As a result, they have no willingness and abilities to bargain with dealers." Aside from prioritizing their own interests, co-op officials also confessed that they were unable to change the farmgate price of vegetables if the market price in general was not favourable. The almost total lack of an organized cooperation mechanism allowed other agents in the market to squeeze farmers.

Difficulties in Service Organization

Young farmers who engage in agriculture are short of agricultural services such as agricultural technique service and financial service. Since Lin town farmers began to cultivate vegetables in the early 1990s, they have expanded the scale of production but without consideration for the necessity of crop rotation. Farmers are swayed by the agricultural capitalization path of improved varieties of seeds, fertilizers, and pesticides. The consequences can be disastrous. The soil accumulates a large amount of bacteria and vegetables become vulnerable to disease. If you plant cruciferous vegetables, clubroot may appear. In 2017, the disease appeared on one-third of the farms in Lin town. Under such circumstances, it is essential to have access to agricultural technology services and relevant training that can prevent or treat such diseases. Young farmers are generally aware of the difficulties in obtaining agricultural technology services; the government focuses these resources on large households and enterprises rather than protecting smallholder farmers who rely on agriculture for their livelihood. After marketization, the services that had been available in the villages and the town collapsed. Individual peasants' agricultural knowledge is insufficient in a climate where imported seeds and various fertilizers and pesticides are dominant.

Peng from MZ Village told us:

> Policies now support those who do nothing rather than who farm. For example, many companies for land circulation get compensation from the government. The ordinary people get nothing ... The government spent more than 700 million yuan to prepare for the cooperative a large cold storage, which can accommodate more than 100 tons of vegetables. When the market is not good, cooperatives can collect vegetables at a low price, storing them in it. And they sell them out when the price rises again. What do the farmers gain?

The young farmer Li from QL Village planted more than 300 walnut seedlings last year and more than 100 died. "Nobody taught you. And I don't know who I can go to ask for help. The walnut seedlings got leaf curl virus. I sprayed pesticides, but there were still more than 100 seedlings dying."

Zhao Xin, a 19-year-old from MZ village, took a chance and planted schisandra, a herb used in Chinese medicine. Zhao had trouble locating the seedlings and his search took him online. "Baidu is the most commonly used channel. Internet is not always a helpful assistant. I bought fake seeds online and it cost more than 3000 yuan." Zhao learned this hard lesson from his own failure. His aim was to encourage villages to join him in starting a Schisandra Park, but this ambition conflicted with the plans of the village committee and the town government. The town supports vegetable cooperatives in the village because nine village cadres are shareholders, and they did not want to develop other industries outside of vegetable cultivation. As a result, the village committee compelled other farmers not to cooperate with Zhao.

With the deepening of the market economy, local governments have been gradually withdrawing from the supply of public services in agriculture such as agricultural machinery services and marketing service shifts from collectivization to levitation. Farmers with their own ideas and interests want to develop these independently. However, the individualization of peasants is also preventing them from self-organizing. Despite the changing environment, farmers retain their expectations of the local government to take a lead in service delivery. Most of the interviewees told us that they feel a strong sense of powerlessness when dealing with the ever-present market risks. Their view is that individuals alone cannot fight this struggle and believe that the government should take responsibility for organizing farmers and providing social services in agriculture.

Young Farmers' Innovations in Sustaining Rural Livelihoods

Confronting the structural risks of agriculture and the dilemma of organizing production, it is difficult for most young farmers to shift away from traditional agricultural production and management. The rural social stratification in the countryside is also an obstacle for young people in their efforts to access diverse resources such as information and finance. In this context, some young returnees have been able to combine their non-farming and migration experiences to innovate in agricultural production in an effort to modernize and sustain their livelihoods.

Broadening Agriculture-Based Multiple Job Holdings

Local development policies oriented towards industry integration provide an important base and environment for young farmers to innovate. The industrial integration in Lin town is focused on merging agriculture with tourism. There are two advantages for such industrial integration in Lin town: beautiful landscape in Lin town and the local government's efforts to push a "combination of agriculture and tourism" against a backdrop of agricultural supply-side reform. The unique climate and natural landscape made the Zengjia Mountains in which Lin town is located one of the top ten summer destinations in China. Every year, about 50,000 tourists escape their urban homes to spend their summer vacation in the area; the most popular period is June to September. These visitors pay to stay with locals in their homes. Some villagers saw this as an opportunity and renovated their homes as guesthouses. To stimulate farmers' enthusiasm for engaging in agrotourism and to accelerate the tourism economy, in 2012, the government advertised tourism in Lin town through its official social media account and built samples of guesthouses for farmers to follow and imitate. Guesthouses operated by farmers gradually developed in this context, with the numbers increasing from 1 guesthouse in 2012 to 102 by the summer of 2017. Most of the newly married young farmers renovated their new houses that were gifted by their parents to include neat and comfortable rooms for urban guests.

The opening of tourism in the area also afforded young farmers the chance to adjust their agricultural production. Some young farmers deepened their "grain and vegetable" cultivation to the mode of "grain+vegetable+service." Vegetables, grains, meat, and eggs are all local food to serve tourists. Vegetables that are often worthless in the market can make good returns when they are provided to guests as "rural food." Guests pay for the meals cooked by farmers and sometimes buy fresh vegetables when they leave. This is not limited to cabbage—tourists are keen on all garden vegetables and grains produced locally. This interest has increased the value added of agro-products for these households and lessened part of their burden in terms of vegetable sales. Other young farmers have broadened their operations since 2016. When food safety became one of Chinese society's central concerns, urban residents increasingly welcomed locally produced, quality food from the countryside. The entrepreneurial activities of Zhao Jun, a 26-year-old from YL Village, is a typical example.

Zhao only attended seven days of senior middle school before he had to abandon his studies and join the workforce. He relocated to Qingdao city in Shandong province where he trained as an apprentice, learning vehicle repair and cooking. He also ran a fast-food restaurant with his brother in Linyi county, Shandong province. After returning home five years ago, he took a contract position as a salesman with a telecom company. Later, he registered a small company with his brother and together they engaged in business outsourcing for the telecom company. After working outside of the village for many years, Zhao was obsessed with making money and he had become proficient at seizing money-making opportunities. After this telecom outsourcing business encountered bottlenecks, Zhao took advantage of the new tourism market and returned to his home village. He invested all of his and his parents' savings to construct and decorate the family home as a guesthouse. Baiyun Guesthouse can accommodate more than 50 people at a time. Tourists arrive daily and the fee for accommodation and three meals per day is CNY 80–100 per person.

Zhao's parents planted 5 mu of land in 2017 and raised more than 100 chickens and two pigs. Even before the guesthouse opened, Zhao had over 100 followers and friends on WeChat. He took advantage of this situation and began to advertise the sale of free-range chickens and local pork via social media. During Spring Festival in 2016, he took pre-orders and then collected 12 pigs from around the village to fulfil these orders from friends around the country. The price was CNY 20 per catty, which was higher than the families would receive if they sold the meat at the local market in Lin town. This successful experience of selling agro-products via the internet gave Zhao a boost of confidence.

Young Farmers' Embedded Entrepreneurship and Its Socioeconomic Impacts

One characteristic of farmers' grassroots entrepreneurship is embeddability, which comprises three aspects. First, young farmers depend on local natural resources and their families' social capital, the latter of which is a distinctive feature of rural entrepreneurship. In Lin town, for example, the local climate, terrain, and environment directly determine the direction and progress of entrepreneurship. Family assets, such as houses and land, also play an important role. The pioneering actions of young farmers are deeply embedded in the rural society. Second, farmers' entrepreneurial actions depend upon family support and family farming. Zhao's successful agro-product enterprise would not have been possible without his parents' farm or their family home. Farmland maintained by parents and unpaid family labour support all contribute to young farmers' livelihood innovations. Thirdly, the entrepreneurial actions of young farmers are deeply embedded in the "acquaintance society," which refers to a reliance on interpersonal relationships. In Zhao's case, he accumulated abundant social capital during his time working away from home; he developed a market in acquaintances, both friends and former customers. Most of the visitors to his family's guesthouse are referred by Zhao's friends.

Embedded entrepreneurship is most prevalent among young farmers who have recent migration experience. With their monetary savings, social networks, and new skills accumulated during migration together with their family's agricultural foundations in the village, young entrepreneurs are able to deepen and broaden small-scale family farming in order to sustain their livelihoods and those of their families. On the micro level, the primary change is the increase in farmers' incomes, which in turn encourages more young people to return home. This reverse migration has also reinvigorated the countryside, slowly remedying the problem of the "left-behind" family members as well as starting to fill the gaps in the agricultural labour force. Virtually all of the operators are young people. The local government has applied strict standards to regulate food safety and service quality to protect the tourism industry, and young people have more easily adapted to this new environment than the older generation.

Thirty-eight-year-old Yang cultivates 2 mu of vegetables, but it is not enough to support her family. She and her husband left the village for three years to find employment. Their youngest son, who was 12 years old at the time, was a left-behind child who remained at home to attend school. When the market for migrant workers declined, Yang and her husband followed the tourists back to their home village. They have registered a guesthouse business licence with the county's Industry and Commerce Bureau and are renovating their house. Yang was very happy to return. "Now I can take care of my son. He's no longer a left-behind child. He's going to enter into junior middle school. Our staying will make him more concentrated in study. All our family can stay together."

Moreover, their entrepreneurship has positive impacts on regional rural development. It reflects the re-grounding of agriculture in an agroecological sense. These young farmers have managed to establish themselves in a new market—catering to urbanites' desire for healthy agricultural products. The farmers began to consciously reduce their use of fertilizers and pesticides, adopting the concept of ecological farming as much as possible. This important conversion is also one step towards solving the aforementioned soil bacteria problems that accompanies extended monocropping. Their entrepreneurship also facilitates urbanrural integration. As the migrant workers return to their home villages and more and more urban residents visit the countryside, the relationship between rural and urban is morphing into one that is increasingly organic. Rational communication and human interaction start and accelerate the process of urban-rural integration.

Conclusion

Among the fruitful studies in Chinese academia on agrarian transition and farmers' dynamics, perspectives of young farmers and their experiences are mostly overlooked. Although some researchers have paid attention to young farmers and examined the youth perspective, their focus was on elite or special groups, such as elite young peasants in Liu (2017), landless farmers in Xu's study (2008), and young farmers related with gang in Deng's (2011) study. With regard to young farmers' entrepreneurship, existing research concentrates on young farmers' economic foundations earned from non-farming activities rather than their agricultural activities.

In this context, this chapter explored the characteristics of young farmers engaged in agriculture from a more general perspective. Through in-depth interviews with 31 young farmers engaged in industrialized vegetable cultivation in western China, this study found that there are significant differences in mindsets of passive young farmers and of returning migrants. They should be distinguished in researchers' studies and the process of policymaking. However, when they stay together in the countryside and live by agriculture, both groups of young farmers need to confront the market's structural risks and predicaments of organizing production. These have negatively impacted the livelihoods and development of young farmers. With the advancement of urban-rural integration and the changes in rural society, young farmers, especially those with migration experiences, have begun to actively utilize environmental advantages and entrepreneurial resources to innovate in agricultural production and management. This is a breakthrough against the aforementioned risks and predicaments. Young farmers' entrepreneurship has an apparent embeddability. It provides them with a relatively stable market, minimizing risks for young farmers in the early stages of their venture. Most young farmers who return home do not expect riches but do for livelihood reasons. Their adaptation has generated obvious economic and social impacts, providing substantive ideas for improving farming, young farmers' situations, the agricultural sector, and the problems that plague rural areas.

References

Deng, Wanchun. 2011. Getting rich together: Participation of grey young farmers in new countryside construction ["ju zai yiqi facai": xin nongcun jianshe zhong de qingnian nongmin de "huihua" canyu]. *China Youth Study [Zhongguo qingnian yanjiu]* 9: 29–33.

Feng, Xiao. 2014. The local logic of farmers' professional cooperative alienation: Based on capital to countryside packaged by cooperative [Nongmin zhuanyehezuoshe zhidu yihua de xiangtu luoji: yi "hezuoshe baozhuang xiaxiang ziben" wei li]. *China Rural Survey [Zhongguo nongcun guancha]* 2: 2–8.

Gao, Yong. 2003. Improvement of the farmers' organization level to enter the market: Investigation and consideration on countryside in Chengdu [Dali tigao nongmin jinru shichang de zuzhihua Chengdu: dui Chengdushi nongcun qingkuang de diaocha yu sikao]. *Rural Economy [Nongcun jingji]* 11: 1–4.

Liang, Dong, and Huifang Wu. 2017. Dynamics and impacts of the feminization of agriculture on gender relationships in rural China: First-hand research in villages in the provinces of Jiangsu, Sichuan and Shanxi [Nongye nvxinghua de dongli jizhi ji qi dui nongcun xingbie guanxi de yingxiang yanjiu: ji yu jiangsu, sichuan, shanxi san sheng de cunzhuang shidi diaoyan]. *Journal of Chinese Women's Studies (Funv yanjiu lun cong)* 6: 85–97.

Liu, Youfu, and Xiangping Li. 2017. Counter-urbanization or pseudo-urbanization: Consideration on returning college students and farmers and discussion with Shen Dong ["ni chengshihua" haishi "wei chengshihua": fansi daxuesheng, nongmin "licheng fanxiang" wenti jian yu shen dong shangque]. *China Youth Study (Zhongguo qingnian yanjiu)* 6: 24–30.

Liu, Tongshan, Feng Mao, and Xiangzhi Kong. 2015. Study on the effect of young farmer elite among new agricultural management subjects: Based on examples of Henan province [Xinxing nongye jingying zhuti zhong qingnian nongmin jingying de zuoyong yanjiu: yi henansheng wei li]. *Rural Economy [Nongcun jingji]* 9: 104–109.

Xia, Zhuzhi. 2017. Farmers returning and entrepreneurship in countryside: Sketches on 38 returning entrepreneurs in H Town [Qianru xiangcun shehui de nongmingong fanxiang chuangye: dui H zhen 38 li fanxiang chuangyezhe de shenmiao]. *China Youth Study (Zhongguo qingnian yanjiu)* 6: 5–11.

Xu, Yong. 2008. To prevent vagrant trend of landless young farmers [Fangzhi wudi qingnian "nongmin" youminhua]. *Exploration and Free Views [Tansuo yu zhengming]* 3: 8–10.

Ye, Jingzhong. 2012. One reaps no more than what he has sown when he is in the market economy [Yifen gengyun weibi you yifen shouhuo—dang nongmin shuangjiao zhanzai shichang jingji zhi Zhong]. *Journal of China Agricultural University (Social Sciences Edition) [Zhongguo nongye daxue xuebao]* 29 (1): 5–13.

Zhao, Quanmin. 2005. Farmers' subject position: Basic point for Farmers' Organization [Nongmin de shichang zhuti diwei: shixian nongmin zuzhihua de jidian]. *Theory Journal [Lilun xue kan]* 7: 76–79.

7

Young Farmers in a "Cucumber Village": A Different Story of Family Farming in Agricultural Specialization from Hebei Province

Lu Pan

Introduction

This village came to our attention by accident. Given the rural labour migration situation in China, especially in the mid-west, it was difficult to find a village with many young farmers. We came across this village, which is well-known in its adjacent areas for its specialization in cucumber cultivation and marketing. When reviewing the site's agricultural development, the groups of young farmers gradually came to stand out. Huang village, which is a pseudonym, is located in Hebei province, about 430 kilometres south to Beijing and 10 kilometres from the county seat. The research team conducted fieldwork in the village in December 2017. We lodged in rural households during our fieldwork. We used both

L. Pan (✉)
College of Humanities and Development Studies (COHD), China
Agricultural University, Beijing, China
e-mail: panlu@cau.edu.cn

© The Author(s) 2024   **183**
S. Srinivasan (ed.), *Becoming A Young Farmer*, Rethinking Rural,
https://doi.org/10.1007/978-3-031-15233-7_7

qualitative data, conducting semi-structured interviews as well as quantitative data collected via a survey. We also engaged in supplementary work in June 2018.

We sampled young farmers[1] using a combination of purposive and snowball sampling methods. Our household host provided the names of our first few interviewees and we were able to expand the sample through introductions by our interviewees. Most interviews were completed in greenhouses where cucumber farmers spend their days. We conducted the interviews while working with the farmers to gain an intimate experience of their daily lives and the hard labour required to keep the farm operating. We also interviewed village leaders, market managers, village technicians, shopkeepers of agro-inputs, and other key informants. Among the 48 interviewees under 45 years old, 27 were male and 21 were female. Most of them (39 interviewees) have only 9 years of schooling and only 7 finished 12 years of schooling. The average age of interviewees was 35.8 years old. Half of them were under the age of 35. Most of them started farming independently in the area in their twenties. They are indeed quite young compared to the general demographic situation of agricultural labourers in China. This chapter will reveal the atypical story of the village and its young farmers.

Community Profile of the "Cucumber Village"

Huang is a small village with a total population of 1109. Among its 267 households, about 96 per cent are involved in cucumber production. The total farmland area in the village is 1246 mu[2] (about 83 ha), among which the 800 cucumber greenhouses occupy over 1100 mu, leaving the few remaining mu as farmland for corn and grain crops. Cucumber production has brought considerable profits to rural households. In 2016, the daily average trading volume of cucumbers was over 40,000 tons with an annual turnover of CNY[3] 400 million. The average per capita income

---

[1] All farmers' names are pseudonyms in this chapter.
[2] 15mu equals 1 hectare.
[3] 1 CNY equals about 0.15 USD at the time of writing

in Huang village during the time of our interviews was about CNY 30,000. In 2018, the national average per capita income was CNY 39,251 for urban residents and CNY 14,617 for rural residents.

It took over 30 years for the village to develop its specialization in cucumber cultivation. Household responsibility land reform took place in 1984 in the village. Due to rapid population growth and limited farmland, land allocated to each villager was no more than 1 mu. With such a tiny plot of farmland, villagers continued to live in general poverty. Village leaders encouraged people to switch to cash crops in order to increase their incomes. They tried apples and apricots, among other crops, before they homed in on cucumbers. The village's climate and environment is suitable for growing and the high demand of labour input required also matched well with the person-land ratio. In order to encourage villagers to plant cucumbers, village cadres and Chinese Communist Party (CCP) members took the lead to construct the first batch of 46 greenhouses in 1988. This first generation were small and simple structures constructed with bamboo and timber with 0.6–0.8 mu of floor space. Cucumber production in the late 1980s rapidly increased producers' incomes. There was a saying in the village at that time that "one greenhouse and one small field, 300 yuan one year yield." During the 1990s and early 2000s, the number of cucumber greenhouses increased from 40 to over 300 and, on average, each household kept one greenhouse. Although cucumbers generated higher profits than grain, due to the high labour demands and the low level of mechanization (it took an hour to manually roll up the greenhouse shutters), most households could only plant one greenhouse. The annual income generated from one greenhouse in 2007 was about CNY 20,000.

In 2007, cucumber production in Huang village encountered its first bottleneck. The price of cucumbers had declined and the greenhouse could not sustain the same number of labourers. Many young people gave up farming and travelled to the cities to work as migrant labourers. One retired village cadre recalled that "cucumber production of our village was in danger that time and it's possible that cucumber production would fail after almost 20 years' efforts." The newly elected village leader during that period actively sought solutions to this crisis. In 2009, Huang villagers received a technological boost with the introduction of greenhouse rollers in the county. This piece of machinery dramatically reduced shutter rolling time from an hour to a few minutes, increased illumination time in the greenhouse, and relieved a portion of farmers' labour burden. It was such a significant change that it allowed farmers to increase cultivation. In 2012, the village committee applied for support from the county government's agricultural poverty alleviation and development fund for agricultural transition. The village planned two large vegetable plantations and applied for a subsidy for 50 additional greenhouses. Thanks to generous subsidies and an upgrade to steel frame greenhouses, there were over 100 households submitting applications to the village committee in the first year alone. In 2017, to encourage increased production, the village committee applied for more funding from the county government to subsidize steel costs for newly built greenhouses. The village leader also applied for CNY 2 million loan from the Rural Credit Cooperative to help villagers develop cucumber production. All of the above economic and technical support has motivated villagers to sustain cucumber production. Since 2013, more and more villagers have returned from the city to resume an agricultural livelihood. Prior to 2018, there were only about 30 young people working outside of the village as migrant workers. Most of the villagers are smallholder farmers with 1 or 2 greenhouses, several big land holders have more than 10 greenhouses, but the average is around 4 greenhouses.

Becoming a Young Farmer

When compared to tens of thousands of villages in China, Huang village is unique due to the number of young farmers in one village. In the context of a hollowed-out countryside, Huang villagers can make a moderate income and be prosperous in their agriculture endeavours in their home village by virtue of their own labour, supportive policies, and favourable market conditions. Their farming experiences reflect the characteristics of young people in rural society.

Entering into Farming: Gender Differentiated Self-Choices

The traditional way of entering into farming in rural China can be called "natural employment" that there was no occupational qualification requirement for age, educational level, and so forth to be a farmer. Farming skills were usually passed to offspring from the older generation without formal training. Young people automatically became farmers once they worked on the land. There was no retirement system accordingly for farmers. The young generation becomes accustomed to farming from a very young age and is then able to take over the farm when they are older. For young farmers in Huang village, taking up farming is not natural employment. It is not completely strange to them, nor are they familiar with it. Most young farmers have certain childhood experiences from when they helped their parents with farm work. Most of them moved to the city when they finished junior middle school and have accumulated rich migration experiences. They were a part of the so-called new generation of migrant workers and their involvement in farming is not a natural process.

For 32-year-old Likai, her first experience with farming was at age 10 when she helped her parents with weeding. She is the family's eldest daughter and had to help her parents with farm work even if she was reluctant. She migrated for work at age 15 and her first job was packaging preserved ducks in a food factory in Baoding city, Hebei province. Two years later, she moved to a spinning factory in Shouguang city in Shandong province and then to a food processing factory where she worked for several years. She returned to the village at age 22 to marry and stayed. It is very common for young women like her to remain at home after marriage.

Wang Tao, 27 years old, only has a primary school education. She did not have the opportunity to attend middle school due to her family's poor economic circumstances and her parents' rejection of her education. She first worked in an ice cream factory in Tianjin municipality near Beijing and then at a clothing factory in Yantai city in Shandong province; her friends helped her to find and secure the latter job.

Thirty-year-old Hu Ting had a very similar experience to Wang Tao. Ting also migrated for work after primary school. She worked in a restaurant in Handan city in Hebei province for years before her marriage. Although she is a farmer now, she told us that her migration experience had a very important influence on her farming career. "We interacted with many different people and experienced many different things, and have become more outward. It is also a very useful experience when we deal with other people as a farmer."

Young women farmers in Huang village are usually the returned migrants. In accordance with village conventions and parents-in-law's willingness, they quit migration post-marriage and stay in the village of their husbands to take care of children. Young women do not have contract land in the village of their husbands and sometimes assist their parents-in-law with farming responsibilities. While young women stay in the village for family care, their husbands often continue working in cities. This situation and these women's experiences reflect the influence of traditional gender norms on rural women's occupational development. In contrast, the reasons for young men's return are more diversified and proactive. Family is only a minor element in their reasoning as most returned to the village to pursue farming as a result of the push-pull effect in rural-urban society, that is, being pushed away by the marginalized migrating experience in the city and pulled back home by the prospect of cucumber production in the village.

Wang Zhichao, age 38, was a migrant worker many years ago. In 2001, he worked in a machinery store in Handan city in Hebei province where he sold combines for grain harvesting. His salary was around CNY 500 per month and this work only lasted for six months. Wang is a National Basketball Association (NBA) fan, and during the finals one year, he left work to watch the game without asking his boss' permission. When he returned to the job site, his boss scolded him and said that he would dock his salary for half a month. Wang was very angry and could not abide by such humiliation. He decided to return home to farm.

Now I've been farming for over 10 years. I will not change my job, it's very difficult. For young people in our village, we don't have special skills nor a high level of education. We can only do hard labour worker in the city and earn 3000 to 4000 yuan a month. Planting cucumbers at home is also very hard work but we can enjoy the freedom and don't need to be disciplined. One summer a few years ago, I went to a construction site for short-term work when my greenhouse fallowed. It was burning hot at noon in August, but we still need to work in the sun. It's very impressive for me. Cucumber production requires intensive work, but I can be my own boss and have a bit higher income than migration. It's more comfortable than migrant work.

When we met him, Wang was planting three cucumber greenhouses. He had various migrant labour experiences on construction sites, in the oil field, and in different factories, among others. He lasted only one or two months at any one job. "No skills or knowledge, it's difficult for us to settle down in the city." Wang returned to the village to plant cucumbers after careful thought and consideration. He had three key reasons for doing so: one, an income earned from agriculture in the village is better than he could earn via migration; second, he can live with his family; and third, after decades of development, cucumber production has stabilized in the village.

The return of young men to the village is the driving force for family farming and agrarian transition. This reverse migration is also a key factor in reshaping young women into vocational farmers as the joint work of couples has organizational advantages that fit well with the intensive labour demands of cucumber production. This is not to deny the possibility of independent young women farmers. There are some left-behind women who farm successfully when their husbands are working in the city. General agrarian transition in the village provides favourable conditions for farming. However, in spite of women and men's differing reasons for entering into farming, it is the joint commitment of young couples to farming that enhances the foundations of family farming and further brings about agrarian dynamics in Huang village.

Household Split: Establishment of New Farming Households for Young People

The return of young men has accelerated household split, which occurs at the same time as the reproduction of a new farming unit. In rural society, household split indicates, on one hand, the birth of an independent son's family and, on the other hand, the distribution of household wealth and means of livelihood among sons to allow for the reproduction of each son's household. In traditional agricultural society, the means of production in agriculture were under the full control of the father, with the son obtaining his share of means of production via a household split. Along with social transformation, the one-off mode of household split prevails in rural society instead of the multiple mode of household split. The timing of household split has also shifted to earlier in a son's life. Usually a son would split with his parents immediately after his wedding. In regions that generally produce low agricultural revenues, household split has more social and cultural implications than economic purpose. Low agricultural revenues provoke young people's migration, and regardless of any household split, it is usually the elderly who remain in the village to run the farm, hence the phenomenon of "grey agriculture." In Huang village where agricultural production is more profitable, the household split has equal cultural and economic significance for young people. Young couples become an independent farming unit after a household split and need to plan and organize agricultural production by themselves. Very often they can only get a few means of production from their parents. Some young farmers are unlucky and do not acquire any land from a household split.[4]

The number of greenhouses in Table 7.1 for the 13 cases is from one greenhouse to nine greenhouses. Some young farmers can inherit one or two greenhouses through household split or build the greenhouses with

---

[4] The average family size in Huang village was 4.15 in the time of investigation. According to the seventh National Census in 2020, the national average family size is 2.62, dropping from 3.1 in 2010 and showing the tendency of miniaturization. In Huang village, as in many rural regions of China, couples in their 30s and 40s usually have two children. If the first-born was a daughter, they were allowed to have a second one in the time of birth control policy. The nuclear family with two children is common in rural area. Married adult son may split with parents in householding but still live in one single yard.

Table 7.1 Farming scale in some households of young farmers

Interviewee	Number of greenhouses	Greenhouse through household split/built with parents' assistance	Greenhouse built by young couple
Hu ting	3	1	2
Cui Weiying	9	0	9
Li Caixia	1	0	1
Yu Kun	2	0	2
Dai Qiuyan	1	0	1
Wang Jing	3	2	1
Wu Xia	7	0	7
Han yan	4	2	2
Cai Hong	4	2	2
Wu Xin	7	1	6
Yao hui	2	1	1
Chen Honge	6	1	5
Xiao Qing	4	1	3

parental assistance. For others, they must secure a greenhouse through their own efforts. Inherited greenhouses are usually constructed in the old style—smaller buildings with bamboo frames. In all cases, young people have managed to expand their scale of production due to their hard work; some even grew to be large holders in the village. In our research, we found that the average household land area for a young farmer is 2.76 mu. For the 21 young farmers who received land as part of a household split, the average area of inherited land is 1.09 mu. This land, however, may be too small and scattered to be able to construct a greenhouse. In such situations, the young farmers have to rent land. It is important to note that the intergenerational transmission of land and greenhouses constitutes a necessary foundation for young people to launch their own farming careers. However, what is more central is their labour and accumulation and acquirement of community resources.

Wang Weichun and his wife operate five greenhouses. They are both 38 years old. Their first greenhouse was constructed in 2005. After their marriage, Wang split received 1.6 mu in farmland from his parents as part of a household split. The young couple wanted to build a greenhouse but had no money. The wife borrowed some money from her family and the couple built their first greenhouse on the land that Wang's parents provided. The second and third greenhouses were built in 2007 and 2009, respectively. For convenience, Wang rented land from his neighbours near his first greenhouse. He can take care of the three greenhouses with minimum transport costs. The first three greenhouses were bamboo frame structures, which meant that the investment cost was minimal—it costs about CNY 30,000 to build a bamboo greenhouse. With two years' revenues from the first greenhouse, Wang was able to accumulate extra capital to expand the scale of the couple's production. The fourth and fifth greenhouses were constructed with steel frames at a cost of CNY 70,000 each. Although the investment increased, it was not difficult for the couple to build the last two greenhouses given the realized accumulation from earlier phases of production.

In our fieldwork, we discovered several explanations for the limited intergenerational transmission of agricultural resources in Huang village. First, parent generations do not have abundant resource to dispose of, especially land. They also face production scale limits. Before the cucumber production boom in Huang village in 2013, each household had only one or two old-style greenhouses. Given the scarcity of land when compared to the population, there were limited production resources that the older generation could pass down to their children in a household split. The second important factor is the special life cycle of the parent generation. As young farmers in Huang village are in their thirties and forties, their parents are mostly in their sixties, an age with a moderate capacity for labour and in urgent need to prepare for their later life. Rural citizens can voluntarily join the New Rural Social Pension Insurance, a policy launched nationwide in 2009 to guarantee older adults above the age of 60 a regular pension. In Huang village, the pension is about CNY 80 per month. In rural society, adult children usually do not provide economic support or living expense for their parents if the latter still have the capacity to work. Therefore, the older generation has to rely on agricultural production to maintain their livelihood and save as much as they can for later life. The high profits from cucumber production mean that older people cannot easily give up their land and pass it on to their children. Although a parent's household and a son's household are separate independent calculation units in agriculture, it does not exclude their intergenerational reciprocity in specific production activities, especially intergenerational support from the older generation. Production scale is often smaller in the parent's household. When they have finished their own farm work, parents often help out with routine chores at their children's greenhouses or collect their grandchildren from the village school. Strong family cohesion and solidarity remains after a household split, which is the foundation of Chinese culture and peasant agriculture.

Yao Hui is 30 years old and has an eight-year-old daughter and a five-year-old son. She has been living in her home village since she married. The couple stayed with one set of parents in the first several years after returning to the village. They didn't need to worry about daily expenditures but lacked economic autonomy. When Yao's daughter was three, her in-laws proposed a household split. Huihui's husband and his brother each inherited one small greenhouse from the parents and jointly share their parents' living expenses. After the split, the whole family were engaged in cucumber production. As the older brother planted three greenhouses and Yao's family only had one, her parents-in-law provided assistance to their older son. If Yao and her husband were very busy, the parents and older brother would come to help. "Although we are not in one household (with the in-laws), we still have very close interaction, just as before and our relationship is very good."

Acquirement of Knowledge and Skills: Active Learning Through Limited Ways

From migrant worker to cucumber producer has been a huge occupational transition for young farmers in Huang village. To become a farmer requires skills, experience, and knowledge that forms the basis for their own farm work after they settle into their new household. There are four major paths for young people to acquire knowledge and skills in cucumber production, which also reflect this group's different demands for technology as production evolves. The first channel to acquire basic knowledge in production is usually within the family. As Huang villagers have 30 years of cucumber production experience, knowledge transfer to young farmers was initially an intergenerational transmission within family. Parents, parents-in-law, and sometimes spouses were farmers' first teachers.

Thirteen years ago, Wu Xia married into a family that lives in Huang village and began cucumber cultivation with her husband. Wu's own parents own cucumber greenhouses in an adjacent village therefore she came to the marriage with production skills. Cai Hong, on the other hand, became acquainted with cucumber production only after she married and moved to the village. She gradually learned the necessary methods and skills from her parents-in-law. Her husband also provided instruction. "It's not difficult. Almost all of the villagers are planting cucumber here, you just need to observe and follow your parents, then you know how to do it."

Secondly, young farmers can receive technical support in everyday production through communication at the community level. Although young people can learn basic production technologies from their parents, they may encounter various and changing technical problems in pest control, seedling management, and so forth in specific production that require outside expertise. The three decades of cucumber cultivation in the village have provided a favourable and supportive environment for young people. Many of them told us that "many villagers have rich experiences in production and they could be technicians beyond our village." After setting up their own greenhouses and as they engage in production, young farmers face new problems that they cannot resolve with internal resources or for which they need new technical inputs. For this reason, communication with other farmers and producers is very important to young farmers. They frequently visit neighbouring greenhouses to provide help and also learn from each other. There are six village technicians who are experienced producers and grassroots experts; these "cucumber doctors" provide technical support to local farmers. From neighbours and local technicians to seedling raisers and agricultural material suppliers, there are diversified sources of technical support that young farmers can draw on at the community level.

The third way to gain technological expertise is from public training services that the village and government provide. The local government has given the village and its vegetable production special attention as a model of cash crop plantation. Inviting external experts to conduct training for the villagers is one form of public service. There is, however, a gender difference in the way in which young farmers acquire this knowledge. For a young couple, it is usually the husband who participates in public training and later shares the information with his wife.

While farmers welcome training provided by the village or local government as necessary, it cannot substantially meet farmers' technical demands. As many young people commented: "the external experts usually have more theories than practices. When it comes to farming practice, they even don't know more than us. Generalized theories are not always applicable in our regions." When compared to generalized abstract training, young farmers prefer to learn technologies and information suitable for their own needs from the internet or via smart phone apps. Most of the young farmers, including many young women, have installed various technology and marketing apps on their phones. Some listen to podcasts on technology in the greenhouse when they are working. The internet, mobile phone, and other information and communications technology (ICT) are very common and popular among young farmers who use them to access flexible and customized information. This special channel in knowledge acquirement also separates young farmers from their parents in terms of farming and marketing methods.

In Yao's family, she and her husband use their mobile phones to access the internet for training and other knowledge acquisition, including a technology program accessed via WeChat.

> This technology account has a lot of information that suits for the production condition in my family. We all plant cucumbers in the village, but each greenhouse is different in terms of location, soil quality, illumination, ventilation, species, etc. So we should not fetish the so-called experts and have to choose the information and knowledge that really suits us.

Dai Qiuyan is 34 years old. Her husband is a fish wholesaler in the provincial capital and she takes care of the greenhouse at home. She is very active in increasing her knowledge about cucumber production. When watching television or reading news on her cell phone, she is constantly on the search for related news. "I have many apps on my phone. I can search for the price fluctuations on my phone for the national market. Sometimes, different regions have different marketing prices. What we got from the internet is just a guideline, you have to have your own judgement."

Previous research showed that there is significant gender differentiation in the acquirement and application of agricultural technologies. Even on farms where women provide the majority of labour, they are still passive in learning new technologies and subordinated in technology extension activities. Many rural women are confined to communication with family or neighbours and reluctant to try new technologies. In Huang village, in contrast, we found that both young men and women farmers are highly motivated to gain new knowledge and modernize their operations. Unlike their parents, they are accustomed to the ease of finding knowledge via the internet. Young farmers agreed that when compared to their parents, they are pursuing a kind of modern agriculture that requires increased adaption and pursuit of knowledge and technology.

Labour Division for Young Farmers: "Man Outside and Woman Inside" in the Production Sphere

Cucumber cultivation is very labour intensive during a season that in Huang village runs from September to June of the next year. Cucumber harvest is 50 days after planting. In autumn and winter, farmers need to pick cucumbers every other day, and in spring, they will pick the vegetables daily. In addition to harvesting, farmers need to adjust the seedlings' height and water and apply chemical fertilizers and pesticides, all of which require a lot of manual work. The farmers told us that "every day there is work waiting for us and we're tied to the greenhouse." In winter, farmers work in the greenhouse from 8:00 am to 4:00 pm. There is a two-hour break at midday for lunch and a rest. Farmers close the greenhouse at 4:00 pm, but related farm work is often done with a head lamp after dinner. In the summer months, due to the hot weather, farmers work from 5:00 am to 11:00 am and then continue from 4:00 pm to 8:00 pm. "We don't have spare time. Even during Spring Festival we need to work in the greenhouse. On the first day of new year, we visit family and

Table 7.2 Daily work in the greenhouse

Task	Frequency (times/day)
Rolling shutters	2/1
Harvesting	1/1–2
Marketing	1/1–2
Falling the seedlings	1/1–2
Applying chemicals	1/2
Mist spraying	1/6–7
Watering	¼–5

relatives in the morning with new dress and then in the afternoon we change to working clothes in the greenhouse. We can only rest for two days even during the new year's holiday" (Table 7.2).

During the busy production season, young couples divide their labour. Daily management of the greenhouse requires watering and mist spraying of chemicals; these tasks are labour intensive, but the latter requires certain technical expertise. Falling seedlings is the most tedious work in a cucumber greenhouse. In a 1 mu area, there are about 4000 seedlings. When they fall, the farmer needs to unlace the rope that supports the seedling, pull it down, and then retie it. This process takes between two and three days to finish in one greenhouse. It is usually the women who take care of falling seedlings and apply chemicals to flowers, while men are in charge of mist spraying, watering, and other chemical applications. Harvest and pick-up are joint tasks by the couple. When the cucumbers are packaged in boxes, the husband will transport them to the village market for sale. In such scenarios, men's labour is largely technically oriented and outward focused, while women's labour is more labour intensive and inward focused. Women work longer hours in the greenhouse than men. Such labour division in cucumber production to a great extent duplicates the conventional family labour division of "men outside and women inside," which ascribes men as breadwinners and women as housekeepers. The labour division in the production sphere echoes traditional gender roles. Most male farmers described their work in watering, mist spraying, and chemical application as "managerial work," while a woman's job is basically "labour work." For men, managerial work is key to a greenhouse's production success.

However, for many women farmers, this kind of labour division in agriculture is not a complete duplication of the "men outside women inside" model in the family sphere. On one hand, women equally contribute to production and revenue. Women's responsibilities are not unpaid work but an important role in the production chain that determines the quantity and quality of cucumbers. In Huang village, women farmers have built strong self-recognition towards their contributions in family farming. As they told us: "I'm doing my part and he (the husband) is doing his part. We have different tasks. It's nonsense to say who is stronger and who is more powerful in agriculture … it's not competition but cooperation. Our works are complementary, and both of our work are very important. The production will be affected if either of us didn't work well." We should also consider that increased access to technology and relevant information needed for farming success keeps the gender labour division flexible. Women can and often do take on the roles that men typically hold.

Huihui and her husband take responsibility for one greenhouse each. They work in the greenhouses separately and work together when harvesting and applying chemicals. Huihui said, "I do not work less than him at all. The only difference is that he has more experience in planting cucumbers than me. We're equal in agricultural production."

In Cui Weiying's family, she and her husband Wang Zhanling discuss important family affairs and take decisions together. Although it is Wang who usually takes on the technical work, it does not mean that Cui has to rely on him. "I can dispense the chemical as well. I can do all of the work by myself. It's not difficult. When he's busy or he's not around, I would do the technical work."

In households that manage large-scale production, women farmers' involvement is not limited to greenhouse production. Due to high labour demands, large holders need to hire labour to work in their greenhouses. It is usually women farmers who are in charge of organizing and managing hired labour. As most of the hired labourers are middle-aged women from neighbouring villages, women farmers have a gender advantage in communicating with these workers.

Haixia and her husband plant seven greenhouses. They cannot manage the workload by themselves and regularly hire three or four labourers to support production. In the busy season, this expands to five or six. Finding stable and responsible workers is the biggest challenge for Haixia, especially in the busy month of March. "In our experience, the market price in March is very good. All of the farmers would like to take the chance to sell as many cucumbers as possible. We have seven greenhouses and cannot miss the timing. Sometimes I need to put aside my work in the greenhouse to search for hired labourers in the villages."

Women farmers' involvement in agriculture and the family labour division in production impacts on labour division in the domestic sphere. Due to long hours working in the greenhouse, women farmers usually do not have the time or the energy to take care of housework. Many households do not cook lunch at home. In order to have more time in the greenhouse, they buy fast food from a village convenience store. Husbands cannot ask their wives to be a traditional "housewife." Moreover, given this shared workload in production, many wives ask their husbands to share equally in the housework. In both production and reproduction spheres, the formerly rigid gender boundary has been diluted for women farmers.

Cui Weiying, for example, usually asks Wang Zhanling to help with chores around the house. He does not complain even when his wife affectionately mocks his efforts. All year round, the couple eat breakfast and dinner at home and have some bread in the greenhouse as lunch. At the end of a working day, they return home to prepare dinner together.

Community-Based Agrarian Transition and Its Implications for Young Farmers

Huang village is a very special case in terms of agrarian transition in contemporary China in that there is no space for urban capital in the community. Family farmers organize all agricultural production, although a few of them are larger in scale and hire labour. There is no horizontal or vertical concentration by agricultural enterprises from outside of the village. Family farming, especially young farmers, is the major (or only) form and driving force for agricultural development in the village. This is due in part to the nature of the village's specialization in cucumbers, which is highly capital-labour intensive and cannot be mechanized for larger-scale production. The critical variable in this case is the role of collective community, which provides important support to young farmers to guarantee their subsistence and a space for development. The collective community has been actively involved in resource distribution, land rent moderation, and innovation extension in cucumber production. The patriarchal role of the collective community partly explains the tendency of repeasantization in Huang village alongside its prosperous commodification. Community's value and obligation to safeguard livelihood of smallholders make it possible for young people to get access to land and involve in agricultural production in a more pleasant way—work on family-controlled land with family labour for the family well-being.

Land Transfer: Breaking Through Household Limits for Young People

Access to land is the most important pre-condition for young people to start farming. It is true for both men and women, but especially for women. Rural women face many difficulties in accessing land. The land contract system[5] defines a rural household as a unit of land contract; however, the household head is usually the male and the land rights of female members in the household are often overlooked. It is common for women's land rights to be violated due to changes in their marriage situation. Because farmland is contracted in the unit of household, when a rural woman marries and settles into a new household, she will lose use right for the land contracted by her native family. According to a national All-China Women's Federation survey in 2010, about 21 per cent of rural women did not have access to land, among which 27.7 per cent lost their land after marriage. This unfair situation has passive impacts on their position in both family life and agricultural production. Landless women feel deprived and marginalized in their husbands' households. The

---

[5] Rural farmland is collectively owned by the community in China. Households get their land contracted from the community based on headcounts. Current contract period is 30 years and size of land contract cannot change during this period.

landless situation also increases livelihood pressure for poor households. In Huang village, prior to 2005, some land was set aside to meet demographic changes and to offer to young women who married into the village. There was no communal farmland to redistribute at all after 2005 and this practice did not continue.

Young men in the village also face land scarcity. Land contracted from the village collective is minimal and scattered; 2 mu of contract land could be scattered over three locations. Most of the contract land runs from north to south. However, land suitable for greenhouse production should run east to west in order to maximize the availability of sunlight. In the 1990s, the limited land area meant that villagers built smaller greenhouse. There were limited land transfers among villagers. A larger piece of land suitable for greenhouse construction sometimes involved land rights of several households, and successful transactions meant communication and negotiation with different households. It was not an easy procedure for individual households, which strained the expansion of the scale of production in the village.

In 2013, the scattered small-scale household production was changed by the community-led land transfer. As farmland in the countryside belongs to a village collective, village committee in Huang village made an overall change on major plots. While keeping farmers' land use right intact, the village committee withdrew land from individual households to consolidate into larger plots for greenhouse construction. After the land has been consolidated and infrastructure (including roads, pipes, etc.) installed, villagers who would like to plant cucumbers apply to the village for a greenhouse site. If it is approved, the producer obtains land use rights for the site and pays rent to the village. For households whose whole plot consists of contract land, their land contract rights remain intact and they pay rent annually as compensation. Village committee-led land was meant to promote land transfers that sustain and enlarge cucumber production. It simultaneously accelerated production as the consolidated land is more suitable for greenhouse production in a geographic sense. In general, greenhouses with more acreage have higher yields.

The villagers welcomed land consolidation as it reduced their transaction cost in land transfers and improved farmland infrastructure. As a resource allocator, the village committee applies related measures to ensure that land distribution is fair and effective.

1. It regulates the standard rent, which in Huang village is CNY 1000 per mu, which has been applied for both land transfers among villagers themselves and those between villagers and the village committee. This rent is much higher than in neighbouring villages where rent usually falls between CNY 600 and 800 per mu. This reflects the higher revenues associated with cucumber production while also protecting the interests of the original land contractor, especially the elderly and those who do not cultivate cucumbers.

2. It moderates rent collection. Due to the tiny area of contract land held by each household and the quantity of land transfers in the village, it is very difficult to calculate land rent for a single household. The greenhouse site that one household might contract from the village committee may involve contract land of several households. The producer would need to separate CNY 1000 per mu of rent by different households and then contact them in order to pay the rent. Some households may not pay their rent on time. To avoid such problems, the village committee set a fixed date in spring to collect land rent. All related households need to pay their annual rent to the village committee on that day. The village accountant then prepares the payments for those who rented land who can collect their rent payment the following day.

3. The village committee also regulates the term and period of land transfers. A land transfer contract is made with the village committee, renting producer, and the original contractor. A basic transfer period is 10 years. The producer has to guarantee that the land will be used for agricultural production; the village committee will reclaim the land if it is not properly used within two years of the transfer's signing. During the contract period, to protect the producer's interests, the original contractor cannot break the contract since greenhouse construction involves a substantial capital investment. To protect the

interests of the original contractor, if the producer wants to terminate the contract, he can sell the greenhouse to another villager with the village committee's prior approval or return the land to the village after restoring the landscape. These measures aim to protect the interests of the different actors involved and avoid land desolation.

4. It guarantees fairness in land distribution. When the village committee consolidated the land for greenhouse construction in 2013, it was divided into 100 plots for greenhouses with an average size of 1.5 mu, and it was numbered for selection. Households who applied for a contract paid a CNY 5000 deposit for each greenhouse and drew lots to see who would receive which plot of land. Land consolidation continued in 2015 and 2016, and all of the land sites were distributed through public lottery. The villagers recognized the value of this transparent procedure and it prevented potential conflicts in the competition for good land.

In the last decade and especially since 2013, more and more villagers have been able to secure land through land transfer to start or expand their cucumber production. Since 2013, farmers have built about 400 greenhouses thanks to community-led land consolidation. The village collective has surpassed the household as the primary allocator and moderator of land. It is an extraordinarily important process, especially for young people who do not have much inherited land.

Wu Xia and her husband operate seven greenhouses. Before she married and moved to Huang village, Wu had 1.5 mu of land in her home village, which that village's committee reclaimed after her marriage. Fortunately, she was able to secure 0.7 mu of farmland in Huang village after her arrival. Her immediate family—Xia, her husband, and two children—hold 2.6 mu in the family, by which they built their first greenhouse. In 2013, when the village committee started to plan for more greenhouses, the couple was very excited. In that year, their two children were in kindergarten and primary school and did not need as much care. Haixia and her husband decided to put all of their efforts in cucumber production. They contracted five greenhouses and later bought the seventh greenhouse from another villager.

Wang Zhanxue is 38 years old. He operates three greenhouses with his wife. His two children are in middle school. In the household split with his parents and his younger brother, Wang received only 1.5 mu of land. The land he uses now for his three greenhouses is 8 mu—all transferred from the village. His own contracted land was transferred to other villagers. He is very content with the family's current production of scale. "The three greenhouses are just right for me and my wife to work without using hired labour. With our own labour input, we can earn CNY 100,000 each year. It is not bad as a farmer."

Land transfer in particular provides the chance for young women to access land and pursue farming independently. For villagers, land consolidation and land transfer already blur the boundaries of land. In most cases, farmers work on land transferred from others and have their own contracted land transferred to other people as well. Whether a young man or a young woman, having his/her own share of contracted land does not matter very much in a young person's pathway to becoming a farmer. Family position for young women is not influenced just because they don't have land in their husbands' village because they are also entitled to transfer land from the village committee. There are cases in which young women farmers transfer land by themselves and pursue farming independently.

Forty-year-old Wu Xin operates two greenhouses on 3 mu of land. She rented the land three years ago from villagers for CNY 1250 per mu. It was higher than the "official" rent in the village because there was some greenhouse infrastructure for cucumber cultivation on the land already. There is another 3 mu of land in her family for maize and wheat, which is her husband's contract land. Her husband has been engaged in migrant work in the city and only returns home once or twice a year.

> He (the husband) doesn't have time or efforts to take care of things in the village, transferring land and planting cucumbers is all my own idea and decision. I manage the whole procedure. I'm now very capable of cucumber production. If there is land available in the village and a subsidy policy, I'd like to build another greenhouse and make more investment. I have confidence to manage three greenhouses by myself.

Agricultural Programmes on Finance and Technology

After land, finance and technology are the other important elements for young farmers who are engaged in developing their own cucumber production unit. For small-scale family farming, initial capital for cucumber production usually comes from family savings. With any profits, farmers are able to maintain simplified reproduction or expand the scale step by step. This is the general economic logic for family farmers who have tried to minimize risks and external dependence. When the local government promotes rapid agrarian transition, young farmers face both risks and opportunities in expanding production. Financial support from the village and local government is critical to reconfigure young farmers' careers. In 2012, the village committee received a subsidy from the county office responsible for poverty alleviation and development in support of the construction of 50 modern greenhouses. The subsidy's distribution was based on villagers' voluntary application. Investment for a standard modern greenhouse with an area of 1.5 mu was about CNY 80,000–100,000 at that time and the subsidy was CNY 60,000. In spite of the favourable policy, villagers were not very active in applying due to a funds shortage and the fear of associated risks. To encourage young people to expand and upgrade production, a village cadre tried to secure free loans from the county for them. With this support and inspiration, several young farmers took the lead and applied for the programme, which also established a foundation for them to later become large holders.

Wang Zhanling and his wife operate nine greenhouses and are one of the large holders in Huang village. They operated 11 greenhouses for some time but sold 2 greenhouses as they were too busy). The couple did not have any greenhouses when they separated from his parents. They received only 2 mu of land that was too barren to plant grain. They started planting cucumber in 2002. They constructed the greenhouses one-by-one with their savings and some money that they borrowed from relatives. With profits from the previous two years, they were able to build the next greenhouse every other year. During the aforementioned 2012 subsidy programme opportunity, the couple wanted to expand their production but were afraid that there would not be sufficient land in the village in the future. The resulting financial burden also caused hesitation. They planned to build five greenhouses, but under the village leader's persuasion, they finally applied for seven greenhouses. To build them all in one year, they applied for a CNY 300,000 loan for one year. As the yields and market for cucumbers in that year were good, the couple was able to repay the loan with profits from their 11 greenhouses.

In addition to the 2012 subsidy program, the village committee continued to apply for financial support for local villagers. In 2015, households could receive CNY 2000 subsidy for newly constructed greenhouses; in 2017, it was CNY 1000. As it is difficult for individual villagers to apply for bank loans, the village cadre applied for CNY 2 million in loans from Rural Credit Cooperatives in the name of the village cooperative to facilitate villagers' demands for finance.

The village committee has also been a promoter of technological innovation. Many sectors of cucumber production have been mechanized, including ploughing, shutter rolling, and chemical spraying. In 2009, the shutter roller for greenhouses was initially promoted in the county when there were only 2000–3000 such machines in the whole county. The village committee introduced the shutter roller to the village with a CNY 1600 subsidy for each machine. Technical innovation in cucumber production has had a very positive influence in improving labour productivity and working conditions for young people, especially since cucumber production is highly labour intensive, unlike grain production. With the promotion of the rotary tiller, shutter roller, and mist sprayer, farmers' manual work has been significantly relieved. It also has emancipatory implications for women farmers as they can more easily take up farm work independently with the support of machinery.

Ciu Weiying told us: "Vegetable cultivation is more laborious than grain farmers because wheat and maize cultivation is fully mechanized. It is good that the farming style has been innovated and we don't need to work as hard as before."

Wu Xin also recalled the challenges: "I remembered how hard it was without the shutter roller. I was standing on the top of the greenhouse to roll up the shutters by my hand. It almost took a whole day for me to do that. Now I only need to switch on the engine and it takes 10 minutes to do the work."

Qiuyan recounted for us: "We used manual pesticide barrels before. We carried the barrel and walked through the greenhouse. Because we don't have much strength, it usually took more than one hour to finish the work. With the mist sprayer, I only need 10 minutes to finish one greenhouse."

The introduction of technology has had a positive impact on equalizing gender positions among young farmers. Women farmers can engage in most farming sectors independently or replace the role of male farmers in production. With the application of such machines, women farmers also have more positive self-recognition of their role in agricultural production. As they told us: "We can do what men usually do on the farm. It is not difficult for us women to do farming."

Providing Public Goods: Development of the Local Market

The local cucumber market provides an important and stable outlet for young farmers' production. It has taken over 20 years for the village to develop its own marketplace. In the 1980s, due to the lack of a marketing channel, cucumber producers had to go street by street with a three-wheeled bicycle to sell their product. In 1993, several small marketplaces formed at the entrance of the village and some "cucumber brokers" emerged. Farmers no longer needed to sell their product in the street. Instead, cucumber brokers would collect cucumbers from households and then sell them to wholesalers from the city. In 1996, the village built a marketplace with an area of 3 mu, which was the only cucumber transaction market in the adjacent area. Every day, farmers sent their cucumbers to the market directly from their farms. Cucumber brokers would organize and moderate the transactions in the market. In 2013, through the village committee's organization and investment, the marketplace was further expanded from 3 mu to 40 mu. There is not only space to accommodate trucks and lorries, but also a warehouse, refrigeration storage, and shops for production materials to serve farmers and wholesalers. The daily transaction volume is about 200 tons, a five to six times increase than before the expansion. Specialized production has an associated stable source of marketing. Current wholesalers come from Beijing and different cities in Hebei, Henan, Shandong, and other provinces. The village market has provided a stable marketing platform for cucumber farmers and farmers from neighbouring villages who also take advantage of the market.

The village committee not only built the market but also played the role of market manager. The three staff members working in the market as cashier, accountant, and gauger are retired village cadres or village party members. They weigh the cargo, calculate the trading volume, and make payments to farmers. The transaction is made through cucumber brokers who are local villagers with rich experience and social networks for marketing. Cucumber brokers organize individual farmers and contact different wholesalers to fix the price. Wholesalers need to pay 3 cents per half kilogramme to the village committee as a management cost. One per cent of this is the cucumber broker's commission, 0.1 cents is for the accountant and cashier as their salary, and the remaining 1.9 cent is for the village committee to pay land rent for the market and to maintain the infrastructure. At the initial stage of transaction, wholesalers transported the cucumbers first and then set the price when they sold the vegetable on the urban market, which meant that farmers did not receive the money immediately. Their payment was delayed and sometimes shrank due to changing market prices. To protect local farmers' interests, the village committee has regulated that wholesalers have to set the price at the time of transaction according to the price in Beijing and Shandong. Payment to farmers cannot be delayed to the next day. The market now has a formal mechanism for transactions. The market opens daily at noon. When wholesalers arrive, they prepay for the goods to the market's managerial staff. Farmers deliver their products to the market and await a price. If the price is good, they will proceed with the transaction. Farmers receive their payment from the wholesaler immediately. The market's regular operation via the village committee is key for the farmers. Without such a convenient and stable marketing channel, young people would not have the confidence to start farming. Many young people told us that this was the reason that they decided to become involved in farming. Wang Zhanxue shared:

The market in our own village is very convenient for us and it saved a lot of transportation cost. For farmers from other villages, when they come to the market, they have to accept the price no matter how low it was because they could not transport their cucumbers back. However, for us local people, we make transactions every day and have become very sensitive to the price. We can have information about the price anytime and gradually have grasped some rules. When the price was low, we're not in a hurry to sell. We can put cucumbers in storage for a few more days to wait for a better price. For example, when it is cloudy, cucumber production is lower; therefore the price is in an upward trend. It's also easier to store cucumbers in cloudy weather. So we can wait for several days to have a better price. There are uncertainties and fluctuations as well; everything is not so definite. We need to consult the price from different channels and pay attention to weather changes.

The local market is not only a platform for cucumber transactions, but also an important public space for villagers to socialize and exchange with the outside world. As cucumber production is highly commodified, the concrete marketplace and abstract market mechanism have become central to farmers. The local marketplace is the place almost every farmer visits every day. "Our daily activity is from home to greenhouse, from greenhouse to the market. We don't go to other places. We also don't have the time to go to other places." Therefore, the cucumber market has become an important place for villagers to socialize and interact. Through involvement in market transactions, the busy farmers who work all day in separate greenhouses have the time to talk to each other. They discuss technical or management problems in the greenhouses, exchange experiences, and talk about deals in other provinces to expand their knowledge of cucumber production and marketing. With this information collected at the market, they can adjust their production or plan accordingly. The market is an especially active space for young farmers to build and enhance their social networks. After their market transactions, young farmers often gather in a restaurant with the money that they earned that day to share dinner and continue conversations about production and rural life. These social interactions, however, are largely confined to male farmers. As marketing is termed as men's work in a rural household, women usually do not go to the market to sell their product unless the husband is busy or away. While the village market is open to all farmers, women farmers have yet to actively integrate into this aspect of farming life.

Agency and Challenges for Young Farmers

Young Farmers' Perceptions of Farming

With the technological upgrades in Huang village in 2009, young farmers gradually became the main body of agricultural development in the village. While many older farmers are still working in the first generation of greenhouses, young farmers are all working in modernized greenhouses, in technical and managerial terms.

Wu Zhiying said that young people are also more careful about farming. "We put a lot of effort and work into greenhouses. We pay much attention to pest control and disease prevention, and we have more sources of information to learn. That's the major difference between us and our parents' generation. Their way of production is very extensive."

Wang Zhanxue told us that while they are modern farmers, they are not producing an organic product:

We're pursuing modern agriculture because our farm work is highly mechanized. Although cucumber is very unique that it cannot be totally mechanized, we have had many machines in the greenhouse. But we're not real modern agriculture because we still highly rely on pesticides and chemicals. It's not organic farming. That is the way of agricultural development, but currently we cannot realize it. Without chemicals, the yield won't be guaranteed.

This is a paradox in young farmers' perception of farming. On the one hand, they are contented with the income from agriculture and proud of their freedom and autonomy. "Freedom," "living with family," and "stable" are the words that they most often use when asked about the merits of family farming. On the other hand, those words are just another expression of their social exclusion in cities. "Stable" also means young people are bound by the hard work required to make a greenhouse successful. If they could secure decent employment in the city, some of them, especially the younger farmers, would migrate. Their perception and assessment of their own farming is based on a compromise of imbalanced rural-urban development and their social exclusion as migrant workers. For all of the young farmers that we interviewed, they do not want their children to become farmers. Wang Zhichao does not want his son to be trapped in farming:

I don't want my son to continue farming. Being a farmer is a symbol of (being a) loser. To do farming is very boring. I really admire those migrant workers. They have another kind of freedom to try different things. They can dispose their time after work. If I have a good job in the city, I definitely will quit farming. But now everything is settled. My life is almost fixed in this trajectory. I can't change it.

For Wu Li, their son's engagement in farming would be a last resort:

My son is 18 years old and he's doing migrant work in Handan city in Hebei province. If he doesn't have a good job or could not make his own family in the city, maybe in the future I will pass the greenhouses to him. There is stable income with greenhouses, at least he can sustain himself. But that is the last resort. As long as he can make a living in cities, I don't want him to come back. Nobody would like to be a farmer. We don't want to be farmers neither, how can we expect our children to be farmers just like us?

Better profits, technological advances, and labour-saving machinery have not removed the social stigma of being a farmer or the rural-urban barriers for villagers. Most of the young farmers want their children to finish high school and secure a good job in the future. In their imagination, a good job should be in cities, stable and less toilsome than farming. Farming is the bottom line for their livelihood security.

Young Farmers' Differentiation in Production

Young farmers in Huang village are not homogenous. They have diversified in terms of production and life pursuits. According to the farmers' own definition, combining the greenhouse scale and use of hired labour, large-scale farmers in cucumber production are those with seven or more greenhouses, a medium-scale farmer has four to six greenhouses, and a small-scale farmer has three or less. Large-scale farmers have a huge capital investment in greenhouses and regularly use hired labour in production. In addition to the farming couple and regular hired labour, they need to hire additional help to harvest during the busy season. For medium-scale farmers, family labour is the mainstay though they may need to hire labourers occasionally during busy times. Small-scale farmers rely solely on family labour. During our fieldwork in Huang village, there were only four large-scale farmers, while medium-scale farmers make up 25 per cent of the farming population, and the rest, about 70 per cent, are small-scale farmers. Differentiation in agricultural production is not significant in Huang village. Production scale has gradually stabilized over the last 10 years and there is no tendency of proletarianization for small-scale farmers. Although hiring labour is very common in Huang village, there is no labour exploitation between larger holders and small holders because there is no agrarian overpopulation in farming households. Hired labour is generally from neighbouring villages that don't plant cucumbers and have a surplus labour force. When comparing small-scale and medium-scale farmers, the government favours large holders for policy intervention and subsidy opportunities. Large holders become models for the local government to propagandize, and they are in turn able to secure policy preferences and subsidies from the government. It has also helped large holders to accumulate additional social capital when dealing with government officials. They also become ideal promoters and advertisers for agro-businesses to sell their agrarian inputs.

Large Holder: Zhai Jizhong

Zhai Jizhong is one of the large holders in Huang village. He is 37 years old and operates nine greenhouses with his wife. His annual land rent is CNY 30,000. The couple were formerly employed as teachers. Zhai graduated from senior middle school in the 1990s, which was very unusual for a child growing up in a rural community during this period. After graduation, he was a village primary school teacher. The job's low salary pushed Zhai to start planting cucumbers in the late 1990s. His wife was a teacher in an urban school for migrant children. In 2006, when their first son was born, Zhai's wife returned to the village to care for their child and help her husband. Due to his capacity for social interaction, Zhai Jizhong developed a broad social network and became a cucumber broker in the market.

Zhai hired seven workers for his nine greenhouses, but he is not as careful a manager as small holders. Cucumber disease in his greenhouses is more serious than in those of small holders. His focus is profit and as long as his yield is high, the quality is less important. When compared to small-scale farmers, Zhai is more sensitive to technological innovation. He invited an agronomist from Henan province to conduct some biological experiments on one of his plots of land.

Small Holders: Wu Xiaoen

Wu Xiaoen and his wife Diao are 33 years old. They have two greenhouses. Cucumbers in their greenhouses grow much better than those of their neighbours, thanks to the couple's intensive management. There are small bricks hanging in the greenhouse to adjust the shape of cucumbers. They use reflective film to make sure that all of the cucumbers have sufficient sunlight. They apply medicine to each flower to prevent botrytis. As Diao told us: "managing a greenhouse is like raising your children. It will give you more reward if you take care of it very well. Greenhouses managed by different farmers would be very different." Indeed, Diao is very careful about her cucumbers. After she learned during a training session that some ingredients in makeup can impact cucumber growth, she had decided not to wear any in the greenhouses.

Small Holder: Yan Liang

Yan Liang is 28 years old and operates two greenhouses with his parents. He did not split the household with his parents because his older brother and sister are working in the city. He told us:

> Cucumbers in our greenhouse are better than other villagers' because the land in the two greenhouses is ploughed by my father by hand, not using the tilling machine. When I harvested cucumbers I thought it's good to plant a greenhouse, at least it's stable. Each greenhouse can generate at least 30,000 yuan, it's not bad to live in the village with that income. In the fallow time of June and July, I can do seasonal migrant work in the city. Our life is not bad. Although my brother and sister work in the city, to be honest, I do not admire them. I have my own house and two greenhouses, and I already have two children because I married earlier than them. Now I can live with my family and parents and enjoy the family life. My brother and sister could not have feelings as I do. My sister is already 31 years old but not married yet. My brother is just engaged. They live in the city, but they face more pressure than me.

These cases of young farmers at different scales of production show their internal differences in farming modes. Large-scale farmers are more capitalist in their efforts to maximize production with hired labour, while small-scale farmers are more like peasants with their priority on quality using their own labour. Profit for a single greenhouse is higher for small holders than for large holders as labour-driven intensification is central to cucumber production. For large holders, in order to reduce wage costs, they try to diversify production varieties and use some of their greenhouses to plant tomatoes and beans, which are less profitable but require less labour input. For young farmers at different scales, the structure of cucumber production in Huang village is stable and most of them would like to maintain their current scale. Key constraints for scale enlargement include access to land and high labour inputs. Stable cucumber production implies that there is no more land available for expansion. Farmers who want to extend their production can only transfer land in neighbouring villages, which would increase transportation costs and is not convenient. Scale enlargement beyond a family's capacity requires more hired labour and would increase wage costs. Moreover, all farmers face the risks of rising material costs and uncertain market situations. Given the current market price for cucumbers, depeasantization will not occur among small holders. This explains both the existence and limited differentiation of family farming led by young people.

Challenges for Young Farmers

Looking at family farming in Huang village as outsiders, its agricultural specialization and its core of young family farmers is quite dramatic and impressive. In this case, the village community played a strong and protective role in supporting the village's overall agricultural development whereby young people could have the opportunities and resources to enhance their family farming operations and maintain their livelihood. The village's collective economy has, in turn, strengthened to be able to provide more infrastructure (roads, environment improvements, etc.) and promote rural development. The county has awarded Huang village with model village status and it receives many visitors who come to learn about its experiences. However, underneath this "successful" case, the village and its young farmers are facing similar challenges to other agricultural villages and producers in China.

1. Fluctuations in the Hegemonic Market

The wholesale market in the village provides a unique opportunity and convenience for farmers but also becomes a baton for farmers' production. Farmers could go to the market every day and sell their product with the help of local brokers, however, they are highly involved and dependent on the market just as farmers elsewhere. Farmers cannot foresee or cope with market fluctuations, nor do they have a say in market pricing. They can only delay selling for a few days to wait for a better price. If the market price keeps declining, the farmers have to accept the price because they don't have other outlets. It is possible for young farmers to break through the limits of the local market with quality products of new species, or by looking for space in retail markets further from home. However, there is a gradual path of dependence among young farmers due to the convenience of the market. Its demands have determined local farmers' production structure—individual farmers cannot bear to take risks and bear the costs of market exploration.

Li Baoxiang shared with us these concerns:

> Our market has been framed, and it's just within this (village) scope. We've seen fruit cucumber in other places, which are more expensive than the species we planted. But in our area it's unusual in the market. Species in small scale would not have a place in the market. No dealers would like to buy for such small quantity. It's impossible for us to change our species.

Wu Li reiterated Li's concerns:

> The village market is our own way to sell cucumbers. We don't have other channels. There are no other wholesale markets in this area. You may have a higher price if you take the cucumbers to the county or to other urban areas but it's very difficult. You have to enter into a new market and compare the prices by yourselves. It's too difficult.

All of the young farmers that we interviewed are sensitive to and worried about the price fluctuations in recent years. As the government promotes agricultural specialization nationwide, vegetable production has been extended in many other villages in north China. Beyond the traditional vegetable production base in Shandong and Hebei provinces, those in the northeast are actively promoting vegetable production. The county government asked the Huang village committee to facilitate the construction of 100 greenhouses in a neighbouring village in one year. With the promotion of cucumber production and the rapid increases due to yields, young farmers in Huang village will face serious market competition if they do not innovate in production and marketing terms.

2. Ecological and Environmental Threats

The declining ecological conditions and resource base is another challenge for sustaining cucumber production in Huang village. The decades of monoculture have generated many problems in the soil.

Farmers apply manure to the soil after ploughing to increase soil fertility. However, insufficient fermentation of manure increases soil hardening. Cucumber production requires significant irrigation but frequent watering also speeds up moss growth on the surface. In addition to common cucumber diseases such as botrytis and downy mildews, severe smog in winter and early spring is a new climate threat for cucumber cultivation. Heavy smog in winter reduces sunlight for cucumbers and aggravates sprout rot. Due to the village's specialized production, the ecological and environmental problems of individual farmers have a cumulative effect for the community. In July and August, manure adds to air pollution and induces a "disaster of flies." In the fallow season of June and July, farmers cut down all of the cucumber vines. However, there is no organized plant waste disposal plan and vines are dumped in a remote area of the village. The vine piles are combustible in hot weather, and bacteria in the plants pollutes surrounding land and the river. Environmental problems are primarily the outcome of large-scale monocropping. As there are very few grain plantations in the village, farmers do not have enough straw to mix in with the manure for fermentation. Some farmers travel to neighbouring villages to procure it, but not all farmers make the effort to do so. As the environmental problems of large-scale production impact all farming households, it becomes an issue of village governance that requires collective action.

3. Lack of Social Organization and Insulation from Social Life

For labour-intensive cucumber production, young farmers in Huang village show their tough spirits by taking up hard work. However, they spend most of their time in the greenhouses, which has, to a great extent, isolated them from the social life of the community. Most of young farmers do not have much spare time to spend with their families, let alone join in leisure activities and socialize in the community. Wang Zhichao compared life in the village to two points in one line—between home and the greenhouse:

We're always moving between these two points and even in a hurry when walking on the road. Take me as an example. My two greenhouses are a bit far from each other. When I have finished work in one greenhouse, I need to rush to the other one. When you see villagers on the road in the daytime, they're always busy running to different greenhouses, just like me. We don't have time to chat or talk on the road. We work until the last day of the year in the lunar calendar. We only rest on the first day of the new year and then continue work from the second day. Previously, when we visited relatives during the new year's holiday, we spent several days to visit each relative. Now we only take half a day to visit them and spend a few minutes with each relative. Due to the busy farm work, I feel like our relationships with relatives, friends, and neighbours are very loose and estranged. Everyone is busy. It's very sad but we can do nothing to change it.

Young farmers' complaints about their social isolation are common. Young women farmers and farmers in their thirties have a strong link to the world outside the village and would like to broaden their views (and those of their children) via travel and sightseeing. However, farm work is their priority. Young women farmers complained to us: "we don't have the chance to wear high heels and beautiful dresses. Every day we work in the greenhouse just like a country woman." They also do not have time to participate or voice their opinions in the public affairs of the village. Village cadres are mostly those with off-farm work and have the time to work for the village. There are no social organizations, informal or formal, among young farmers to take the lead in innovation or to address common marketing problems. Over the past decades, it has been the village committee that has steered agrarian transition and innovation. In the balance between innovation and stability, young farmers usually lean towards the latter. To break through their path dependence on the village committee, young farmers need to organize themselves and take an active role in social change with encouragement and institutional support from the local government.

Conclusion

The stories of young farmers in Huang village are peculiar, not only to other countries but even to many regions in China. Is it too unique to represent the commonalities of young farmers in China? If we consider their stories in a different light, the case of Huang village also proves that it is possible for young farmers to settle in agriculture and sustain their households. In this case, the key to foster young farmers is the collective agency of the community. When small holders are involved in commodification and capitalized farming and in the face of the risks from infinite markets, it is widely believed that the best counteraction is the existence of organizations or collective actions among farmers. This is the reasoning for farmers' cooperatives and associations as well as various social movements. In the Chinese context, the natural form of rural people's organization is the village, especially their home village. The village is not only territory where people live in a compact and socially interconnected space, but also an interface between the government and the farmers. Through villagers' recognition of village authorities and mutual trust within the community, the Huang village collective was able to reallocate internal resources (land) to allow for the entrance and expansion of young farmers as well as strive for and channel external resources (e.g. subsidies and machinery) to strengthen the economic competitiveness of family farming. It is not the individual agency of single young farmers that facilitate the village's vibrant situation, but its collective agency as a whole. Considering the rural labour migration and increasingly atomized rural livelihoods, it is difficult for many villages and rural communities to organize for collective action in agricultural development. However, the case of Huang village once again confirms the importance and viability of such collective action. In this sense, it's a unique case with common implications (Map 7.1).

Map 7.1 Study area. (Source: Ministry of Natural Resources Map Technical Review Center)

8

The Youth Dividend and Agricultural Revival in India

Sudha Narayanan, M. Vijayabaskar, and Sharada Srinivasan

Introduction

Fifty-four per cent of India's population is under 25 years of age and, by 2019, the median age of Indians was estimated to be 29 years. As per the 2011 Population Census, close to 34 per cent of India's rural population belonged to the age group 15–34. In 2012, an estimated 56.6 per cent of

---

This chapter is a revised and updated version of Vijayabaskar, M., S. Narayanan, and S. Srinivasan. 2018. "Agricultural revival and rural dividend." Review of Rural Affairs, *Economic and Political Weekly* 53, no. 26–27: 8–16.

S. Narayanan (✉)
Indira Gandhi Institute for Development Research, Mumbai, India

International Food Policy Research Institute (IFPRI), New Delhi, India
e-mail: s.narayanan@cgiar.org

M. Vijayabaskar
Madras Institute of Development Studies, Chennai, India

S. Srinivasan
Department of Sociology & Anthropology, University of Guelph, Guelph, ON, Canada
e-mail: sharada@uoguelph.ca

S. Srinivasan (ed.), *Becoming A Young Farmer*, Rethinking Rural,
https://doi.org/10.1007/978-3-031-15233-7_8

221

rural youth in the age group 15–29 years continued to rely on agriculture, forestry, or fishing as a source of livelihood (GoI 2013a). While the presence of a sizeable young population is believed to offer a demographic dividend, policy efforts to realize the dividend have not met with success as is evident in reports of jobless growth and the poor quality of employment generated outside of agriculture. Poor prospects for livelihoods within agriculture, its declining importance as a sector in the national economy, and aspirations of rural youth and their parents to find futures in non-farm sectors suggest that, like elsewhere, agriculture today is an unlikely option for the young in India.

While youth as a distinct social and demographic category has come to occupy a significant place in recent policy imagination in India,[1] and agriculture continues to occupy policymakers, the two are rarely brought together in research and in policy. The purpose of this chapter is to bring the question of youth in agriculture into focus. What do we know about young people in farming in India? Despite a large share of rural youth involved in farming, there is limited research or policy attention on the issues and challenges that they face around farming, non-farm opportunities, succession, and intergenerational transfer of resources and knowledge. This is reflected in data that are not always available by age, making it challenging to draw inferences specific to young farmers, even more so with respect to young women farmers.[2] We draw on statistical data and scholarly material to examine the situation of young farmers in India. Although the paper implicitly understands a farmer as someone with access (ownership, shared, renting, etc.) to land (or a productive resource)

---

[1] Since 2000, several policies have been directed at "solving" the youth problem on the one hand and "utilizing" the youth potential for national economic goals on the other. The first National Youth Policy was formulated in 2003, followed by the 2014 National Youth Policy. In 2008, the Ministry of Youth Affairs and Sports split into two separate departments—the Department of Youth Affairs and Department of Sports. Under the 12th Five Year Plan (2012–2017), the Planning Commission appointed a Steering Committee for Youth Affairs and Sports with an emphasis on skills, employability, and addressing socio-psychological issues among youth.

[2] Apart from the Census of India, the major source of nationally representative data on farmers is the National Sample Surveys (NSS) 59th Round in 2002–2003 and 70th Round in 2013 that collected data on farm incomes. The other NSS rounds cover employment profiles of people and the quinquennial agricultural census collects information on operational holdings; none of these focus on farmers as such. Evidence on young women farmers is even more scarce, since most data are at the household level, with the male head of the household presumed to be the "farmer."

who invests a large part of their time and labour in farming, actual definitions are quite varied.

Consistent with the focus of the research on which the chapter is based and of this collection, we adopt a youth perspective to understand the generational dimensions of the social reproduction of rural communities, the lives of young people within the agrarian economy, their paradoxical (apparent) turn away from farming in this era of mass rural un(der) employment (Cuervo and Wyn 2012), and youth subjectivities. In doing so, the chapter also engages with developmentalist and policy discourse that views the movement of people out of agriculture as a transitional imperative (Chenery 1979), even as global sustainability discourses place the family farm as a bulwark against incursions of industrialized and corporatized agriculture (McMichael 2008; Food and Agricultural Organisation n.d.). Despite the realization that conventional routes of labour transition out of agriculture are not available to many, policy initiatives to make agriculture attractive for youth livelihoods have been few and far between. The purpose of the chapter is not to argue that all (rural) youth undertake or remain in farming, but it is to make a case for improving the livelihood prospects within agriculture in a context of changing youth aspirations. We argue that a clearer understanding of the issues is essential to frame a nuanced approach to support the role of youth in agriculture and the role of agriculture in youth livelihood strategies.

The chapter is organized as follows. In the following section, we draw on national farm household surveys to map the demographic profile of (young) farmers in India. Section "Staying in, Exiting, and Entering Agriculture: A Review of Evidence" draws on the literature on farmer exit, entry, and continuation in agriculture, to provide a nuanced understanding of the generational crisis in Indian farming. In section "Structural and Policy Issues Within Agriculture," we turn to the structural problems that youth in farming confront in securing "decent" lives. Section "What Next for Young Farmers?" reflects on the policy crossroads and section "Empirical Study of Young Farmers" sets up the next two chapters that are based on original interviews with young farmers in Tamil Nadu (TN) and Madhya Pradesh (MP).

Socio-Demographic Profile of Farmers in India

Agriculture and allied sectors on which 54.6 per cent of India's workforce relies have registered a rapid decline as a share of national income, accounting for only 16.1 per cent of the Gross Domestic Product (GDP) in 2014–2015.[3,4] Evidence from two NSS rounds suggests that over the decade spanning from 2002–2003 to 2013,[5] the median and the mean age of the head of an agricultural household have increased by around two years to 39.3 years for the mean and to 38 years for the median, indicating that heads of agricultural households are now older. However, the change does not seem rapid (Vijayabaskar et al. 2018). Further, they find that a greater proportion of youth among the scheduled tribes (STs) are likely to farm than those from the scheduled castes (SCs)/other backward castes;[6] young people from other general castes are comparatively much less likely to be farmers. These differences seem to disappear among the older cohorts, but only beyond 65 years. Gender gaps exist, and the proportion of women who participate in farming is consistently less than those of men in farming. It seems that while the generational crisis in farming is not yet evident in terms of the average age of a farmer, there is a distinct pattern of rural youth, even in farm households, being disproportionately disengaged from farming (Table 8.1). This is apparent from other data sources such as the Census and NSS Employment-Unemployment Surveys.

---

[3] The figures are for the share of the agriculture and allied sector in total employment as per the Census of India, 2011 (GoI 2016, 35).

[4] At 2011–2012 constant prices (GoI 2016, 4).

[5] The 2002–2003 survey data were collected for "farmers" and the 2013 data were collected for agricultural households. For the 2013 survey, NSSO defined an agricultural household "as a household receiving some value of produce more than INR 3,000 from agricultural activities (e.g., cultivation of field crops, horticultural crops, fodder crops, plantation, animal husbandry, poultry, fishery, piggery, beekeeping, vermiculture, sericulture etc.) and having at least one member self-employed in agriculture either in the principal status or in subsidiary status during last 365 days" (GoI 2014b, 3). The income cut-off was not applied as a criterion for sampling in 2002–2003. Further, the definition used for "farmer household" in 2002–2003 made possession of agricultural land as a necessary condition for inclusion whereas it was dispensed with in the 2013 survey's definition for an "agricultural household."

[6] These are categories recognized by the Constitution of India to denote historically disadvantaged social groups.

Table 8.1 Selected indicators of farmers and agricultural households

	Madhya Pradesh	Tamil Nadu	India
Data from Agricultural Census (2015–2016)			
Operational holding (million)	10.0	7.9	**146.45**
Operating area (million hectares)	15.7	6.0	**157.82**
Average size of land holding (hectares)	1.57	0.75	**1.08**
Proportion of small and marginal holdings	71.46	91.74	**86.08**
Proportion of semi-medium and medium holdings	27.54	8.05	**13.35**
Proportion of operational holdings that are large	1	0.21	**0.57**
Data from National Sample Survey Situation Assessment Surveys (2012–2013)			
Share of rural youth in agriculture, forestry and fishing (Distribution of workers aged 15–24 years by NIC 2008 classification according to usual principal status approach (ps) for each State/UT [rural])	**76.1**	**33.1**	**56.9**
Agricultural households (as percentage of rural households)	71	35	**58**
Agricultural households (millions)	6.0	3.2	**90.2**
Share of agricultural households engaged in animal husbandry	56.5	60.3	**62.7**
Proportion in each demographic group engaged in cultivation as a principal activity			
Men (over 45 years)	77.7	72.8	**69.5**
Men (18–45 years)	70.2	37.5	**54.3**
Women (over 45 years)	42.8	40.7	**35.6**
Women (18–45 years)	48.9	37.9	**33.1**
Mean age among cultivators (years)	37.3	44.3	**39.3**
Median age of cultivators (years)	35.0	44.0	**38.0**
Proportion of cultivators who have not completed primary school	62.1	44.9	**50.8**
Land ownership by agricultural households	1.4	0.8	**0.9**
Proportion of agricultural household leasing in land	7.1	11.3	**16.4**
Proportion of agricultural households leasing out land	1.9	2.5	**3.2**
Sources of income (share from different sources in percentage)			
Income from wages/salary	21	42	**32**
Net receipt from cultivation	65	27	**49**
Net receipt from farming of animals	12	16	**12**
Net receipt from non-farm business	2	15	**8**
Income (INR per month)	**6210**	**6980**	**6246**
Expenses (INR per month)	**5019**	**5803**	**6223**

In terms of education, in 2012–2013, it was less likely that a farmer would be illiterate or completed just primary school or less and it was more likely that a farmer in 2012–13 had educational attainment of high school or beyond, relative to 2002–2003 (Vijayabaskar et al. 2018). This might reflect a general trend of a greater number of people who are now studying more, so that farmers in 2012–2013 are on average more educated than they were a decade earlier. This trend seems to undermine conventional understandings about Indian agriculture that attributes its relatively lower productivity to farmers' lower literacy levels.

There is also an indication that there is a lower preference for formal training in agriculture among youth (Fig. 8.1). Among the younger cohorts, technical training in agriculture accounts for the lowest share of all those with technical degrees, while those for engineering are much larger among the younger cohort relative to older age cohorts. The current preference for training in engineering over training in agriculture is likely a reflection of the declining importance of agriculture. While this pattern is the same for men and women, the difference between cohorts in the proportion trained in agriculture relative to engineering is larger for men. The gender gap appears larger for agriculture than for other disciplines, including engineering. At the same time, even in the absence of data, it would not be hard to guess that graduates from agricultural universities rarely enter farming, choosing instead to either work in sectors unrelated to agriculture or work in downstream agribusinesses or agricultural financial institutions.

Staying in, Exiting, and Entering Agriculture: A Review of Evidence

An oft-cited statistic from the NSS 59th Round Survey of Farm Households (2002–2003) is that as many as 40 per cent of respondents said they would quit farming if they had a choice.[7] Although the survey did not focus on youth, it suggested that, in general, low profitability and

---

[7] The NSS 59th Round data, a nationally representative survey of farmers, is unique in recording if farmers are content being farmers. The survey asks: "Do you like farming as a profession?"

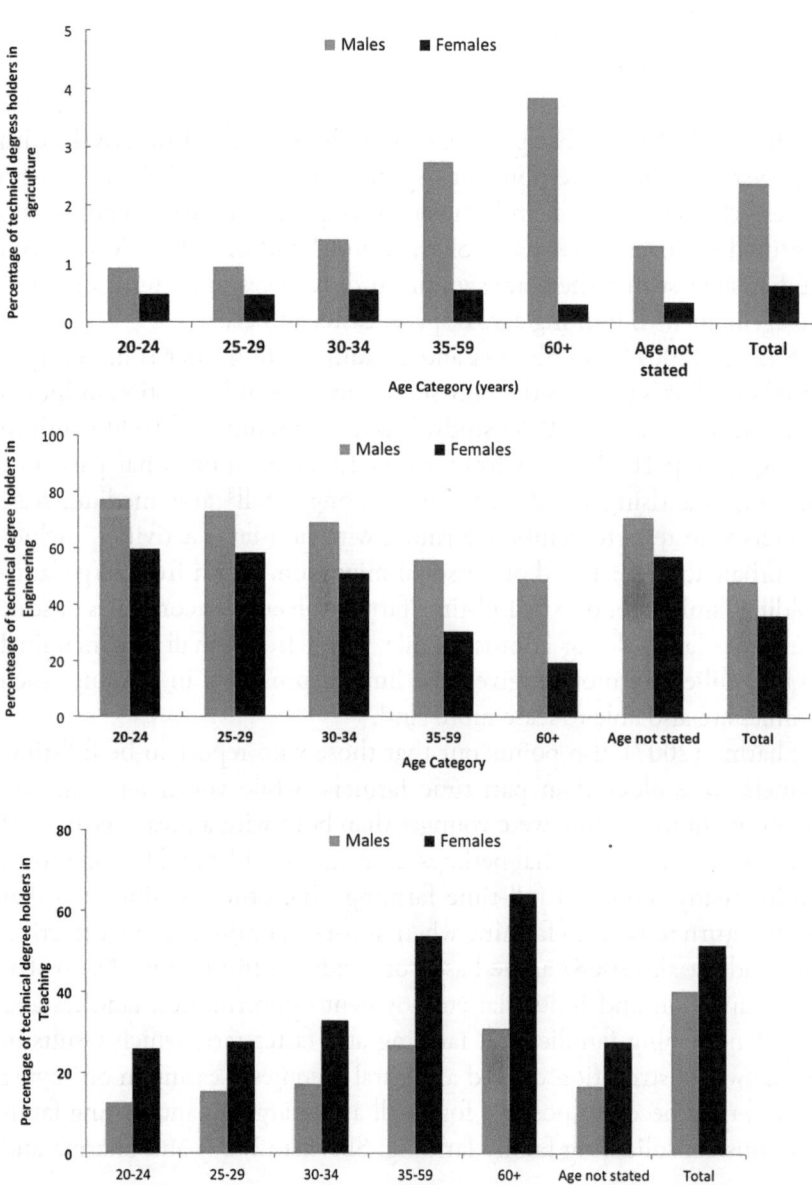

Fig. 8.1   Technical education in agriculture, engineering, and teaching compared (Source: Computed from data of the Government of India (GoI) 2016)

risk associated with incomes were the main reasons cited for the preference to exit from farming. Researchers have noted that this preference is higher among resource-constrained farmers (Agarwal and Agrawal 2017; Birthal et al. 2015). Exit preference was also correlated negatively with the age of the farmer-respondent (Agarwal and Agrawal 2017). But who leaves, who stays behind, and who enters is quite complex and not always captured in macro-level data (Sharma and Bhaduri 2009). Micro-level studies suggest that there are significant differences in patterns of youth engagement with farming across space, caste, and class.

Sharma (2007) and Sharma and Bhaduri (2009) offer some insights based on what is perhaps the only survey on the youth question in Indian agriculture. Sharma's (2007) study, based on a sample of 1609 youth in the age group 18–30 years from across 13 states, found that part-time farming is a rising trend, especially among small[8] and medium-scale farmers who tend to combine farming with non-farm activities, including urban activities based on seasonal migration. Youth from large landholding families tend to be full-time farmers given the economies of scale that large landholding affords. While youth from small and marginal farm families are mobile, given the limited prospects in farming, such families are also able to lease more land.

Sharma (2007) also points out that those who report to be full-time farmers were older than part-time farmers, while youth reporting no involvement in farming were younger than both with a mean age of 24.4 years. This could imply that perhaps as one grows older and has one's own family, many return to full-time farming. The other possibility is that youth return to take up farming when non-farm options are unattractive. Djurfeldt et al. (2008) argue based on evidence from Tamil Nadu that with education and industrial employment opportunities, landless and large landowning families exit farming at a faster rate, which results in less skewed distribution of land and rural incomes. Leasing in or buying of land then becomes possible for small and marginal landowning families, thus consolidating family farming. Sharma (2007) and Sharma and

---

[8] Marginal farmers are those who own less than one hectare of land, small farmers have one to two hectares, medium farmers have two to four hectares, and large farmers have over four hectares of land.

Bhaduri (2009) suggest that part-time farmers and youth not involved in farming are generally from higher castes, have a higher number of years of schooling, and have more employable skills. These youth are also generally from villages close to urban areas, indicating the impact of urbanization on deagrarianization (see also Djurfeldt et al. 2008). These patterns seem to be stronger in regions with a low value of agricultural production per capita and in villages close to towns. While proximity to markets is a key factor affecting returns to farming and in retaining youth in rural areas, it also has the effect of enabling youth to take up more non-farm activities. As Krishna (2017) demonstrates, villages that are at a distance of more than five kilometres from a town or a city tend to be relatively poorer than those that are located closer to urban settlements.

At the individual or household level, the pattern is more evident among castes higher in the social hierarchy, the better educated, and youth with non-farm skills. Interestingly, small, marginal, and large landholders show an inclination to withdraw from farming. While small and marginal farmers are perhaps, at least in part, being pushed out of farming, large farmers appear to take advantage of non-farm opportunities, being better off in terms of education and access to financial capital.

In Bundelkhand, in northern India, Narain et al. (2016) found that marginal farmers are more likely to want to exit farming than the medium landholding size class. Somewhat differently, in Gujarat, Patel (1985) studied the aspirations of youth to emigrate and found that neither the rich and secure nor the dismally poor showed a propensity to emigrate, albeit for different reasons; it was people in the "middle" who were mobile. She attributes this to pressure on land. Given the difficulties of land reform, the pressure on land made the surplus population restive (Patel 1985). Given that the study is somewhat dated, it is possible that the profiles of who wants to leave and who stay are today different from the 1980s.

Jeffrey (2010) in his ethnographic work in Uttar Pradesh describes the emergence and experiences of the "educated unemployed," a generation of youth from rural landowning families. Better-off landowning families increasingly send their children away for urban education and jobs, a phenomenon also noted by Balagopal (2011) in the context of coastal Andhra Pradesh in the 1980s. Many of these youth cannot find jobs and, given their newfound (educational) status, are reluctant to engage in farming. At the same time in relatively developed states such as Tamil Nadu and Punjab where youth withdrawal from agriculture may be occurring at a faster pace than in other states due to urbanization and other related processes, we are beginning to witness a small stream of well-educated, urban middle-class youth turning to farming as a lifestyle choice or as an enterprise (Shandal 2016).[9] Within agriculture, field research shows that youth tend to find certain activities more attractive than others (such as dairy, poultry, orchards, and horticulture); these are areas where returns are relatively higher. However, youth in rural areas believe that cultivation of field crops is the least difficult to enter, given that one does not require costly investments upfront if land is available (Umunnakwe et al. 2014). Studies on contract farming and contemporary supply chains suggest that, on average, younger farmers are more likely to participate in new marketing forms (Singh 2012). Overall, it appears that certain sub-sectors within agriculture appeal more to youth than others, but access to such avenues may be limited.

Village studies also lend support to the entry of segments of lower castes into farming. For example, Rao and Nair (2003) conclude that in Andhra Pradesh, the landownership pattern among caste groups has undergone a significant change—while the dominant caste has lost land, the backward and scheduled castes are reported to have gained land. Sharma (2007) notes that in Bihar, the traditional farming castes like the Bhumihars were selling land, and backward caste groups such as Yadavs were increasingly acquiring land. While such land transfers can seem socially progressive, the low returns to agriculture, particularly in relative terms and the growing crisis in the sector (Vasavi 2012; Deshpande and Arora 2010), may warrant a different reading of this phenomenon, wherein the lower castes are trapped in low-return occupations. Movement out of agriculture is also tied to non-economic aspirations. Agricultural labour is ascribed low status in the caste-based division of labour, historically associated with scheduled castes and other castes lower in the caste hierarchy. Upward mobility, as Tilche (2016) notes in her study of the

---

[9] See Karthik (2017) and Raju (2017). Tamil Nadu, for example, has a vibrant organic farmers' movement.

Patidars in Gujarat, is therefore associated with movement out of such manual work. Farm work may therefore not be appealing.

Structural and Policy Issues Within Agriculture

Existing studies thus identify several recurring themes that emerge in the context of youth entry and continuation in agriculture, some better understood than others. A few of these can be characterized as structural conditions associated with agriculture. Unremunerative agriculture constitutes one of the strongest push factors prompting exit from the sector. Research has confirmed the negative effects of the Green Revolution,[10] such as depletion in quality of soils, increase in use of purchased inputs, and extensive extraction of ground water through private investments (Reddy and Mishra 2009), has led to a process of capital intensification of agricultural production without commensurate increases in yields and/or returns. Accompanying these agro-ecological factors are a series of policy shifts, such as reduced public investments in research and development, and a lack of technological breakthroughs in rain-fed and drought-prone agriculture, which account for 60 per cent of cropped area. For much of the post-reform period, terms of trade were against agriculture except for the period from 2004–2005 to 2010–2011 when high world prices led to agricultural produce prices remaining higher relative to non-agricultural produce (Dev and Rao 2015).

Unviable Size of Holdings

The shrinking size of landholdings has been a major structural factor contributing to smallholder vulnerability. The average size of a landholding has declined by half, from 2.28 hectares (ha) in 1970–1971 to 1.16 ha in 2010–2011 (NABARD 2014). There has also been a steady increase in the share of marginal and small landholdings at the national level, and at

---

[10] The Green Revolution refers to the introduction of high yielding varieties of wheat and rice introduced in India in the late 1960s and early 1970s that required intensive use of fertilizers and water.

present, this segment accounts for 85 per cent of all operational land-holdings in the country, although accounting for only 44 per cent of total area being cultivated. Marginal landholdings increased from 9 per cent of lands cultivated in 1970–1971 to 22 per cent in 2010–2011. Trends indicate that within each farm size category—marginal, small, medium, and large—the landholding size has declined, implying that there has been no consolidation of holdings in any size category.

This reduction in operational land holding size has been partly driven by a successive division (sub-division) of inherited land in the country-side. Other factors such as distress sales that we discuss later have also been observed. Notwithstanding the evidence that smallholders in India might be more productive or efficient (Gaurav and Mishra 2015, for example), there is ample evidence that small holdings in India are smaller than the threshold size and hence unviable, a point that the Government of India recognizes explicitly: "The results of the 70th Round NSS show that positive net monthly income—i.e., difference between income from all sources and consumption expenditure—accrues only to farmers with landholdings of more than 1 hectare" (GoI 2016, 15).

While the continued non-viability of small-scale farming and of frag-mentation of land push children from such families to move out of farm-ing in search of urban employment, they pose an obstacle even to those (youth) who might be inclined to farm. Entry options into farming among lower caste youth that we noted earlier may not necessarily con-stitute upward mobility in a phase of relative decline in incomes from agriculture.

Rural Land Markets and Land Use

An important factor that contributes to reproduction of marginal land-holdings and hence to agrarian distress is the nature of emerging land markets. While unviable landholdings are constraining, there is little evi-dence of land consolidation due to either buying or leasing. A major fac-tor that may have prevented owners of unviable landholdings (or for new entrants into farming) from accessing additional land is the rise in costs of rural land, especially in relation to returns from agriculture. As

Chakravorty (2013) demonstrates, there has been an increase in the levels of activity in rural land markets since the late 1990s, followed by a tremendous increase in rural land prices during the last 10 years or so. Rising values of land due to growth in real estate activity as a consequence of higher incomes and demand for real estate from overseas Indians attract buyers who invest in land and keep prices high. Investment of black money is another major source of demand for land (GoI 2012). The expansion in credit for housing in post-reform India has also increased effective demand for land and given the inelastic supply of land, generated price increases. As a result of such demand, Chakravorty (2013) contends that rural land prices in states such as Punjab are higher by 20–30 times (one of the highest in the world) when compared to prices that would reflect agricultural productivity. Rural land values are, therefore, determined more outside of agriculture. Under such conditions of financialization of land, active land markets may not always generate outcomes that are welfare enhancing for marginal and small farmers (Vijayabaskar and Menon 2017). One consequence of rising land prices is that farmers have limited capacity to expand their farms, and young (and new) farmers are put at a huge disadvantage. These entry barriers are even more acute for women, who typically do not have access to land of their own. Although laws provide for inheritance, it seems to be the norm that women do not stake a claim in order to preserve their relationship with their brothers, often justifying their stand by rationalizing that if they did stake a claim, the already small landholdings would become non-viable (see Agarwal 1994, for instance).

In the absence of proper insurance markets and in anticipation of rising prices, land is seen as an important hedge against risk and hence property owners do not want to sell, even if their own capacity to invest in land to improve returns is limited. More than 60 per cent of Sharma and Bhaduri's (2009) respondents reveal that while complete withdrawal from farming was high on their agenda, selling land was the last option. The ties to land are maintained possibly because one cannot completely rely on non-farm opportunities, but also because of social meanings ascribed to owning land, apart from expectations of land price hikes. More than a third of their young respondents mention that they would like their children to continue farming, not only because there was a lack of opportunities elsewhere, but because that is what they had done for generations. In these instances, land does not pass to more efficient farmers; it is not the case that its sale offers an exit option for farmers. Demand for land is therefore not tied to desire to pursue farming as also pointed out in a study of rural Telangana (Jakimow et al. 2013).

In extreme cases, however, in the absence of effective policy interventions to address price and production risks, farmers end up relying on distress sales as micro-level studies of rural land markets reveal (Krishnaji 1991; Sarap 1995, 1998). Farming households also respond to risks by diversifying their livelihood options. Rather than invest in land to improve or stabilize returns from agriculture, they may consider investing in their children's education or to access non-farm employment, and hence a possible future career outside agriculture. Even before the onset of agrarian crisis and relative decline in agricultural incomes vis-à-vis incomes from other sectors, agriculture surplus was being invested outside agriculture rather than towards expansion in agricultural investments (Balagopal 2011). Diversification has seldom meant economic mobility or reduced vulnerability for most rural youth.

Diversification Sans Mobility?

NSSO's Situation Assessment Survey of Agricultural Households for the crop year 2012–2013 indicates that 57.8 per cent of households have at least one member who is self-employed in farming. Although a large share of households continue to rely on agriculture, many do not rely exclusively on agriculture; 68.3 per cent reported farming to be their main source of income in that year.[11] On average, agriculture accounted for only 60 per cent of the income for farm households (NSS, 70th Round). While income from crop cultivation and animal farming account for 48 per cent and 12 per cent of income, respectively, as much as 32 per cent of income in the household is derived from wages, working on others' farms as well as off-farm (computed using data from the NSS 70th

---

[11] Given that these are agricultural households, one would expect it to be 100 per cent; it is telling that only two-thirds say it is their main source of income.

Round). These suggest that rural is no longer synonymous with agriculture.

Over the past two decades, the contribution of the non-farm sector in rural GDP has grown significantly—from 37 per cent in 1980–1981 to 65 per cent in 2009–2010—accompanied by a marked increase in the share of non-farm employment over the same period (Papola 2013; Reddy et al. 2014). However, the quality of employment outside agriculture has been poor, marked by either poor wages or incomes. In 2009–2010, salaried employment constituted only 20 per cent of all jobs in the non-farm sector (Himanshu et al. 2013). Sectorally, the bulk of employment generation has been in the construction sector, which accounted for 35.74 per cent of all jobs created between 1990–1991 and 2015–2016 (Bhattacharya 2018). Two aspects of the employment boom in construction are worth noting. First, it tends to employ men in larger numbers and relatively more mobile men at that. Second, employment is insecure and casual for most jobs. Thus, while the rural non-farm sector is no longer a "residual" employer, it offers "decent" exit options only for a few (Jodhka and Kumar 2017). Studies also suggest that occupational mobility is lowest in agriculture and allied occupations, and half of all children of farmers end up being farmers themselves (Motiram and Singh 2010). While the ratio of non-agricultural productivity to agricultural productivity has increased from 3.97 per cent to 5.83 per cent from 1983–1984 to 2011–2012, the construction sector has a labour productivity that is only 58 per cent higher than that in agriculture.

As a way out of agriculture, rural households are investing considerably in education. According to the All India Survey on Higher Education (AISHE) 2014–2015, 24.3 per cent of youth in the age group 18–23 years are in some form of higher education compared to 19.4 per cent reported in 2010–2011. Such investments have, however, not been backed by adequate openings in the job market. Despite having registered one of the highest growth rates since 2000, the growth in India continues to be accompanied by concerns of joblessness (GoI 2018),[12] especially among the educated and those from rural households.

---

[12] See also Paroda et al. (2013) for proceedings of a symposium focused on youth in agriculture in South Asia.

According to a Ministry of Labour and Employment, Government of India survey (GoI, 2013, 43): "Every 1 person out of 3 persons who is holding a graduate degree and above is found to be unemployed based on the survey results… for the age group 15–29 years. In rural areas the unemployment rate among graduates and above for the age group 15–29 years is estimated to be 36.6 percent whereas in urban areas the same is 26.5 percent."

This clearly indicates an emerging crisis in employment with available employment opportunities not commensurate with rural youth aspirations (Cross 2009; Jeffrey 2010; Jeffrey et al. 2005a and 2005b; Jeffrey and Young 2012). Young men from rural farm backgrounds often engage in "timepass," that is, passing one's time for its own sake, and enrol in one course after another waiting for their preferred employment to materialize (Jeffrey 2010). This is also tied to quality of education and first-generation learning in the absence of social networks in landing them jobs (Jakimow et al. 2013). Apart from the inferior status assigned to farm work as discussed earlier, the desire to move out of the rural is, therefore, also tied to lack of access to quality education and to networks that facilitate access to better non-farm options. Such aspirations are belied by a lack of commensurate employment for the educated, continuing to be in farming in a context of growing income differentials between agriculture and non-agricultural sectors. In this context, micro-level studies (such as Anandhi et al. 2002; Srinivasan 2015) point to a growing crisis of masculinity among rural young men, who, unlike older generations of men, are not able to assert their identity based in farming. The unattractiveness of farming is further fuelled by the desire of rural women to marry out of farming (Bourdieu 2008; Srinivasan 2015). Overall, youth aspirations in rural areas are often not built around farming but around strategies for a way out of agriculture.

What Next for Young Farmers?

The discussion so far pieced together information from secondary sources highlighting that there is scarce attention to young farmers in policy and research. With an agrarian crisis, an ageing farm population, and a youth bulge, can youth revive the prospects of agriculture in India? And can agriculture revive hopes of youth? The agrarian crisis, precipitated by the non-viability of small-scale family farming (low productivity, poor market returns, low soil fertility, water scarcity, high levels of indebtedness), lack of public investment, and the continued dependence of a significant share of the population on agriculture for their livelihoods, is in reality also a demographic crisis as (rural) youth have not been able to effectively move out or move into agriculture in economically secure ways. If India is to reap dividends from the demographic youth bulge, the revival of quality rural employment—in particular of prospects in agriculture—will be crucial. Likewise, prospects in agriculture cannot be revived without addressing the youth question.

A youth or generational perspective demonstrates that we do not know much about youth in agriculture—their aspirations, variations across regions, how they access resources (land, knowledge, and skills), challenges they encounter, and so on—information that is necessary to offer workable strategies. The discussion highlights the need for not only greater visibility of young farmers in research and policy, but also more importantly an intersectional approach to reviving agriculture, tackling rural poverty, and youth livelihoods.

Agarwal and Agrawal (2017) note that governments tend to assume that farmers would be better off in cities while emergent farmers' movements presume that all farmers want to farm. The evidence on farmers' preferences for exit is clearly more nuanced. Further, rural households are already showing through their adaptation strategies what may be viable. Increasingly, households are combining incomes from self-cultivation with incomes from non-farm employment and business. Declining employment elasticity in agriculture (Majumdar 2017) also implies that households can undertake agriculture without much labour expenditure, allowing pluri-activities. Creating non-farm employment in rural areas would enable youth to forge livelihood pathways in the countryside and in turn, contribute to agriculture's revival (Chand et al. 2011). Similarly, ruralization of manufacturing as noted by Ghani et al. (2012) may also contribute to a "high road" to rural diversification. Efforts are necessary to quell the growing rural-urban disparities in access to quality healthcare and education that further accentuate vulnerabilities emanating from the agricultural sector.

Possibly in response to the realization that all is not well with the non-agrarian economy in terms of employment, the government launched a new project in 2015, "Attracting and Retaining Youth in Agriculture" (ARYA), supported by the Indian Council of Agricultural Research (ICAR) and implemented by Krishi Vigyan Kendras (KVK), a public institution meant to provide technical support to agriculture.[13] The National Commission of Farmers (NCF), constituted in 2004, was tasked with recommending measures to address agrarian distress. One of the sub-tasks was to suggest strategies to attract and retain youth in agriculture. In each of the six reports that the NCF submitted between 2004 and 2006, there is an explicit recognition of youth aspirations to move out of agriculture. The commission, however, restricted itself to suggesting a role for youth employment in custom hiring and skilling for animal rearing.

A sectoral and an economistic approach to integrating youth into farming may not work given the complex set of factors that render the agrarian rural inferior. The challenge may also involve revalourization of agricultural work without valourizing caste. While improved returns may provide some incentives, in the absence of a reversal of social norms around labour in agriculture, such policies may be socially regressive. In addition, the gender-neutral category of youth implicitly refers to young men.[14] This often leads to the neglect of young women in policies directed towards youth. Inheritance laws and social norms around land rights also marginalize young women from policies that focus on youth participation in farming. The family farm as conceived in the conventional sense cannot be the unit of organizing production; a flexible arrangement that can transcend sectors but is spatially located in the rural will have to be envisaged. Further, exploring new forms of collective organization of the agrarian economy may potentially weaken caste hierarchies, status, and patriarchal relations that undergird the family farm.

---

[13] See also Paroda et al. (2013) for proceedings of a symposium focused on youth in agriculture in South Asia.

[14] We undertake a detailed analysis of the situation of young women farmers in India elsewhere (see Narayanan and Srinivasan Forthcoming).

Finally, there is a strong push from youth themselves to revive farming as evident, for example, in a growing number of urban youth embracing farming out of their own volition. Political activity around access to land has also witnessed a rise recently, for example, in Jignesh Mevani's land to Dalits agenda (*The Outlook* 2018) and the Land March in Maharashtra (Dhawale 2018).

The problems that youth face in agriculture must be given more serious attention than has been the case in recent research and policy debate. This would entail a move away from viewing agriculture not merely as a source of surplus labour, but as a sector that generates social values around land and work that cannot be reduced to monetary valuations. To accomplish these, we need to understand better the choices that young farmers, both men and women, make in terms of choosing to be or become farmers. In order to do so, youth have to be a priority in policy and research—in the ways that questions are framed, data are collected, and solutions sought.

Empirical Study of Young Farmers

Inspired by the current (lack of) focus on young farmers in research and policy in India, the authors set out to gather information based on primary research on the lived realities of young farmers in two different contexts: the highly urbanized southern state of Tamil Nadu (TN) and the primarily agrarian central state of Madhya Pradesh (MP). Despite both states contributing significantly to the agricultural GDP of India and being major producers of several crops, the contrasts between their agrarian prospects for youth are stark. In the remainder of this section, we discuss salient aspects of the state's agrarian context that shape youth's involvement in farming. Similar to the discussion thus far at the national level, while there is a lot of material on agrarian and non-farm contexts, there is limited literature on rural youth and young farmers per se (Image 8.1).

Image 8.1 Sites for the becoming a young farmer study in Madhya Pradesh and Tamil Nadu, India

Tamil Nadu

Tamil Nadu represents one end of the national spectrum with probably the best parameters in terms of what can be referred to as structural transformation (Tamil Nadu Human Development Report 2017, TNHDR 2017 hereafter). It has the second lowest share of income and employment from agriculture among the larger states, the highest share of employment in manufacturing among major states, and a vibrant services sector. The diversification of economic structure and employment has been accompanied by one of the highest levels of access to tertiary education and urbanization (with 50 per cent of the population living in urban areas).

An important feature of its urbanization is that it is relatively better diffused among other urbanized states with a wider spread of small and medium towns in the state. This spread of urbanization is likely to have shaped the aspirations and opportunity structures for rural youth. Our contention is that these factors have not only significantly shaped youth aspirations and their dispensation towards farming but also generated economic and social incentives that shape such aspirations.

Although agriculture accounts for less than 8 per cent of the state's income, 33 per cent of rural youth continue to be employed in the sector. This 33 per cent is the second lowest in the country among major states, indicating a higher diversification of livelihoods. Looking at the break-up of those dependent on agriculture into cultivators and agricultural labourers, there has been a steep fall in both the share and absolute number of cultivators since 1991 (TNHDR 2017), suggesting a movement away from working on one's own land among this section of the workforce. Nearly 92 per cent of operational landholdings are in the marginal and small category and the share of total land under such operational holdings has actually increased from 55.6 to 60.6 per cent between 2000 and 2010–2011. The Agricultural Census 2015–2016 also slots Tamil Nadu with one of the lowest average land holding sizes (0.75 ha.) among the major states in the country. Normally such a decline is attributed to fragmentation due to partible inheritance apart from the development of capitalist relations that leads to differentiation within the cultivators. In Tamil Nadu, given the fact that it has witnessed a rapid decline in fertility levels (TNHDR 2017), the role of sub-division and fragmentation due to inheritance is likely to have played a lesser role. There is some evidence of the role of regional political mobilization in paving way for other modes of land transfers (Jeyaranjan 2020).

Although the share of agriculture in the state's Net State Domestic Product is half that of the national average, yields for most of its major crops including paddy, sugarcane, and horticultural produce is one of the highest in the country. The rapid diffusion of Green Revolution technologies in the state paved the way for the emergence of a strong farmers' movement in the 1970s and 1980s that mobilized around better prices and subsidies for an input-intensive production regime. The demands led to the provision of free electricity, which in turn led to intensive production and surplus generation through extraction of ground water using electric pumps. Even as it enabled a set of dryland farmers to enhance their livelihoods, excessive extraction of ground water over time, declining water tables, and growing capital intensity have led to an agroecological crisis and the undermining of agricultural livelihoods. Extraction in nearly two-thirds of the ground water blocks in the state exceeds the replenishment levels. Despite improvements in yields, the median annual income for a farmer in Tamil Nadu in 2012–2013 was less than INR 20,000[15] (net of cultivation costs) and was one among the 17 states with such poor returns to agriculture. Such poor incomes from agriculture are particularly striking in a state that has the third highest per capita income among major states in the country (TNHDR 2017). This means that significant sections of cultivators are not in a position to reproduce themselves solely through farming. It also implies that for the next generation, there are incentives not only to diversify but also to move out of agriculture. Rural households in Tamil Nadu are some of the most diversified and least dependent on agriculture in the country (Vijayabaskar and Balagopal 2019)

Income from wages and non-farm businesses account for higher incomes than that from cultivation. Importantly, income from dairy is one of the highest among all Indian states. Diversification out of agriculture has been an important livelihood strategy. While the construction sector has accounted for bulk of employment outside agriculture in the last decade and a half (TNHDR 2017), the state has also witnessed considerable investments by households in higher education. According to report of the 2014–2015 All India Survey on Higher Education, 45.2 per cent of Tamil Nadu's youth in the age group 18–23 years are engaged in some form of higher education. The highest by a sizeable difference among all major states. Importantly, the rate, although lower than that of the overall category, is also relatively higher for the marginalized scheduled caste youth, suggesting a more broad-based increase in investments in education across castes. A lack of adequate employment opportunities for the educated in the non-rural sectors, other than in construction and low-end services (drivers, security guards), implies that the exit options for large sections of cultivating households, despite investments in higher education, are fraught with uncertainties.

---

[15] In 2012–2013, the exchange rate was approximately INR 54=USD 1.

Another important dimension has been the relatively strong social security net in Tamil Nadu. Although agricultural wages have increased across the country, it is one of the states where the real wages started increasing much earlier. In addition, the implementation of an effective public distribution system (PDS) that allowed for households to access 20 kilogrammes of free rice and other provisions means that labouring households do not have to depend exclusively on their wage incomes for access to food. This state-proffered food support has meant that agricultural households are free to seek other employment. The Mahatma Gandhi National Rural Employment Guarantee Act (MGNREGA) has also been effectively implemented in the state, allowing for increases in real incomes (Kalaiyarasan and Vijayabaskar 2021). Accompanied by better access to education and rural-urban linkages, diversification into non-farm employment is believed by some of the farmers to have undermined their prospects in agriculture. Such aspects of the regional economy play an important role in the processes of entry into farming, being in agriculture, and perceptions of what it is to remain in agriculture.

Madhya Pradesh

Madhya Pradesh (MP) is a contrast to Tamil Nadu in a number of ways. The fifth most populous state in India, MP has a population of 72.6 million, of which nearly 72 per cent live in rural areas. Around 21 per cent of the population is tribal (belonging to scheduled tribes, i.e., STs), relative to 8.6 per cent in India. An estimated 15.6 per cent of MP's population belongs to the scheduled castes (SCs), a share compared to the country average (16.6 per cent). MP has been one of the more backward states within India. Unlike Tamil Nadu, it lags behind on several key human development indicators and has high rates of infant mortality (54 per 1000) and child malnutrition (57.9 per cent of those under five years are underweight) and lower life expectancy—64.2 years versus the Indian average (67.9 years). The rates of rural poverty are high as well at 35.74 per cent compared to 25.7 per cent in rural India as a whole in 2011–2012 (Bhanumurthy et al. 2016); rural poverty among the historically disadvantaged communities is even higher at 55 per cent and 41 per cent among STs and SCs, respectively. Being landlocked and with 29 per cent of the land covered in forests, MP does not possess some of the location advantages of its neighbours, or indeed Tamil Nadu, in terms of access to ports and large cities. It continues to be primarily agrarian with as much as 34 per cent of the state's GDP coming from agriculture (in 2013–2014) and over 70 per cent of the workforce still dependent on agriculture. In the 2012–2013 survey of agricultural households—in contrast to Tamil Nadu where agricultural households derived only 43 per cent from cultivation and animal rearing—in MP, this figure was 76 per cent, suggesting that diversification out of agriculture is fairly limited (Table 8.1).

In agriculture, however, MP has stood out among the major Indian states for having the highest growth rates in agricultural GDP over the past decade. It is among the largest producers of wheat, soyabean, maize, gram, canola, and mustard (Government of India 2017). Commentators note that this spectacular increase in agricultural GDP in the context of a widespread crisis in Indian agriculture is attributable to expanding irrigation (both major and minor irrigation, such as canals and wells) and a government procurement system that assures a minimum price to farmers, notably for wheat (Gulati et al. 2017). Some point out that due to better infrastructure in terms of road access, farmers have now been able to monetize their produce more easily (Gulati et al. 2017). For both wheat and soyabean, MP has emerged as an important source of produce for processors of flour and solvent extractors, respectively. Recent years have also seen the emergence of several not-for-profit initiatives by firms as part of their corporate social responsibility obligations.[16]

Notwithstanding this success, agriculture in MP faces challenges. Only about half the land is under cultivation and less than a third of cultivated land is cultivated more than once a year. According to the Agricultural Census of 2015–2016, there are 10 million operational holdings in the state covering 15.67 million hectares. The average size of operational holdings is therefore 1.567 hectares, much higher than in Tamil Nadu. In 1970–1971, the average size of operational holdings was over four

---

[16] In 2014, India implemented a Corporate Social Responsibility (CSR) provision in the Company Law Act requiring certain companies to spend at least 2 per cent of their average net profits made in the preceding three years on certain eligible activities for the larger social good.

hectares and it is evident that landholding size here is declining rapidly like in most other Indian states, mainly on account of sub-division. Between 2010 and 2015, MP registered among the largest increases in India in the number of operational holdings (Government of India 2020). As much as 71.46 per cent of the operational holdings counted in the Agricultural Census of 2015 are small or marginal (under two hectares). The presence of a large ST population adds another dimension to understanding MP's agrarian context. The Constitution of India mandates that land in tribal areas (denoted as Schedule 5/6 areas) cannot pass hands to non-tribals, as a safeguard to prevent land alienation. MP's forests are also governed by the Forest Rights Act, which shapes access to forests and exerts an influence on both agriculture and livestock practices.[17]

Tamil Nadu and Madhya Pradesh thus offer interesting contrasts from the perspective of youth in agriculture, shaping their decisions to become and be farmers. Based on in-depth interviews with 98 young farmers (18–46 years old), the next two chapters analyse the experiences and lived realities of becoming and being young farmers in these two states. Given that there is considerable variation in the agricultural contexts within each state, our findings are not intended to offer generalizations. What they do allow us to do by privileging young farmers' voices is to add much-needed insights into young farmers' experiences, the ways in which supply, demand economic factors, and sociocultural factors (norms around gender roles, marriage, inheritance, aspirations) interact in shaping the lived realities of young farmers, and hopefully inform further research, policies, and programmes to support becoming young farmers.

By way of concluding this chapter, we offer a few insights from the research conducted in the two states. The biggest challenge facing young farmers and the future of farming more generally is the declining land size and soil and water depletion. How climate change will further impact farming prospects needs to be studied urgently. While the proportion of the population dependent on farming varies as exemplified in the choice of the two states, it seems that availability of viable non-farm livelihood opportunities will be key to the continuation of farming. The extent of

---

[17] The Act is officially called The Scheduled Tribes and other Traditional Forest Dwellers (Recognition of Forest Rights) Act, 2006

urbanization and the development of non-farm, urban sectors for employment generation offer (or not) exit options as evident in the contrasting cases of Tamil Nadu and Madhya Pradesh. While we do not focus specifically on young women farmers (we do this elsewhere as it warrants a thorough focus), the challenges posed by sociocultural norms and high levels of gender discrimination become immediately evident in studying their farming experiences. Finally, the interviews offer much-needed evidence to counter the misconception that young people leave farming, never to return. Farming is a process marked by years in school, disinterest, urban aspirations, marriage, family responsibilities, different livelihood and income strategies, and migration.

References

Agarwal, B. 1994. *A field of one's own*. Cambridge: Cambridge University Press.

Agarwal, B., and A. Agrawal. 2017. Do farmers really like farming? Indian farmers in transition. *Oxford Development Studies* 45 (4): 460–478.

Anandhi, S., J. Jeyaranjan, and R. Krishnan. 2002. Work, caste and competing masculinities: Notes from a Tamil village. *Economic and Political Weekly* 37 (43): 4397–4406.

Balagopal, K. 2011. *Ear to the ground: Selected writings on class and caste*. New Delhi: Navayana.

Bhanumurthy, N.R., H.K. Amar Nath, S. Bose, P.D. Adhikari, and J. Arkajyoti. 2016. *Madhya Pradesh, state millennium development goals report 2014–15*. New Delhi: UNICEF and National Institute of Public Finance and Policy.

Bhattacharya, P. 2018. Which are the top sectors that generate employment in India? *Mint*, April 12. https://www.livemint.com/Politics/7XXmUWyxkSEGKoWXJqUuHM/Which-are-the-top-sectors-that-generate-employment-in-India.html.

Birthal, P.S., D. Roy, M.T. Khan, and D.S. Negi. 2015. Farmers' preference for farming: Evidence from a nationally representative farm survey in India. *The Developing Economies* 53: 122–134.

Bourdieu, P. 2008. *The bachelors' ball*. Cambridge and Malden: Polity.

Chakravorty, S. 2013. *The price of land: Acquisition, conflict, consequence*. New Delhi: Oxford University Press.

Chand, R., P.L. Prasanna, and A. Singh. 2011. Farm size and productivity: Understanding the strengths of smallholders and improving their livelihoods. *Economic and Political Weekly* 46 (26–27): 5–11.

Chenery, H.B. 1979. *Structural change and development policy.* Baltimore: Johns Hopkins University Press.

Cross, J. 2009. From dreams to discontent: Educated young men and the politics of work at a special economic zone in Andhra Pradesh. *Contributions to Indian Sociology* 43 (3): 351–379.

Cuervo, H., and J. Wyn. 2012. *Young people making it work: Continuity and change in rural Places.* Melbourne: Melbourne University Publishing.

Deshpande, R.S., and S. Arora, eds. 2010. *Agrarian crisis and farmer suicides.* New Delhi: Sage Publications.

Dev, S.M., and N.C. Rao. 2015. Improved terms of trade for agriculture: Results from revised methodology. *Economic and Political Weekly* 50 (15): 19–22.

Dhawale, A. 2018. *The Kisan long march in Maharashtra.* New Delhi: Leftword.

Djurfeldt, G., et al. 2008. Agrarian change and social mobility in Tamil Nadu. *Economic and Political Weekly* 43 (45): 50–61.

Food and Agricultural Organisation. n.d. Family farming knowledge platform. Accessed August 10, 2019. http://www.fao.org/family-farming/background/en/.

Gaurav, S., and S. Mishra. 2015. Farm size and returns to cultivation in India: Revisiting an old debate. *Oxford Development Studies* 43 (2): 165–193.

Ghani, E., A. G. Goswami, and W. R. Kerr. 2012. *Is India's manufacturing sector moving away from cities?* Harvard Business School Working Paper 12-090. Cambridge, MA: Harvard Business School. https://www.hbs.edu/faculty/Pages/item.aspx?num=42436.

Government of India (GoI). 2012. *Black money.* New Delhi: Central Board of Direct Taxes, Department of Revenue, Ministry of Finance, Government of India.

———. 2013. *Report on youth employment-unemployment scenario (2012–13).* Vol. 3. New Delhi: Ministry of Labour and Employment, Labour Bureau, Government of India.

———. 2014a. *Key Indicators of land and livestock holdings in India.* Report no. NSS KI (70/18.1). New Delhi: National Sample Survey Organisation, Ministry of Statistics and Programme Implementation, Government of India.

———. 2014b. *Key Indicators of situation of agricultural households in India.* Report no. NSS KI(70/33). New Delhi: National Sample Survey Organisation, Ministry of Statistics and Programme Implementation, Government of India.

————. 2016. *State of Indian agriculture, 2015–16.* New Delhi: Ministry of Agriculture and Farmers Welfare, Department of Agriculture, Cooperation & Farmers Welfare, Directorate of Economics and Statistics, Government of India.

————. 2017. *Policy for Kharif marketing season 2017–18.* New Delhi: Commission on Agricultural Costs and Prices, Ministry of Agriculture and Farmers Welfare, Government of India.

————. 2018. *Economic survey 2017–18.* New Delhi: Ministry of Finance, Government of India.

————. 2020. All India Report on Agriculture Census 2015–16. Agriculture Census Division, Department of Agriculture, Cooperation and Farmers' Welfare, Ministry of Agriculture and Farmers Welfare, Government of India.

Gulati, A., P. Rajkhowa, and P. Sharma. 2017. *Making rapid strides-agriculture in Madhya Pradesh: Sources, drivers, and policy lessons.* Working Paper 339. Delhi: Indian Council for Research on International Economic Relations

Himanshu, Lanjouw P.A., R. Murgai, and N. Stern. 2013. Non-farm diversification, poverty and economic mobility, and income equality: A case study in village India. *Agricultural Economics* 44: 461–473.

Jakimow, T., C. Tallapragada, and L. Williams. 2013. A future orientation to Agrarian livelihoods: A case study of rural Telangana. *Economic and Political Weekly* 48 (26 and 27): 129–138.

Jeffrey, C. 2010. *Timepass: Youth, class, and the politics of waiting in India.* Stanford: Stanford University Press.

Jeffrey, C., and S. Young. 2012. Waiting for change: Youth, caste, and politics in India. *Economy and Society* 41 (4): 638–661.

Jeffrey, C., P. Jeffery, and R. Jeffery. 2005a. Dalit young men and formal education. In *Educational regimes in contemporary India*, ed. R. Chopra and P. Jeffery, 256–275. New Delhi: Sage Publications.

Jeffrey, C., R. Jeffery, and P. Jeffery. 2005b. When schooling fails: Young men, education and low-caste politics in rural North India. *Contributions to Indian Sociology* 39 (1): 1–38.

Jeyaranjan, J. 2020. Tenancy Reforms in Tamil Nadu: A Study from the Cauvery Delta Region. In *Rethinking Social Justice*, ed. S. Anandhi, K.R. Manoharan, M. Vijayabaskar and A. Kalaiyarasan, 255–83. Orient Blackswan.

Jodhka, S., and A. Kumar. 2017. Non-farm economy in Madhubani, Bihar social dynamics and exclusionary rural transformations. *Economic and Political Weekly* 52 (25 and 26): 14–24.

Kalaiyarasan, A., and M. Vijayabaskar. 2021. *The Dravidian model: Interpreting the political economy of Tamil Nadu.* Cambridge and New York: Cambridge University Press.

Karthik, S. D. 2017. Techie to quit job and turn organic farmer. *The New Indian Express*, March 18. http://www.newindianexpress.com/states/tamil-nadu/2017/mar/18/techie-to-quit-job-and-turn-organic-farmer-1582753.html.

Krishna, A. 2017. *The broken ladder: The paradox and potential of India's one-billion.* New Delhi: Penguin Random House.

Krishnaji, N. 1991. Land market–On dispossession of peasantry. *Indian Journal of Agricultural Economics* 46 (3): 328–334.

Majumdar, K. 2017. *Labour mobilization from farm to non-farm: Contemporary structural transformation of Indian economy.* Presented at the 5th National Conference of the Network of Agrarian and Rural Studies (NRAS), Bhubaneswar, Odisha, October 27–29.

McMichael, P. 2008. Peasants make their own history, but not just as they please.... *Journal of Agrarian Change* 8 (2–3): 205–228.

Motiram, S., and A. Singh. 2010. How close does the apple fall to the tree? Some evidence from India on intergenerational occupational mobility. *Economic and Political Weekly* 47: 56–65.

Narain, S., A.K. Singh, and S.R.K. Singh. 2016. Perception of farming youth towards farming. *Indian Research Journal of Extension Education* 15 (2): 105–109. http://agritech.tnau.ac.in/farm_association/orgfarmtn.pdf.

Narayanan, S., and S. Srinivasan. forthcoming. No country for young women farmers: A situation analysis of young women farmers in India.

National Bank for Agriculture and Rural Development (NABARD). 2014. Agricultural landholdings pattern in India. *Rural Pulse* 1: 1–4.

Outlook, The. 2018. Give every Dalit 5 acre land, 70 percent are landless, says Jignesh Mevani. *Outlook*, January 18. https://www.outlookindia.com/website/story/give-every-dalit-5-acre-land-70-are-landless-says-jignesh-mevani/307019.

Papola, T.S. 2013. Employment growth during the post-reform period. *The Indian Journal of Labour Economics* 56 (1): 1–13.

Paroda, R., I. Ahmad, B. Mal, Y.L. Saharawt, and M.L. Jat. 2013. Regional workshop on youth and agriculture: Challenges and opportunities. In *APAARI, Proceedings and Recommendations*, 23–24. Islamabad: Asia-Pacific Association of Agricultural Research Institutions (APAARI) and Pakistan Agricultural Research Council (PARC).

Patel, N. 1985. Youth aspirations vis-à-vis national development: Participate or emigrate? *Sociological Bulletin* 34 (1–2): 39–48.

Raju, K. 2017. Software to farming: Engineer ready to reap rich yield. *The Hindu*, July 26. http://www.thehindu.com/todays-paper/tp-national/tp-tamilnadu/software-to-farming-engineer-ready-to-reap-rich-yield/article19361528.ece.

Rao, N. 2017. Assets, agency and legitimacy: Towards a relational understanding of gender equality policy and practice. *World development* 95: 43–54.

Rao, G.N., and K.N. Nair. 2003. Change and transformation in rural South India: Findings from villages studies. *Economic and Political Weekly* 38 (32): 3349–3354.

Reddy, D.N., and S. Mishra, eds. 2009. *Agrarian crisis in India*. New Delhi: Oxford University Press.

Reddy, D. N., A. Reddy, N. Nagaraj, and C. Bantilan. 2014. *Rural non-farm employment and rural transformation in India*. Working Paper Series No. 57. Patancheru: International Crops Research Institute for the Semi-Arid Tropics.

Sarap, K. 1995. Land sale transactions in an Indian village: Theories and evidence. *Indian Economic Review* 30 (2): 223–240.

———. 1998. On the Operation of the land market in backward agriculture: Evidence from a village in Orissa, Eastern India. *The Journal of Peasant Studies* 25 (2): 102–130.

Shandal, M. 2016. Is post-green revolution agriculture in Punjab, India an example of Karl Polanyi's double movement? MA thesis, University of Guelph.

Sharma, A. 2007. The changing agricultural demography of India: Evidence from a rural Youth perception survey. *International Journal of Rural Management* 3 (1): 27–41.

Sharma, A., and A. Bhaduri. 2009. The 'tipping point' in Indian agriculture: Understanding the withdrawal of the Indian rural youth. *Asian Journal of Agriculture and Development* 6 (1): 83–97.

Singh, S. 2012. *Modern food value chains India*. New Delhi: Samskriti.

Srinivasan, S. 2015. Between daughter deficit and development deficit: The situation of unmarried men in a South Indian community. *Economic and Political Weekly* 50 (38): 61–70.

Tamil Nadu Human Development Report. 2017. Tamil Nadu State Planning Commission, https://spc.tn.gov.in/tnhdr2017.html.

Tilche, A. 2016. Migration, bachelorhood and discontent among the Patidars. *Economic and Political Weekly* 51 (26–27): 17–24.

Umunnakwe, V.C., V.K. Pyasi, and A.K. Pande. 2014. Factors influencing involvement in agricultural livelihood activities among rural youth in Jabalpur District of Madhya Pradesh, India. *International Journal of Agricultural Policy and Research* 2 (8): 288–295.

Vasavi, A.R. 2012. *Shadow space: Suicides and the predicament of rural India.* Gurgaon: Three Essays Collective.

Vijayabaskar, M., and G. Balagopal. 2019. *The politics of poverty alleviation strategies in India*, UNRISD Working Paper, No. 2019-7, United Nations Research Institute for Social Development (UNRISD), Geneva.

Vijayabaskar, M., and A. Menon. 2017. Peripheral agriculture? Macro and micro dynamics of land sales and land use changes in the 'rural' economy of Kancheepuram. In *Political Economy of Contemporary India*, ed. R. Nagaraj and S. Motiram, 205–229. New Delhi: Cambridge University Press.

Vijayabaskar, M., S. Narayanan, and S. Srinivasan. 2018. Agricultural Revival and Reaping the Youth Dividend, *The Economic and Policitcal Weekly* 53: 26–27, 30 Jun, 2018. https://www.epw.in/journal/2018/26-27.

9

Becoming/Being a Young Farmer in a Fast-Transitioning Region: The Case of Tamil Nadu

M. Vijayabaskar and Radha Varadarajan

Introduction

This chapter focuses on the pathways into farming among youth in select locales in the state of Tamil Nadu, southern India and what being in farming means for their lives and livelihoods. In doing so, we offer some interpretations about youth experiences of being a farmer, what they perceive as constraints in securing decent livelihoods, and the institutional context in which such processes and perceptions are embedded. We point out that regional political economy shapes the institutional context in which youth develop aspirations and dispensation towards farming. The extensive diversification into the non-farm and the broad base of education in the farming communities in the state, we argue, have deeply

M. Vijayabaskar (✉)
Madras Institute of Development Studies, Chennai, India

R. Varadarajan
University of Madras, Chennai, India

© The Author(s) 2024
S. Srinivasan (ed.), *Becoming A Young Farmer*, Rethinking Rural,
https://doi.org/10.1007/978-3-031-15233-7_9

253

influenced how farming is located in relation to other occupations. We elaborate on these details beginning with a discussion of data collection including our study sites and the sample of young farmers.

Data Collection

Despite a relatively more diffused rural-urban continuum, inter-regional differences persist across Tamil Nadu in agricultural and human development. To understand how such regional differences shape the processes of becoming and being a young farmer, we chose to interview young farmers in three different locations: (i) western Tamil Nadu or what is referred to as *Kongunadu* has been the centre of Green Revolution in the state as well as a region that has diversified extensively based on investments of agrarian surplus into industry and services. The region also has relatively larger farm holdings compared to the rest of the state, especially among the Kongu Vellalas, an agrarian caste that has also extensively diversified out of agriculture. Here, we conducted interviews among farmers across four adjoining districts: Coimbatore, Erode, Tiruppur, and the Nilgiris. While the first three are typical of the region, the Nilgiris is a hilly region with a vibrant horticulture. (ii) The districts bordering Chennai also have a strong agricultural economy traditionally based on tank irrigation and paddy cultivation. Urban expansion and poor tank management have eroded this economy in part, but expansion of urban demand for fruits and vegetables has led to agricultural diversification and intensification on the periphery. Landholding size on an average is, however, smaller than in western Tamil Nadu. We conducted interviews in Kancheepuram district bordering Chennai. (iii) There are also districts like Tiruvannamalai that are predominantly agrarian with relatively poor transport and employment links to the urban and low levels of development. Tiruvannamalai is one of the most backward districts in the state with a large Dalit population who have limited access to land. As elsewhere in the country, caste differences underlie access to land. The lowest castes, referred to officially as Scheduled Castes (Dalits is the term that caste members prefer), have historically been landless agricultural workers, and even when they own land, they are mostly marginal and small farmers.

We interviewed 58 young farmers to capture regional and caste divergences. The largest number of farmers—42—that we interviewed are from western Tamil Nadu, out of which 36 young farmers belong to the Kongu Vellala caste, three were young Dalit farmers, and three were young farmers from the Badaga caste in the Nilgiris. We interviewed eight young farmers each in Tiruvannamalai and in Kancheepuram district bordering the Chennai metropolitan region. In total, we interviewed 16 women farmers, with 11 from Coimbatore and the remaining from Tiruvannamalai. The Tiruvannamalai young women farmers are Dalit and are members of a collective that a civil society organization coordinates to practice organic farming. We also interviewed six older farmers, two organic farmers who have moved from high-paying non-farm jobs into agriculture, two members of organic farmers' movements, and two officials associated with the Tamil Nadu Agricultural University.

The young farmers in western Tamil Nadu tend to have more land on average (four to six acres) when compared to farmers in Tiruvannamalai and Chennai regions, with the largest landholding size being 20 acres. In the latter region, young farmers mostly owned one to three acres of land with the largest landholding size being five acres. Importantly, across the state, farmers can commute to a nearby town to take up non-farm work. Crops grown varied from tree crops such as coconuts, horticulture and floriculture, paddy, turmeric, and tobacco. Farmers have often responded to changing physical and market conditions through shifts to different crops.

In terms of age profile, the average age of our sample of respondents was 36, with six of them below age 30. In general, our sample farmer households had at least one member employed outside agriculture. Dominant non-farm employment options include working in workshops, garment or textile units, or undertaking petty service provisioning. Non-farm businesses include a woman farmer's husband running a tea shop in the village and a male farmer diversifying into agro-processing and setting up a groundnut oil mill. A few in the Chennai region reported taking part in real estate activities. Although not directly reported, a few households were also engaged in money lending. In some cases, they had siblings working in professional white-collar jobs elsewhere who help these young farmers to financially negotiate the vulnerabilities emanating from agriculture. Among respondents in Chennai and Coimbatore regions, a few worked in factories or a small firm before quitting or losing their jobs and taking up farming. One young male farmer (35 years old; owning 10 acres) continues to work in an IT-related job in Bangalore and manages the family farm with his grandfather's help. One young woman farmer was also an *anganwadi* (government childcare) worker. A few of the Dalit women farmers work as agricultural labourers when they have less work on their own farms.

Pathways into Farming

For all of the young farmers that we interviewed, the pathway into farming was through inheritance, to continue their parents' vocation. Most respondents had on average 10 years of schooling. A few respondents secured a diploma in engineering, or an undergraduate degree, including in engineering. There was one farmer who had a graduate degree. Barring such farmers, many respondents cited a failure to pursue education as the main reason for entry into farming. Importantly, entry into farming among the better educated was tied to a lack of access to quality jobs. This is best represented by Mohan aged 31 years,[1] a farmer who grows vegetables in a village near Chennai. His parents used to grow paddy earlier; a rising demand for vegetables from the city enabled the shift. He completed a diploma in mechanical engineering and found temporary employment in an auto firm near his home. Within a year of Mohan joining the firm, it shut down due to financial losses. After failing to secure similar employment, he entered into farming and continues to work with his parents on the farm. In the case of most young farmers with lower educational qualifications, either a parental inability to sustain their child's education or failure to pass final exams led them to stay home and help their family with farm work. According to Murugan, a 43-year-old farmer in western Tamil Nadu with eight years of schooling: "I didn't study very well. So, I remained stuck in agriculture. Thinking back, I am not sure how I would have done had I been in any other job. I like what

---

[1] We use pseudonyms in this chapter.

I do. I think the only key to success here is constant hard work. I think that is the key to success anywhere. Here, it appears harder—that's all."

Only in three cases, including two from the Nilgiris, respondents decided to take up farming with their family while studying. The two young farmers in the Nilgiris mentioned that they did not aspire for much else as the entire village was involved in agriculture. In the third case, the farmer from western Tamil Nadu was always keen to enter farming, even though he had completed a diploma in engineering. Some of the male farmers also said that their parents insisted that they should acquire a good education, even if they wanted to enter farming. Despite not being the preferred choice at the time of entry, many of these young farmers see this entry favourably in relation to working outside agriculture.

The pathways into farming for women farmers overlap with that of male farmers, but there are differences. Although born in cultivating or agricultural labour households, women's entry into farming is largely tied to their marriage into farming households. As in the case of men, many of these women helped their parents on the farm before marriage while still studying. Young women farmer respondents discontinued their education after 8–12 years of schooling. In addition to factors that young male farmers report for discontinuing education, women farmers state that their parents did not encourage them to travel far for higher education. After dropping out, they tend to work in agriculture but, more importantly, in the non-farm sector such as in the power loom weaving units, garment factories, and cotton spinning mills that dot across western Tamil Nadu. They discontinue the non-farm work once they are married. After marriage, they either find themselves assisting the husband and in-laws on their farms or on occasion, assist the in-laws when the husband is engaged in non-farm work or farm on their own. At times, women also work on their parents' lands if they are older with no sons or support from these sons. This happens due to South Indian kinship arrangements where daughters are usually not married too far away from their natal homes. Still, farming is not something that most young women or their families choose. Excerpts from an interview with Jothi (28 years), who is married to a 36-year-old farmer in Erode district, western Tamil Nadu, reveal this:

I like to study. I finished my 12th. My village is near Arichaloor. But my father had no money. We were two sisters. There was a compulsion that if they spent on my education, they need to spend on her education as well. So, they stopped both our education. I wanted to study after marriage, but since I had a child right after, I couldn't get around to studying further. I used to work for a local garment company before marriage ...My father hated it, if we went to the farm. He wanted us to be inside the house. He would tell me that even if I did not know to read/write or was not interested in studying, I could just go and keep typing at a computer or teach at an elementary school. He was very particular that both his children were actually people who were well educated and who were placed in good jobs.

But married into a farming family she had little choice. Selvi, a 34-year-old farmer, again from western Tamil Nadu, has a similar narrative about her and her brother's entry into farming:

My husband and I both studied up to 10th standard. I was so passionate about studying further. I was a very bright student...I used to cry for several days, asking that I should be allowed to go to school. But, my father wouldn't allow me. There was no one to go from there to the school with me. Actually, I had to change two buses to get to the school. So, it was not safe for me, particularly while returning home. Even for the 10th grade, I had to walk four kilometres. I begged my parents, but then, slowly, I also let it go. It was not their fault, you know! Those days were like that. It was not important for anyone. My brother tried to drag on for another two years. But, after my father passed away, my brother had to get back to farming... Whenever I would go rearing cows and doing farming, if I found any paper, I would read it. My father did feel sad about it and got me lots of books, but then he couldn't help me in getting to school. After I got married, I forgot all about it. Then, I had to convince myself saying farming is the thing I am probably destined to do.

The case of 33-year-old male farmer Subbu, who has an undergraduate degree in commerce, illustrates how entry into farming was often not the first option, but over time becomes the preferred option:

When I grew up, agriculture was rewarding. It was lucrative. But, I did not want to be in farming. I wanted to get out of farming, go to Coimbatore and get a job. But, now things have changed completely. I want to be only in agriculture... Before getting married, I used to work in a garment factory as a staff supervisor in Tiruppur... After getting married, I quit. I was working full-time in farming. Now, recently, since there is no water, I went back to the garment factory job... With the commute, it gets difficult to come back and do anything. But, weekends, I am at the farm, helping out with whatever I can.

Given the importance of family labour to small farms, farms become less viable when labour is inadequate. Another pathway in this context is when parents fall ill or one of them dies leaving the children (often sons) with no option other than to take over farming. Entry into farming, however, allows them to diversify into different livelihoods as we mentioned earlier. Young men's entry into farming, especially after working in the non-farm sector, highlights the vulnerabilities that the bulk of non-farm employment poses. Poor incomes, lack of economic mobility, and job insecurity have all contributed to educated youths' entry into agriculture. At the time of entry into agriculture, the move is regarded as downward mobility. Over time, they prefer to continue to be in agriculture. At the same time, the low status associated with farming also means fears of not being able to find a bride. Farmers we interviewed in western Tamil Nadu often spoke of reluctance on the part of women and their parents in marrying men who are primarily in farming.

Crisis of Agrarian Masculinities

While the heteronormative patriarchal-patrilineal context of farming makes it hard for women to be identified as independent farmers and, more importantly, to own land, which pushes them to look outside farming for better lives, there are also caste markers of manual work. Apart from issues of the viability of livelihoods, young men tend to refrain from entering farming because of a growing fear that they may not find a bride. "I can tell you one thing—no one in agriculture gets a bride!" says Perumal, a 30-year-old male farmer who has been married for a year. Responding to a question about how he managed to get married, he says,

"Only circumstances. Maybe because I own land." Often young women tend to be better educated and prefer not to undertake manual work on the farm. It is, therefore, not merely tied to farming's economic viability and the probability of a decent livelihood but also to a set of values and meanings ascribed to manual work and agricultural work. Although this devaluation of "peasant work" has global resonance (Bourdieu 2008), in Tamil Nadu, it is also tied to the dominant political narrative that conceives social justice as a move away from a caste-determined division of labour (Aloysius 2013).

The views of some of the farmer respondents in the state strongly echo Srinivasan's (2015) observations on the failure of peasant youth to marry within their caste. Jothi provides some insights on this shifting ascription of status to a farmer:

> Parents do not encourage girls to get married to someone in farming. So, what men do is, they take up [non-farm] employment around their 20s and once they get married, they quit their job and then come back into farming. This happens a lot! It is so popular now that the girls insist that their husband should not move into farming after their marriage. A situation has been created where boys find work outside, despite having more than 10 acres to their name... This is only for the past few years. Before that, there was nothing like this. When I got married, 10 years ago, I had no issues working at the farm, or marrying someone who is into full-time farming.

Venkatesan (42 years old), who left a low-paying job to take up farming, points to how rather than income from agriculture, it is the shifting social value around farming that undermines his chances of marrying:

> I haven't got married yet... There are so many men here in the village...even those who do part-time work are not getting married because no woman wants to marry a farmer. No one respects farmers... They expect only a daily/monthly income... In 2007, I was getting 5000 Rupees. I wanted to do what I was interested in. So, I quit. They would not have bothered if that is all I made as long as I had a job. No one is realizing that they cannot get a raise or keep the job these days.

While land ownership still commands and bestows social status, farm livelihoods are not valued when compared to work in the non-farm formal economy. While this is not an issue in the Tiruvannamalai region, farmers near Chennai express apprehensions about the diminishing prospects of marrying. While male respondents that we interviewed are already married, they concede that the prospects for marriage have diminished at present. This desire is also tied to material shifts in the state. Broad-based access to education in Tamil Nadu has meant that members of cultivating castes have managed to access high-end jobs, triggering aspirations that often result in exit from agriculture.

Access to Land

Most of the study respondents own less than five acres of land, consistent with the bulk of the farming population in Tamil Nadu and even India at large. As Sivakami, a 32-year-old woman farmer in western Tamil Nadu, says, "We need at least 10 acres to do agriculture exclusively...especially if you have to pay school fees and other expenses." Her husband works in a workshop in Kangayam, a nearby town, and she too worked in a garment factory before marrying. Few respondents report buying land towards consolidation within agriculture. This inability to access is tied to rising rural land prices across the country (Chakravorty 2013, also see Chap. 8). Chakravorty attributes this price rise to urban actors' speculative investments in land, which is also borne out by a Tamil Nadu-specific, micro-level study (Vijayabaskar and Menon 2018). During our fieldwork, we did observe manifestations of this phenomenon in two of our field sites—around Chennai and western Tamil Nadu. Given the poor returns on farming, the increase in land prices means that farmers cannot afford to buy additional agricultural land to consolidate their holdings.

Dhanapal, a male farmer in his early 40s, explains why it is not possible to consolidate through buying or leasing land in western Tamil Nadu where land prices have been increasing rapidly:

See, he quotes 12 lakhs (INR)[2] for this land, and another quotes 25 lakhs, because his soil is better, and he has access to road. How to consolidate? Particularly, after the agricultural lands are converted to site plots, the same is charged at twice or thrice the rate. How can we farmers buy that and use that for cultivating crops? It is very difficult to get plots next to each other. Once we decide to consolidate and buy the first two or three plots, the news will leak and the people who are in the neighbourhood will not part with their land or quote exorbitant prices that we cannot afford. See, when we bought this land, it was 10,000 (INR) per acre. I bought two acres. Then, I wanted to buy the adjacent plot also. He started quoting 20,000 (INR) without any remorse.

Demand for land from non-farm sectors (real estate, factories, roads, and other urban infrastructure), where returns are higher, generates new barriers to access land for farming. This is true of the Chennai region where land conversions from agriculture and powerful urban actors' purchase of agricultural lands are visible in areas where we conducted our field study. In the Nilgiris, there is hardly any additional land to be bought or accessed. In the next section, we discuss how our respondents have accessed land in this context.

Accessing Land

Among the traditionally patrilineal landowning caste groups such as Kongu Vellalas, land is invariably inherited. There are, however, differences in the modes of inheritance. When there is more than one claimant to inheritance, a formal division of land and title transfer may take place even when the father is alive. For example, after the marriage of one of the sons, the father may choose to formally transfer the land. This is particularly likely when there are tensions between siblings over sharing inherited land. In case of daughters, when they need the land for certain expenses, they are given their share. In most cases, children tend to work on the parents' farms without any formal transfer. Often, the lands are formally in the father's name who, in turn, may jointly own it with his

---

[2] One lakh INR translates into approximately 1288 USD.

siblings. In such cases, access is guided more by customary rights than by formal ownership. The earnings are used or shared by the entire family.

In some cases, farmers sell the inherited land located elsewhere to buy the piece of land on which they are currently farming. Another strategy to access land is evident from the Nilgiris district, where land is jointly owned by several members of the community. One respondent in his late 30s cultivates two acres of land that has been held by his family since the time of his grandfather and perhaps even earlier. The entire village is an epicentre of high-value horticultural production and produce for leading food retail companies. Households share labour responsibilities, relying on relatives for harvesting while they take turns to guard the land from wild animals at night. The absence of clear titles implies that they cannot easily transact lands or access government schemes that require proper titles. In another case, a male farmer works on land that his wife inherited. Marriages in the areas under study are often village endogamous and patri-virilocal, unlike in Madhya Pradesh, where we also conducted fieldwork. Cross-kin marriages persist, and while on the decline, are an option for families with land that do not have sons. Such marriages ensure that the land remains within the family. A landowning family without sons may seek a groom for their daughter with an interest in farming or may shift to a non-farm business. Land management options of landowning families with daughters only is an area that requires further investigation.

The Tamil Nadu government passed the Hindu Succession (Tamil Nadu Amendment) Act in 1989 that allows daughters equal access to share in the ancestral wealth. This does not imply that all women inherit an equal share of the land. At times, as part of a dowry, land is transferred to the bride along with gold and other assets. When the husband is involved in farming, the woman's parents tend to transfer a share of the property as it is critical to their daughter's livelihood. We came across some cases where women farmers inherited or are likely to inherit the land that they are farming because they are the only children. The decision to transfer land to daughters, therefore, rests on a complex set of factors that involves the relative extent to which sons' livelihoods are dependent on the land vis-à-vis that of the daughters as well as the daughters' post-marital residence.

One Dalit farmer in western Tamil Nadu inherited land that his father received under the "*Bhoodan* scheme," a government programme that involved the landlords voluntarily giving land to the landless. His father had then slowly, over time, started buying small parcels of land from neighbouring farmers. Another Dalit farmer bought his initial plot of land from the landlord where he worked as a tenant and then purchased adjoining plots over time before the land prices began to reflect demand from non-agricultural sectors.

Leasing is another means to access land. However, we found very few instances of leasing among the farmers that we interviewed. These are often lands owned by kin who are no longer in agriculture or family members who are not physically able to manage their lands. Forty-two-year-old Murugesan from Coimbatore inherited two acres through this strategy to increase access to land:

> I have leased in about two acres. The owners are rich businessmen who live in Tiruppur....I pay 3000 rupees per annum as the lease amount. I culti-vate vegetables there. They have leased it out to us as we take care of the land. If it is unattended, thorny bushes will spread and people will also start to smuggle soil out of that field. So, it is mutually beneficial... The water comes from our farm. They don't have water source there. That's another reason they leased it to us...

The increase in land prices has helped farming households to sell a por-tion of their land to meet life cycle expenses such as marriage, childbirth, or illness. None of the farmers that we interviewed had bought land for agriculture in the 10 years prior to the interview. Mary, a 42-year-old Dalit woman farmer in Tiruvannamalai, also cites issues with transferring the land to her name. Her husband inherited two acres of land on his parents' death, and she uses this land to work in a women's collective for organic farming. Responding to a question on whether it will help her if the land is transferred to her name, she says: "Since my brothers might come for a share, the land will never be transferred in my name. But my husband is happy with the co-operative farming. So, he allows me to use it the way we want." This raises the issue of the relationship between for-mal ownership and decision-making on the farm, especially through coercion rather than through any formal claim-making.

Titling, Access, and Decision-Making

Access to land may not always imply exclusive rights to make decisions on land use for young farmers, especially when their parents continue to be actively involved in farming. In most instances, farming decisions are not made by the young farmers and are guided more by norms of the active (grand) parent. The transfer of ownership or control over decision-making is also on occasion dependent on the health of the (grand) father. If the parents continue to work, most farm decisions are taken by them, the (grand) father in particular. If they are too old to work, then the son(s) tend to take charge, irrespective of whether they actually own the land.

Perumal, age 33, lives close to his parents in western Tamil Nadu; he inherited 12 acres of land from his father. He is trying to move into organic farming and has diversified into beekeeping. His father does not approve of his new ventures:

> He doesn't like me doing natural farming...He wants to just continue doing what they did—some small amount of maize, have goats, cows, and carry on. This is a point of contention between us. Like when I was getting ready to plant bananas, they sowed maize without informing me. They do things like this. But they have no problem in my being in agriculture. My idea is to grow a forest in this land, other than a small piece of land that can be used for sowing crops that we need. But their idea is different. They are not for it. They want to continue grazing the lands with goats.

Such instances, though not often reported, illustrate the nature of intergenerational tensions that young farmers face in undertaking new farming practices. Young women farmers have to navigate gender and generational challenges (more on this in the forthcoming book on young women farmers).

When the land they are to inherit is a part of their father and his siblings' joint property, decision-making and even investment become difficult. Rani, a 36-year-old female farmer in western Tamil Nadu, illustrates this dilemma. She works on land that her husband inherited from his father, a share of which belongs to his brother. At present, the families rely on one well for irrigation. Given the declining water tables, the water is insufficient to optimally irrigate all of the land. However, since there is ambiguity around sharing the costs to be borne for a new well, Rani's husband has not taken any initiative to sink another well. They are particularly concerned about the implications of formally partitioning the lands. They need to be assured that the new well stands on their share of land if they are to pay this cost. Otherwise, they cannot realize the full benefits of this substantial expense. Issues around joint cultivation surface in other ways, as Subbu illustrates: "Half the land is in my grandfather's name and the remaining half belongs to my father. The borewell is on my grandfather's side of the field. So, it is difficult to get water if we are not on the best of terms with that uncle (father's brother). The electricity service is also shared. So, we have to work jointly." Such relations, however, also enable the survival of the young farmer and the small farm.

Support Networks

Although farmers do not explicitly mention this reality, family labour and kin support are critical to family farming. This has become particularly important in a context of rising agricultural wages and access to paid labour. Paddy cultivation has become highly mechanized but vegetable cultivation and floriculture require considerable labour on a continuous basis. Household labour becomes critical to sustaining such cultivation. In most cases where vegetable cultivation is undertaken, we find that there is often a reliance on labour from the larger family with in-laws and siblings chipping in. "All of us in the family must work. There is no other way," says Siva. "In (an) agricultural family, there is no alternative. We do bring in the ladies to harvest. We also pitch in. We all four have to pitch in." His wife Jothi narrates her work schedule. "In a day, I usually go to the farm at 9 a.m. after sending the children to school. Then, I am there for the whole day. There is one or another task. Finally, in the evening, again, I have to go and usually stay there until late evening to take care of the buffaloes and cows."

In fact, all households in the Nilgiris villages work on each other's farms with little reliance on paid labour. Murugan has two children and runs a grocery store along with his farm. He talks about the role of his family in farm work:

> I have two children. One is in the 11th year of school and the other in 8th. They are also helping on the farm. One son helps me and the other son helps my wife. They help out with all the chores. From weeding to harvesting to rearing cows, to helping the mother in the kitchen and other places, both my sons work very well. They go to a private school in the town nearby. They are very co-operative…(they) help out in the store as well. My wife does most of the work at home and at the shop. She milks the cows and helps out at the farm whenever needed. We all have to pitch in. My father and mother also pitch in. Otherwise, farming is not possible. Balancing with other businesses is also completely impossible. We cannot hire labour for all these activities. It is unviable.

In case they need to leave the village for any personal contingencies, they have to depend on neighbours or relatives to take care of the farm or the livestock. Else it becomes difficult to manage.

The embedding of farming in familial and kinship relations can be both enabling and disabling. While challenges of control over decision-making emerge when farming households access resources that kin jointly own, kin networks also act as a means of support especially during crisis or contingencies. Sudden illness or death in the family is often compensated by such support. Kin relations are expected to not only fill in for managing the farm or related activities but also take on additional responsibilities such as taking care of, educating, and marrying off the children left behind. Women farmers are even more enmeshed in such relations. Women farmer respondents, even those who own land, are embedded in a socio-cultural milieu that does not allow them to make decisions independent of the male head of the household or the extended family. While in several instances we find women contributing substantially to farming and dairy operations, they do not report to being the primary decision makers with regard to farming practices. In the case of the aforementioned Dalit women farmers' collective in Tiruvannamalai, a civil society organization (CSO) enabled their entry into organic farming. However, following the CSO's withdrawal, given the complex relations in which ownership, use, and spheres of reproduction are embedded, it was not appropriate to ask how the women became farmers, even when they owned land. This is also true for male farmers who are not always in a position to take decisions independently as they rely on the larger social networks that shape their cultivating practices. Rao (2017) alludes to the role of relational webs that do not allow for a linear relationship between land rights and women's empowerment. The family farm as a unit of production tends to reproduce patriarchal relations that deter women farmers from exercising sole authority over decision-making on their farms even when they hold land titles. Youth engagement with farming cannot be understood without locating them in this complex web of social relations within which farming is rendered possible in the first place.

Learning to Farm

Contrary to popular imagination, farming is not an unskilled economic activity. Even educated youth employed in high-paying urban jobs who quit and enter into (part-time) farming with the help of kin or friends realize the importance of skills to successful farming. As mentioned earlier, all of the farmers that we spoke to come from families that are traditionally farming households. Often, they grew up on the farm learning by doing, helping their family part-time on the farm. For men, they would either continue working on the farm after completing their education or return to the occupation after working elsewhere. For women, they would work outside agriculture or stay at home helping their parents until marriage. We must note the important caste and class differences in the extent to which women in particular are socialized into farming. For instance, it is common for young women from well-off Kongu Vellala families to pursue a university education, which takes them away from farm work. Even in such cases, many of these young women would return to the land and work on their husband's land after marriage. In two instances, young women farmers told us that they began to farm only after marriage.

Our interviews suggest that most knowledge required for farming is acquired through work on one's family land. Tamil Nadu's agriculture sector is one of the most commercialized in the country, made possible by considerable diffusion of Green Revolution technologies since the mid-1970s. As a result, even the older generation of farmers has been exposed to the use of new technologies, offering learning opportunities for the younger generation of farmers. In addition, young farmers also access other networks in the village or the neighbouring villages through kinship or through schooling. Importantly, thanks to the explosion in internet access, some of these young farmers belong to learning networks directed at organic farming and alternate marketing networks. Farmers use social media to seek advice about crop disease or pest infestations and to share photographs or audio files. Members of social media groups generally support such learning processes.

Apart from sharing experiences and learning from other farmers, social media also offers considerable scope for forging solidarities that transcend the traditional kinship networks. The role of government extension programmes seems to be minimal except in instances when they encourage adoption of new crops through subsidies. One farmer told us that he shifted to mulberry cultivation in response to such an initiative. Agricultural universities also share new innovations with farmers through such networks.

Apart from diversification into more profitable crops like fruits and vegetables, and tree crops such as coconut cultivation to address labour shortages, dairy has become a major source of farm income. Some government incentives, including the promotion of sericulture, contribute to innovation. Apart from such standard diversification, some farmers have also entered into organic farming, in addition to the Dalit women's collective that we discussed earlier.

The state has one of the largest organic farming networks in the country. This diffusion is largely attributed to the missionary zeal of Nammazhvar, an agricultural scientist turned organic farmer activist. Many young farmers who trained under him became new nodes for diffusion of such practices. Two of our respondents became organic farmers through this route. Another respondent narrated his efforts to introduce new machines on his farm by interacting with his friends abroad and trying to get into buying and selling such machinery.

Critical Resources for Being a Young Farmer

Apart from land, water was the most important resource constraint for the young farmers in our study. Inability to consolidate their landholdings, lack of labour, and, on rare occasions, market prices too came up as important constraints to make farming viable. Despite the mention of prices, hardly any of our respondents mentioned the importance of access to markets as a critical resource. The exception was farmers undertaking organic farming who find perishability a concern when consumers are located in distant locations and their transport and marketing logistics are less developed. They point out that the absence of adequate marketing channels is a major constraint for sustaining the shift to organic farming. The water crisis, however, is something that looms large in most farmers' understanding of what is most critical. This is especially true in the Chennai and Coimbatore regions where two important processes are visible. The extraction of groundwater is the primary driver for the commercialization of agriculture in the regions. In Coimbatore, its increasing extraction has led to depletion of water tables and undermined access to an assured water supply. Saravanan, 36 years old, narrates his travails in this regard:

> I have eight bore-wells. None of them have enough water. It's at 1000 feet. I had to spend 1 lakh per bore! That was very painful. As we dug one, it will work for some time. After some time, there won't be any water. Then, I have to get another drilled. This kept on and on. In this dry season, there is no water anywhere.

In the case of villages neighbouring Chennai, irrigation has been largely tank based and supported by open wells. For a range of reasons (see Janakarajan 2004), poor maintenance since the colonial period has undermined tank-based irrigation, compounded by problems posed by urbanization and the emergence of speculative markets for land. Farmers that we spoke to attribute the absence of water to the erratic pattern of rainfall and the decline in amount of rain in recent years. "We have not had good rains in the last three or four years. How will the *eri* (village tank) have water?" Muthusamy, a farmer in a village near Chennai, echoes this common perception in the region. The spread of real estate activity is also seen as an issue as it disrupts traditional channels of flow into tanks and reduces incentives for community management of village tanks. While farmers have adopted drip irrigation practices, particularly in the Coimbatore region, this is still insufficient to ensure water security.

In the Tiruvannamalai region, farmers complained much less about water. Access to better prices for their produce is a more critical issue for young farmers at this field site. A few mentioned the lack of know-how for new cropping practices or physical inputs into farming to be critical. This may have to do with the norm that despite the collapse of public/government extension services, input suppliers such as fertilizer and pesticide firms often double as advisors for farming services. Water, however, assumes critical significance. While these agricultural resources are critical, farming also depends on farm households' income diversification.

The Non-farm and the Young Farmer

The pluri-activity that the secondary data reveals about agricultural households in rural Tamil Nadu (Vijayabaskar 2017) was evident in our conversations with young farmers. Often, household members undertook many non-farm activities and occasionally secure formal sector jobs in the urban economy in the state. What we observe in general is a mutual dependence between insecure farm livelihoods and insecure non-farm diversification options. As mentioned earlier, youth often come into farming due to a loss of employment or the poor quality of waged employment outside agriculture. There are also instances when young farmers were forced to leave agriculture and find off-farm employment due to water scarcity. Selvi explains this process:

> I split my time between the farm and household chores. My husband has now started working at the local grocery store. He has to go at 9 a.m. and comes back to 9 p.m. There was no income in agriculture the past two years because there was no water. The drought was very severe. If agriculture was remunerative, there was absolutely no need for him to go out... My husband and his brother were both forced to find a job outside. We pool our salaries and the income from agriculture and then split the expenses...

At the same time, respondents are also aware of the insecure nature of non-farm employment. "A company job is not permanent. Sometimes, there won't be any job for you. As far as agriculture is concerned, if there is enough rains, there is no problem in making money," says Karthi, a young farmer who also works in a workshop in nearby Kangeyam, western Tamil Nadu. The preference for agriculture also emanates from their perceptions of quality of work. The sentiments expressed by Subbu echo the feelings of many young farmers who have moved from factory or formal sector employment to farming. "I never liked that job. I hated the timing. The monotony of work and then a lot of regulation of when to come in, when to get out. All these were frustrating me. Only because of water [scarcity], I had to go back…"

It appears that for many engaging in farm work also means engaging in a mode of working that involves certain temporal rhythms that they feel comfortable with. At the same time, it is also a rhythm that does not equip them with the world of modern waged work that is more rigid and coercive. It is this absence of coercion—"I don't have to listen to orders from others. I don't have to follow strict work hours"—that appears to be attractive to those who are currently in farming. Such preferences for agricultural work, as discussed earlier, stand in contrast to the aim of many of our younger farmers to leave farming as soon as they completed their education. Nevertheless, most, if not all, farmers, especially those with young children, say that they do not want their children to make their living from farming.

The Family Farm's Future

Though some of the young farmers do feel that farming is preferable to working in non-farm sectors as waged workers, they seldom want their children to continue in farming. As Murthy, a farmer with 20 acres of irrigated land, said of his two daughters: "I want a hi-fi life for them and not like mine." At the time of our interview, we noticed two cars parked in front of his house. With excellent access to canal water, his farm is quite remunerative, and he has no alternate income. Even under such conditions, there is a strong sense that the family is losing out on many things that the urban offers. Another farmer, Madheswaran (42 years old), wants a different life for his son: "As long as I am around, I do not want him to suffer like I did on the fields." This perception of suffering is tied not only to hard work on the farm but also to risk and uncertainty of income and a lack of access to consumption lifestyles opened up in a globalizing economy. Eswari, a 35-year-old farmer in Tiruppur, western Tamil Nadu, sees a move out of farming as economic mobility for her children. "No, I don't want their lives to become ruined such as ours. We had no choice then. But now, the children have choice. They can study and get themselves a job and do well for themselves. I don't think they need to suffer here in agriculture." Another farmer, 39-year-old Durai, talks of how converting farmland into real estate may mean more economic security for his daughter. "...I will convert these [farm land] into residential plots, into sites... With the returns, we will do some business... I will sell the plots and build ten houses instead of agriculture, to be able to safeguard my daughter's future."

Combined with the fact that young men increasingly feel that a life on the farm is not desirable for most potential brides, the reproduction of the family farm in its present form in a patriarchal-patrilineal context seems unlikely. This sense of unviability is also due to the poor quality of public schooling in rural areas and a growing perception that education in expensive urban-based private schools is critical to secure non-farm futures. Murugan provides an insight into the significance of education to farming households:

> I spent about Rupees 80,000 per annum per child for school fees. There is also fees for the bus. My children are not great at studies but since everyone else is sending them to school, I too am forced to. But I also believe they need some basic education—know how to sign, how to read and write, and have some basic English knowledge. One of our neighbours spends Rupees 3 lakhs on education.

Many English-medium private schools in nearby towns engage school buses and vans to pick up and drop children from these villages. Such aspirations, however, are not in line with the predominantly poor quality of non-farm employment that we observe. Under such circumstances, a combination of farm and non-farm incomes is possibly the most viable option in the near future to ensure better incomes and lives. Nevertheless, to most farmers we spoke to, modern education is the only route to social mobility and away from insecure farming futures. This also raises the question of mobilization among young farmers.

Young Farmers in Policy and Politics

Although there are no other specific policies aimed at ensuring better livelihoods for young farmers, the state and central governments have been trying to improve farm livelihoods through various measures. Subsidies for installing drip irrigation is an example. While a share of farmers have installed drip irrigation systems, they often prefer to deploy systems from private suppliers as they feel that those that the government supplies are of inadequate quality. Some farmers are also of the opinion that subsidizing drip irrigation alone is pointless without a secure water source. There are other support services aimed at innovative practices such as the adoption of integrated pest management practices and Systems of Rice Intensification (SRI) that young farmers see as useful. The Tamil Nadu Agricultural University runs a website for organic farming networks and has an incubation centre for innovations to address recent problems such as pesticide overuse or uncertain water access in farming. Unfortunately, while earlier there was a strong relationship between the formal study of agriculture and entry into agricultural support services, a few respondents are of the opinion that many of them enter into such courses in order to prepare for general public service examinations. It is also rumoured that the farmers' lobby is resisting efforts to legislate against excessive ground water extraction as at present ground water is the important source of irrigation for farming.

In the realm of politics, while political parties do appeal to male youth and their concerns, it is primarily in terms of the absence of urban employment. The narrative is increasingly centred around which party sought to promote more employment opportunities through attracting investments or supporting specific manufacturing sectors. Simultaneously, in the realm of agriculture, politics has focused on loan waivers as well as on starting new irrigation schemes. We interviewed farmers, including women farmers, who campaigned for a new scheme to divert waters from a river for irrigation in the Coimbatore-Tiruppur region. Some farmers were also involved in addressing ecological issues in the village such as excessive use of plastics in the Erode district in western Tamil Nadu. A few of them report to use their contacts in political parties to help fellow farmers access government subsidy schemes. Their claim-making is largely through political parties and as members of a particular region or as a farming community/caste rather than as members of a youth constituency. In 2016–2017, many were sympathetic to the large-scale protests in the state around the ban on *jallikattu*, a traditional bull-taming sport popular among farming communities, with a few participating in the protests. They are also aware of farmers' protests elsewhere against methane extraction projects in the Cauvery Delta, which is believed to be detrimental to the region's agriculture. Social media networks are critical resources, not only for learning about farming practices, but also to imagine themselves to be a part of larger community of farmers. The extent to which such networks are leveraged to form an exclusive pressure group for sustaining agriculture is not clear. Given the aspirations for diversification out of agriculture and the association of manual work with low social status, mobilization around reviving agriculture is likely to be combined with demands for better educational standards and "decent" employment options outside agriculture.

References

Aloysius, G. 2013. *Village reconstruction*. New Delhi: Critical Quest.
Bourdieu, P. 2008. *The bachelors' ball*. Cambridge: Polity.
Chakravorty, S. 2013. *The price of land: Acquisition, conflict, consequence*. New Delhi: Oxford University Press.
Janakarajan, S. 2004. Irrigation: The development of an agro-ecological crisis. In *Rural India facing the 21ˢᵗ century: Essays on long term village change and recent development policy*, ed. B. Harriss-White and S. Janakarajan, 59–77. London: Anthem Press.

Rao, N. 2017. Assets, agency and legitimacy: Towards a relational understanding of gender equality policy and practice. *World Development* 95: 43–54.

Srinivasan, S. 2015. Between daughter deficit and development deficit: The situation of unmarried men in a South Indian community. *Economic and Political Weekly* 50 (38): 61–70.

Vijayabaskar, M. 2017. The Agrarian question amidst populist welfare-interpreting Tamil Nadu's emerging rural economy. *Economic and Political Weekly* 52 (46): 67–72.

Vijayabaskar, M., and A. Menon. 2018. Dispossession by neglect: Agricultural land sales in Southern India. *Journal of Agrarian Change* 18 (3): 571–587.

10

"I Had to Bear This Burden": Youth Transcending Constraints to Become Farmers in Madhya Pradesh, India

Sudha Narayanan

Introduction

As part of the study, two districts were chosen in Madhya Pradesh— Sehore and Chhindwara—that fairly represent diverse agricultures and socio-demographic characteristics. Sehore, located 37 kilometres from the state capital, Bhopal, has ample groundwater and canal irrigation with deep black soils that support high yields. Over the past two decades, Sehore's widespread adoption of soyabeans, an important cash crop, has transformed the district's agriculture (Kumar 2016). Bhopal's growth as a city shapes life in this district, especially in terms of employment options. Most of Sehore's farmers are from traditional land-owning castes. With a Scheduled Tribe (ST) population comprising 11.1 per cent of the total

S. Narayanan (✉)
Indira Gandhi Institute for Development Research, Mumbai, India

International Food Policy Research Institute (IFPRI), New Delhi, India
e-mail: s.narayanan@cgiar.org

© The Author(s) 2024   **277**
S. Srinivasan (ed.), *Becoming A Young Farmer*, Rethinking Rural,
https://doi.org/10.1007/978-3-031-15233-7_10

inhabitants (in 2011), Sehore has a few pockets that are predominantly tribal.[1]

In contrast, Chhindwara is overwhelmingly tribal, with 36.8 per cent of the total population classified as being from the Scheduled Tribes; Gonds and Bhils constitute the major tribal groups. It is located amidst rich forests and hills. Its undulating terrain and mostly light stony soils that don't hold water are not ideal for agriculture. Its relative lack of accessibility implies that the cropping patterns are dominated by food crops rather than cash crops, although this reality is gradually changing. In recent years, those with irrigation have been able to grow wheat and gram in the winter season. Soyabeans have recently become an important cash crop here as well and most farmers sell their produce in the government-regulated markets or to private traders in the nearest towns. Wheat, though, is the dominant crop in Chhindwara. Migration to nearby cities within the state has been a way of life for people here since the mid-1990s, given that work in agriculture is restricted to one or at most two seasons during the year. In the absence of a vibrant non-farm sector, migrants work on other farms or in construction.

The cropping pattern in the two districts that we studied are soyabeans and pigeon peas in the rainy season (*kharif*) and wheat and gram in winter (*rabi*). The winter crop is usually grown only when the rains are good or by those with irrigation facilities. Many farmers (men and women, old and young) also migrate for work—usually to work on farms growing soyabeans or in construction—but all of them migrate only for short spells and never travel too far from their village. A majority of households have some livestock—one to four animals, including buffaloes, crossbred and native cows. Although at the state level, livestock is growing in importance as is evident from milk production figures, we found in our sample that livestock ownership had reduced over the decade preceding the study. The reasons most commonly articulated include the shrinking pasture area, limited access to forests due to restrictive use rights, water constraints as well as dearth of labour—there is not enough help available to graze the cattle and mind them. Few households own poultry or goats

---

[1] The Constitution of India recognizes Scheduled Tribes (ST) along with Scheduled Castes (SC) as officially designated groups of historically disadvantaged people in India.

as their ownership is deemed to be a marker of low social status and associated with specific lower castes.

We interviewed 40 young farmers, 11 of whom were women. We also interviewed 11 older farmers, of whom 2 were women. Our entry points into the farming communities were established via two non-governmental organizations (NGOs): Under the Mango Tree (UTMT) and Samarthan. Whereas UTMT has been working with tribal communities to promote beekeeping, Samarthan's work in this area is around human rights and they had recently been assisting farmers to access government programmes. Including young farmers from a range of ages and land size classes was our aim when we selected young farmers to interview, and they are not necessarily representative of all farmers in the district. The farmers that we selected also come from different types of families and from different castes and tribes represented in the village, both traditional land-owning classes and others. A fifth of our sample constitutes women, identified via village self-help groups. Our respondents come from 10 different villages—five each in Sehore and Chhindwara. We ensured that we included villages where our partner NGOs were active and where they were not. While our selection of Sehore and Chhindwara was driven by the contrast that they would likely offer, Sehore's tribal areas resemble Chhindwara more than they resemble the other parts of Sehore.

The majority of our respondents are from tribal communities: 32 are from Scheduled Tribes (mainly Bhils and Gonds), 16 are from Other Backward Communities (OBC), and 6 are from Scheduled Castes (SC).[2] In terms of religious profile, all barring one self-identified as Hindu. The average age of the young farmers in our sample was 35.8 years, with as many as 20 under 30 years of age. The youngest farmer that we interviewed was 18 years old. Most of the farmers owned between 1 and 10 acres, much larger than the national average. While the majority of farmers (old and young) in our sample had little formal education—25 of them had only a primary-level education or were not illiterate—younger farmers, both women and men, had more education than their older counterparts. Eight young farmers had completed tertiary education. In

---

[2] As with Scheduled Tribes, Scheduled Castes and Other Backward Communities are officially recognized as socially disadvantaged.

our sample, most farmers from Chhindwara had no access to irrigation, an issue that will be discussed later as one of the key challenges for farming in general and for young farmers in particular. In Sehore, most farmers do have access to irrigation; their problem is that yields for soyabeans have plateaued and do not bring in same financial returns as in earlier years. In different ways, farmers in general are operating in contexts where farming is challenging. Our interviews with the farmers focused on their pathways into farming as well as the barriers and opportunities that youth in agriculture face.

The chapter is organized as follows. Section "Becoming a Farmer" focuses on pathways into farming and how these differ both within the younger cohort and how they differ from the previous generation. Sections "Barriers Faced by Young Farmers" and "Support for Young Farmers" elaborate on the challenges that young farmers face in agriculture and their support systems. Section "The Future of Young Farmers and Young Farmers in the Future" reflects on the future of these young farmers and the section "Concluding Remarks" concludes the discussion.

Becoming a Farmer

Entry into Farming

The young farmers that we interviewed were farming well before they became farmers in their own right. In this sense, it seems that the line between becoming and being a farmer is unclear. Most men started helping their parents or grandparents when they were as young as 8 or 10 years of age. Those who started helping that young were typically grazing livestock—taking livestock to the pastures or into the nearby forests, minding them throughout the day. One young farmer recollected that he regards the bamboo stick that he used to guide his family's small herd as his first agricultural tool. Almost every respondent was already acquainted with some of the key farming operations by age 12. Usually, the boys assisted adult men in ploughing and field preparation. It was only later, when they had built stature and strength, perhaps even skill, were they able to use the plough independently. One young farmer recollected how at the age of 15, he was frustrated at being unable to handle the plough and fell repeatedly trying to manoeuvre it. All of the young male farmers that we interviewed came from families that owned agricultural land, even if meagre.

Among the young women farmers, there were three who were not from farming families and began farming only after marriage. For young women farmers, their first roles in agriculture were around age eight or nine when they assisted adult women in the family with weeding and harvest. In Madhya Pradesh, as in most of India, farm tasks are historically gendered, with women almost never involved in plough use, while tasks such as sowing, weeding, and harvesting are often considered women's work. Although these norms have diluted over the years, the accounts of young farmers suggest that their entry into agriculture was quite traditional and in keeping with the gender norms of the time.

Overall, we see little evidence of new farmers as such—for reasons that are discussed later—and there are none in our sample. The young farmers in our sample, especially the men, belonged to three broad groups. The first group became full-fledged farmers by force of circumstance, even if some of them entered farming willingly. A second group consists of those who had been able to finish school and/or vocational training or tertiary education, but were unsuccessful in finding jobs in the non-farm sector. They were farming, but constantly looking for opportunities outside the farm sector. The third group of young farmers has little interest in education, and despite their parents' best efforts, these young people looked for an opportunity to drop out of school to take up farming.

Farmers in the first group entered farming after quitting school, often due to family exigencies. Typically, it was because their parents could not afford school-related costs, a family member died, or because they had to take over farming responsibilities from the father, especially if they were the older amongst siblings. For example, one young male farmer, the son of a former village headman, told us:

> My father did not ask me to drop out of school. He wanted me to study. But I saw that he needed help, so decided to discontinue (school). I was the eldest so I had to bear this burden. My mother always used to tell me that

there is no one in the house to look after farming and people were always calling my father in the village as he was the village head. So I had to do it. If I did not, farming would have been adversely affected [42 years old, ST, farms 12 acres, Chhindwara].

Some of them dreamed of an alternate life—wanting to become policemen, engineers, and other professionals—and had little interest in farming. One young male farmer recounted:

I have done an MCom and MBA. I wanted to have a nice job in some good Multinational company But something happened in my family and I had to come back. I was done with my MBA and looking for jobs in Indore in 2006. I got the news that my father was sick and he was paralysed. In 2008, he expired and the entire household burden was then on me. So I decided to stay in the village to fulfil the needs of my family [32 years old, ST, farms two acres, Chhindwara].

Another young man was forced to migrate for non-farm work in order to support the family farm, which he helped his father manage:

I used to look up to different people back then. I wanted to be something. I had big dreams. I wanted to a get a good service (job) but when my mother passed away, everything got affected. Then I left studies and joined farming. At that time, money was needed for farming and for the household, so I started going to Pachmarhi (150 kms away) to work [36 years, ST, farms 12 acres, Chhindwara].

Some young women farmers rue the fact that their parents did not realize the importance of education and pulled the women out of school to work. One young female farmer recounted how she aspired to earn a job as an *anganwadi* (government-run crèche) worker, but did not have the minimum schooling required to apply for the post.

Young farmers in the second group are biding their time farming. Farming to them is a fallback option. Some of the younger farmers in this group are still hopeful of landing a job outside and hoping to exit farming, in equal measure because farming is not deemed to be viable (especially in the context of very small landholdings) and because they aspire

to a different life. Some of them completed courses to increase the probability of securing a job, including vocational training to become a
mechanic, electrician, computer operator, mobile phone repairer, and so
on. One young male farmer in Chhindwara, about 22 years old, completed a bachelor's degree as well as two vocational training courses and
was waiting to see which one would land him a job. A government job,
and the security that comes with it, continues to be the aspiration of many.

Within this group, however, not all of farmers harbour the feeling that
they are "stuck" in farming. Some of them have come to terms with their
own status as farmers and are seeking ways to become successful farmers.
In this process, their engagement with NGOs working in the regions has
energized them into becoming enterprising farmers.

Given the low profitability of agriculture, many parents prefer that
their sons leave agriculture. They invariably see school education as a
pathway out of agriculture. For daughters, while basic schooling is desirable, the education of daughters is not specifically seen as a vehicle out of
agriculture, since the future envisaged for all daughters is marriage. Yet
given that jobs in the non-farm sector are scarce and of poor quality, the
option of a career outside farming is little more than notional. This is
especially true in Chhindwara, where jobs are hard to come by in nearby
towns. One young male farmer, aged 32 and who has never worked outside the village, explained that finding decent work outside was not easy:
"Even if I went out, getting a job would have taken time. I would need
resources at that time for maintaining myself. I had no such means—no
contacts in town who could help me get a job."

For those without such networks or resources, farming is the only
option. For some, the resource constraints are often so overwhelming
that they have no alternative. One young male farmer, aged 32, had completed an undergraduate degree but could not secure a job related to his
qualifications. He is the only son, has no interest in farming the family's
seven acres, and had this to say about his father: "He would scold me and
ask me to come to the field and work. I was not very interested in farming, but I would go to avoid being scolded. We were financially weak so
I had no other option."

The third group consists of those who were either passionate about
farming or those who preferred farming to further education. Most are

now enthusiastic farmers—proactively seeking information from civil society organizations that worked in agriculture and the local government officials associated with agricultural programmes. Most of these farmers are actively experimenting with new techniques (System of Rice Intensification, for example, in eastern Chhindwara or experiments with organic agriculture), new crops, new ways of growing crops, and exploring ancillary activities such as beekeeping, mushroom cultivation, and aquaculture.

In general, among men, the younger generation seems to have greater say in their decision to enter farming than the older generation when they were young. Older farmers recollect that in their youth, there was no question on whether or not they would farm. In most cases, villages did not have schools and going to school was a rare option. Jobs outside of the farm sector were equally rare. There was no room to aspire for a life outside of farming. An older male farmer, age 55 and from Sehore, exclaimed: "What aspirations? My father handed farming to me and said cultivate and feed yourself! That is it. I started farming young and did not consider anything else."

Another older farmer, 55 years, told us: "One is born a farmer." Yet another 52-year-old farmer said: "My father was a farmer. This then passed on to me." Among the older farmers, there seemed to be no room for imagining an alternate livelihood.

For women, it seems that things have not changed very much across the generations. As with the older generation of women farmers, women today are generally socialized into playing a supporting role in family farm operations, rather than the lead. Younger women farmers today have an opportunity to attend school, whereas older women farmers typically did not, but across both groups, schooling is not viewed as a way to secure a non-farm job. That said, at least a few young women farmers that we met did articulate their aspirations—of becoming a schoolteacher or an anganwadi worker, for example. Almost all of these women indicated that they regarded, or rather came to accept, not necessarily out of their own choice, farm and domestic work as being intrinsic parts of their lives and responsibilities.

Generational Shifts and Succession

Across generations, helping with farming did not give young farmers a major say in decision-making, neither in farming operations nor in how to use the proceeds. Most decisions continue to be taken by the family's older adults, including the choice of which crops to grow, required inputs, and seed use (stored/saved or bought). One father-son pair that we interviewed said that they discussed the decisions thoroughly. The son, 18 years old, had his own ideas about digging a farm pond, engaging in aquaculture, and buying a tractor. The father, 45 years, said that he gives his son a free hand to experiment, but ensured that they took small steps that would not risk the survival of their farm enterprise. This limited role in decision-making did not seem to deter the young farmers.

In contrast, young women farmers typically have little say in any decision, not as daughters, not as daughters-in-law. But not all women are in this position. Older women have more agency, unsurprising in the context of patriarchal structures that bestow some privileges based on seniority. One young man, about 25 years old, told us how his grandmother took all of the decisions on which crops to sow and how to care for the plants, noting that no one in their family disputed her knowledge of farming. Another older woman farmer, aged 55, mentioned that members of her family usually discussed decisions on which crops to sow together with the older women in the family.

A young male farmer becomes an independent farmer, fully managing his farm, only when he marries and has a family of his own, partitioned and obtained ownership of farmland, or both. The lack of ownership rights over land was not always seen as a deterrent to adoption of new technologies. A 24-year-old, high school-educated farmer in Sehore with six acres told us: "Ownership doesn't matter. We use the latest technology irrespective of whose name is on the *patta* (the land record)." Another young farmer, aged 30, with a high school education and who operates a four-acre farm in Sehore however said: "I will be more proactive to use new technologies and new farming methods... once the land is in my name."

For young women farmers, neither marriage nor inheritance is likely to offer them full control over the management of a farm unless they are single women or household head. Managing a farm single-handedly, however, comes with its own challenges. Women in these cases invariably lease out land or hire farm managers to handle the operations. We interviewed one young widow in Sehore who said that she relied on a hired manager to operate and sell produce. She received half the produce or proceeds from any sale while the farm manager kept half. The decisions on what to grow and inputs to use were decided jointly after consultation amongst family members.

Doing Things Differently

Across the board, farmers feel that compared with the previous generation, knowledge of farming practices has increased. As one farmer put it, before agriculture used to be *"dekha-dekhi,"* meaning that you did what you observed around you. Today, one can obtain training, knowledge, and rules for performing different farm operations. Today's young farmers rely more on machines for harvesting, sowing, and tilling than their parents' generation. Weedicides have replaced manual weeding on many farms and purchased hybrid seeds are used as a norm rather than relying on saved seeds. Perhaps the most significant shift has been cropping pattern changes; the emergence of the borewell[3] and access to water has resulted in a shift to crops such as wheat and gram. Millet has gone out of cultivation, replaced by maize, soyabeans, and pulses. For a brief while, the new cropping pattern proved to be very lucrative, especially soyabeans in Sehore. However, in recent years, yields have plateaued, prices have remained low, and more and more farmers are switching to maize. Those with ample irrigation are also diversifying into fruits and vegetables, both in Chhindwara and Sehore. In some villages, especially in Chhindwara, farmers started using inorganic fertilizers only a decade ago. Many young farmers have recently become acquainted with agroecological farming and organic farming and have a shared sense that chemical

---

[3] A well drilled vertically into the ground, as opposed to an open dug well.

inputs are unsustainable. "Once you use it, you need to keep using it...and more of it" was most young farmers' refrain. Both NGOs and the government at different levels, national and state, seem to have been instrumental in promoting vermicomposting and in encouraging farmers to apply organic manure. Yet few defined themselves as organic farmers. Most were still experimenting and testing out methods in some plots. Indeed, no farmer that we interviewed was an exclusively organic or agro-ecological farmer, although many were practising several principles of these techniques.

For each of these decisions—whether to go organic or not, to diversify, and in the choice of seeds—the younger generation negotiates with the older generation. One young male farmer in Chhindwara set up what comes close to an experiment, using organic methods in one plot and inorganic in the other to test the relative merits of each. His father had passed on and he was the heir, the sole male member in his family, and family farm manager. This afforded him the space to make decisions that might be harder in multigenerational family farms. In these latter cases, young farmers have to persuade older adults in the household, usually the father, to dedicate a small patch where the young farmer can plant and do with what they want. Young women farmers, in contrast, have virtually no agency. Their exposure to new ideas and techniques is also limited, bypassed as they were by an extension system with male trainers that is oriented to training male farmers, for example.

Young farmers in Chhindwara and Sehore are also seeking to diversify into allied activities,[4] including mushroom cultivation and beekeeping. The 18-year-old male farmer in Chhindwara, from the father-son pair referred to earlier, is an enthusiastic farmer for whom beekeeping is his new passion. An NGO introduced him to it. Many young farmers are looking to do things differently, urged on by government extension work-ers, Corporate Social Responsibility (CSR) efforts (mentioned earlier in the chapter), and peer networks that play a major role, a point that we return to later in the chapter.

---

[4] Allied activities in the Indian context refer to those other than cultivation, such as animal rearing, including livestock, poultry, beekeeping, and processing.

Barriers Faced by Young Farmers

"Farmers Cannot Buy Land"

It is hard to have a conversation with young farmers without a mention of the small size of their holdings. Although Madhya Pradesh has larger average land holding sizes than most other states in the country, successive subdivisions over generations have left holdings too small to be remunerative. Over the years, land prices have also risen significantly, dramatically in some villages, especially in those closer to the city or those that have access to water. This puts the possibility of buying land to expand a farm out of reach for most farmers. "A farmer cannot buy (land)… if there is a businessman or someone with a job, they can buy. But not a farmer," said a 60-year-old male farmer in Sehore, with 10 acres of land and leasing an additional 6 acres. He noted that in contrast to when he was young, "A farmer today cannot dare to buy land."

The farmers who manage to do so are those with non-farm occupations to supplement their income. A few others have bought land using a loan and paying it off by leasing out the very land that they purchased. Farm incomes, we were told, are not large enough to repay such loans in Sehore or Chhindwara. In Sehore, where soyabean farming has been financially rewarding, recent increases in costs of cultivation without a commensurate increase in the price of soyabeans have left many farmers indebted.

Some young farmers had larger than average landholdings, typically because their parents had the opportunity to buy land a couple of decades ago (when prices were still affordable). In almost every village, it seems that "outsiders," that is, individuals not from the village, have bought up land and many were leasing it back to the villagers to farm. At the same time, many farmers also emphasized that those who sold their land only did so due to extreme circumstances. No one, except those in distress, sold land today, they said, in part because it is impossible to acquire any later on and because there was barely enough to subdivide amongst the next generation. More importantly, however, securing employment in the non-farm sector and in nearby towns was so difficult that land served as a fallback option, an insurance of sorts. In the tribal areas in the study, restrictions on land sales to members outside the community meant that it did not make sense to sell land, but it also meant that those in distress often had to sell their land for a paltry sum within their tribe. The larger implication of this is that for those aspiring to become farmers without land of their own, the difficulty in securing their own land posed a formidable entry barrier to farming. This is despite the fact that in the study area, caste was rarely considered a barrier to land ownership.

For most farmers who cannot afford to purchase land, the only way to expand farm operations is by leasing land. Leasing is common and, by all accounts, caste and social identity do not play a role in who leases to whom. Lease rates are high for land with irrigation and leasing poor-quality land simply does not make sense for most smallholders. Leasing is based fully on trust. Some farmers have sidestepped the constraint of land by "encroaching" on forest land. Although most of our examples came from tribal communities in Sehore, this was widespread in Chhindwara as well. We were told that all that the villagers needed to do was to pay forest officials a modest sum to farm forest plots that, in a few cases, they had been cultivating for decades, pre-dating the restrictions on the use of forest land. The state has recently been attempting to regularize these plots. In the larger context of the limited scope for expansion of farms, illegal occupation and usurping of land by powerful interests seems common. A 26-year-old farmer that we interviewed was farming his mother's plot. He stated that although his father had land in his native village, it had been illegally occupied by a powerful family in the village. As his family could not evict them, they had left the village altogether to be able to farm his mother's land. They managed to buy a few acres in his mother's village to make the farm viable.

In the literature on youth and farming, access to land is often regarded as a chief barrier to becoming a farmer. In the context of Madhya Pradesh, as most land is inherited and given the limited capacity to purchase more land, most farmers receive land in their name only when the parent (usually the father) dies. It is typical to see young men and women identify him/herself as a farmer, be fully involved in managing the farm but without ownership of the land, again blurring the boundary between becoming and being a farmer. As long as the father is alive, it is not uncommon to see all of the children farm together on their parents' land, with vary-ing levels of involvement in decision-making. When the father passes on, if he is survived by his wife, the land ownership documents will reflect the names of the wife and his children. It is common for male siblings to then carve out space for themselves and farm separately. Until then, siblings tend to farm together with the parents. On occasion, they might farm collectively but demarcate their individual shares. Marriage is a similarly important life event. When sons marry, fathers often settle the property in their name to ensure that they can farm independently, sometimes even if they live in a joint family. It is common practice to demarcate the land, anticipating future partitioning, even if the formal partitioning is several years in the future. An older farmer in Chhindwara, aged 52, with three sons of his own, spoke of his arrangement with his siblings over the 12 acres that he manages that have still not been formally divided: "We have not done the paperwork, but we each have 12 acres of land based on understanding. We have also demarcated land."

In some families, the demarcation of plots implies that these are now managed separately. In other cases, the demarcation is notional and fami-lies farm jointly. One farmer, around 28 years of age, pointed out: "Father divided the land (16 acres) amongst us four brothers, but we work together and collectively rent in 5 to 10 acres."

Our study is replete with examples of full-fledged farmers who work land that is not held in their name. A young farmer that we interviewed left school to help his grandfather farm after his father abandoned his mother to live with his new wife. This young man has been farming for several years and he had this to say about the prospects of inheriting land: "My grandfather has about 15 acres... it will go to my father and then get divided between me and my step brothers. There are four of them... right now the land is in my grandfather's name. I will get some part of it in future. I would want to buy some more land and prepare for the educa-tion of my children when they are born" (22 years, male, ST, 5th grade, works on 15 acres of family land, Chhindwara).

As far as inheritance of land is concerned, gender plays a key role. Until 2005, daughters did not inherit ancestral land as a matter of course and Indian law deemed sons to be the legitimate heirs to familial prop-erty. From 2005 onwards, daughters were also eligible to inherit land.

Typically, when the male landowner died, his wife, sons, and daughters would each have their name on the land record as heirs. From all accounts, this was the practice followed by government clerks. However, in most cases, ownership records mattered little, and the male siblings would gain full control of the land. In general, most respondents mentioned that sisters typically give up their claims to land. Some suggest that sisters do not want a share of what is already a small inheritance and thereby deprive her brothers of a livelihood. Others suggest that marrying sisters well with a dowry is considered equivalent to inheriting land. Frequently, young farmers point out that sisters often seek and value that support of brothers even after marriage and are willing to give up their share of land to ensure that they have this support. This is recorded in other studies in India too (Rao 2017, for example).

While daughters' inheritance is a subject that is discussed and debated in the family and community, this is not the case for sons. Irrespective of whether sons migrate, quit agriculture, or leave the village, they inevitably partake of the inheritance. In several cases, the young farmers that we interviewed had brothers who worked in non-farm sectors and often were settled permanently in neighbouring towns, leaving the brother who remained to take care of the farm. In many of these cases, they seemed to have a symbiotic relationship—the brother working in the outside would send money to maintain the farmer-brother's household and fund investments on the farm, contributing to land levelling, boring a well, and so on. A share of the food produced would go the other way, from farm to the city. In that sense, even those who exited farming continued to maintain links with farming. Our interviews suggest that in the bequeathing of land, there is no succession discussion, of which brother would take over the farm, and whether more (all) of the land would be allocated accordingly. The following quote reflects this uncertainty, although this is in the context of whether this young farmer thinks that he will inherit land: "I don't know what is there in their (my parents') hearts. We will see; if they wish to give, they will give; if not, that is also okay" (32 years, SC, graduate, seven acres; Chhindwara).

Irrespective of the difference in family situations, property subdivision appears to demand cooperation amongst siblings in ways that perhaps was not required in the previous generation—whether or not to give land to sisters but also how it works between siblings, since earlier, all of them remained in farming: "I have 15 acres in my father's name… When my father is not there, it will depend on what me and my brother decide. If we can't cooperate, then we shall divide it equally" (36 years, ST, sole operator of family land, Chhindwara).

Land Quality and Water Availability

While the size of land is a big constraint, many interviewees also mentioned the quality of land as a significant barrier in the study area. In several villages across Sehore and Chhindwara districts, especially in the latter, the land is of such poor quality—undulating and strewn with boulders—that several parcels had been left fallow for generations. In many cases, farming families had manually removed boulders to be able to farm the land. Older farmers point out that in such circumstances, they could afford to leave many of these plots uncultivated, partly because their earlier unrestricted access to the forests allowed them to collect enough food. One 80-year-old farmer recounted that in the years past, villagers used to buy maize in Chhindwara and sell it in the village. With the growing importance of agriculture, the flow is now reversed. Agriculture expansion has meant that marginal lands are now being brought under cultivation and investments in land are needed to ensure adequate yields from these plots. Corporate Social Responsibility (CSR) organizations, philanthropic entities established by leading private sector conglomerates, soon arrived in the region with the funds and the means to level the land (more on their role later in the chapter). Many young farmers seemed reenergized about farming since these land improvements significantly improved their prospects in agriculture.

Another huge constraint is water availability, especially in Chhindwara. The undulating land and soil structure means that water is not available throughout the year. CSR initiatives have recently focused on creating and reviving village tanks and ponds to address these needs. Without water, farmers can only depend on one crop annually (the *kharif*) during the monsoon. A second crop is feasible only if there is enough moisture in the soil in winter. In contrast, having access to a water source opens up the possibility of harvesting up to three crops a year.

The Imperative of Secondary Occupations

For young farmers, these constraints circumscribe the extent to which they can farm and identify themselves as farmers, despite any preference for farming. Many are forced to seek alternate sources of income. Most young farmers feel that as long as they continued to migrate and take up non-farm jobs to supplement their farm incomes, they would consider themselves as workers rather than farmers. Indeed, for every young farmer that we interviewed, the idea of who would qualify as a successful farmer typically centred on one's capacity to sustain a livelihood solely from agriculture. Most considered someone to be successful at farming if agricultural income alone was enough to support the family. Others articulated related issues: a successful farmer is variously one who can earn profits, has large land holdings, is free from debt, has irrigation (is able to "grow wheat and gram"), has farming knowledge, is industrious, and one who does things in a timely manner. Our respondents linked each of these attributes with profitable agriculture.

For many young farmers, supplementing income from non-farm sources was the only way that they could invest in their farm. On the other hand, both older and younger farmers maintained that youth's needs have vastly expanded: "Today a farmer has a compulsory need for a motorcycle, good quality food at home, a mobile, and other things. For all these, a farmer does not have enough means to fulfil… so has to go out" (32 years, ST, Chhindwara).

A 60-year-old male farmer in Sehore was less charitable: "Today, people want fashion, everyone wants to wear jeans and shoes that cost Rupees 1000. How will agriculture provide that?…Even a child these days wants a couple of pairs of shoes. They don't adjust. Also, education has become expensive these days." Older farmers in Sehore, both men and women, also pointed out that young women farmers in their families are interested in expensive make-up and beauty products.

The state-run employment programme—Mahatma Gandhi National Rural Employment Guarantee Act or MGNREGA[5]—used to be a reliable source of work within the village. It seems that in recent years, however, the MGNREGA was not implemented well. The more enterprising farmers were trying their hand at mushroom cultivation, beekeeping, and so on, but the majority continued to depend on construction work. Several were engaged in petty trade, running tea shops, tailoring units, driving a tractor, working with NGOs in the area, and so on. For many, however, the only option is to migrate for work, usually to the nearest town or city.

Migration and City Life

How do young farmers view work in the city? City life per se is not an aspiration, especially among those who prefer farming. Young farmers, both men and women, assert that if agriculture were prosperous, they would not migrate. Even as migrants, they don't necessarily experience life in the city, given that they tend to be confined to construction sites, with little free time.

Most young farmers feel that being a farmer allows them to consume what they produce and does not leave them dependent on food purchases. Many farmers also value the freedom and flexibility that farming affords them relative to a routine job in an urban area, not to mention the clean environment. Some also associate urban jobs with drudgery: "I feel that in the village, one can do different things and grow diverse crops. In town, the nature of work is the same. The schedule is also the same. Outside, people get up at a time, bathe, and do the same work daily. In the village, I can grow different things and do not have to buy from outside" (22 years, 5th grade, ST, Chhindwara).

He shared little with his friends who had migrated: "They mostly talk about money—we earned so much, or got this or that. This is all they talk about. Sometimes I feel that I too should go, but then I realize that I

---

[5] Implemented between 2006 and 2008, the scheme guarantees each rural household 100 days of unskilled manual work within the village at a minimum wage, which is paid based on the work completed daily.

should not. Here, I work hard… and I don't have to buy food from outside."

Among many of the young farmers, not only is there no aspiration for a life in the city, on the contrary, they seem to associate it with a poorer quality of life in the balance, based on their conversations with migrants. This is especially pronounced in tribal Sehore and Chhindwara. In Sehore, in the non-tribal areas, it was not unusual for at least one member in an agricultural household to be engaged in non-farm activities, often in the nearby city of Bhopal. A 24-year-old male farmer in Chhindwara told us that many young people in the village who migrate tell him about how they feel homesick and miss their families. As a result, he never wanted to leave. Another 32-year-old male farmer in Sehore, who farms seven acres of family land, said that he never wanted to work outside the village: "I hear from people who go out that they work night and day and conditions are very tough."

A young male farmer who used to migrate told us: "Work in Indore was very taxing. We would sleep for an hour or so and then work rest of the time. I got so weak after a while that I had to leave that work and come back to the village to recover" (29 years, ST, farms three acres, Sehore).

Another male farmer, about 30 years old, who wanted to migrate for work for the experience said that he asked other youth who regularly migrated to take him along: "They say 'no, you are better off here.' They tell me I should be happy with what I have."

Another noted that he loved the village: "My family and home are here. The air here is nice. In towns there is so much heat. Our village is so cool and you can go any direction. There is no tension here" (20 years, ST, studying in 10th grade after a break, also farming 20 acres of family land, Chhindwara).

In general, it would seem that many who farm today would not migrate if they had a choice, that is, if their land and the produce they grow provided them a source of livelihood, and they could live free from debt.

Sometimes life events, such as marriage, mark the transition to working full-time on the farm and giving up migration. One young male farmer, about 35 years old, who had migrated extensively each year to work on construction sites, said: "…when I got married, I decided to stay here in the village. I also had enough experience in construction, so I started getting more work locally. Then I stopped going to town."

A young woman farmer narrated that she used to migrate regularly for work even after her marriage. When she had a child, she looked for opportunities to become a farmer full time. That opportunity came when she got a well under a state government scheme. Her rainfed farm could now be cultivated throughout the year, and she could afford to stop migrating for work and engage in farming full time. Another young woman farmer told us:

> It was difficult when we used to migrate because you are working for others, harvesting their produce. We had to listen to others but when you are at home, you are working on your own land, on your terms. That makes all the difference. I don't even feel like migrating anymore because now I have a family and I have to look after them…Farming has also become better with new techniques…now we try to get as much work as possible in village.

Some young farmers migrate out just to keep busy, so that "the empty mind does not become a devil's den." A young male farmer recalled: "I once migrated to Hoshangabad to work on the soybean farms…But this was not out of necessity rather something I did with my peers for going out of the village for 10 to 15 days" (30 years, ST, unmarried graduate, farms six acres, Sehore).

An older farmer, 55 years, noted that migration was common among youth, but added a cautionary note about those who do not fully evaluate the related perils: "Their limbs are working so they can go out and work, but there will be a time when they won't be able to do this. Hence, they should stay here and build a strong foundation for the future. But they are not thinking about the future like that."

Support for Young Farmers

The strongest support system for the youth in our study area is perhaps the NGOs and CSR initiatives active across many of the villages. These include the implementation of watershed works, incubating farmers' groups and producer companies, offering training in new techniques, offering allied activities, and arranging farmer visits to other districts, among others. In some places, NGOs such as Samarthan have identified *Kisan Mitras* (Farmer Friends) that assist farmers in accessing government programmes. Several of them are young male farmers themselves.

In many spheres, the government is supportive, including in the provision of subsidies for drip irrigation, sprayers, seeds, and other agricultural necessities. There are also many functioning programmes for tribal farmers' welfare. Credit is another area where cooperative societies have a role in providing crop loans at lower interest rates, along with cheaper seeds and fertilizers. One farmer told us about a toll-free telephone number to register for tractor rental services at a cheaper rate of 300–400 INR per acre rather than 600 INR.

At the same time, many also express dismay and mistrust of both CSR and the government. Several farmers told us that government programmes are difficult to access. Village social networks that facilitate access and elite capture of government resources are not unknown. One young woman farmer claimed that most of the benefits of government scheme accrue to a network of elite families within the village. In other cases, bureaucracy was a barrier: "There are government programmes, but these programmes rarely help the poor farmer. The officials make a fool of poor farmers. They make him come to the office 10 times for something and the farmer tires and eventually gives up."

Another young male farmer said:

Under the government scheme, soil testing cards were made and I also had my land soil tested. But the farmer has to take initiative here. They have to go to government officials and demand these things as officials just want to sit in office and not work. They fill papers anyway, get any man from village and click his picture to show they are meeting their quota. So the farmer has to be knowledgeable.

A 55-year-old farmer also registered his dismay: "Once a *gram sevak* (village functionary) informed us that if we take a picture of us using tractors in the farm, we will each get 2000 INR. So we did send them a picture, but we didn't get the amount. The officials can also get unreliable at times."

Kisan Mitras had provided a useful resource for many young farmers, helping them to access benefits. As the quotes above suggest, farmer awareness is important. A 50-year-old farmer explained why he and his children stay away from CSR initiatives that seem to benefit so many other farmers: "They are building wells, planting trees on people's land, if they were to ask for money tomorrow, where will I give it from? So I got scared and stayed away."

Some expressed cynicism, saying that cooperation was no longer there in the village and each one was left to fend for himself/herself. At the same time, many young farmers said that they helped each other out, especially during the peak agriculture season when timely operations are crucial. Farmers' groups (usually male dominated) and women's self-help groups are important sources of support; they are often organized by the CSRs. A farmer elucidated the role of these NGO and CSR initiatives:

> They have introduced organic farming, gardening, and maize cultivation to the farmers. They provide good quality seeds to the farmers, which then yields a good produce that fetches above average price in the market. Apart from this NGO, there is a *Khet Pathshala* (farmers group) that is quite active in the village. *Kisan Mitras* (farmer friends) appointment by the government is also associated with it. Basically about 30 farmers in the village meet once in a while and discuss their problems. We are also told about the latest techniques, seeds, chemicals, and schemes. This not only helps the farmers but also encourages them to be enterprising. For example, we grew maize last year in about six acres of land and it yielded 80 to 85 quintals (1 quintal=100 kgs.) of produce (higher than is typical in this area). When we shared this with our peers, the farmers in the neighbouring village grew maize on 25 acres of land and the yield was 500 to 600 quintals. So everyone benefits from it [24 years, OBC, high school, farms six acres, Sehore].

Whatever the form, it seems clear that even in the minds of older farmers, today's young farmers need active support: "I think some support from government or an organization or even people themselves is needed. A farmer alone cannot dig a well. But the community can come together and have some built between a few families. If a blind man is provided a cane then he too finds his way and moves forward slowly" (52 years, ST, farm 12 acres, Chhindwara).

The Future of Young Farmers and Young Farmers in the Future

Most young farmers today, even those enthusiastic about farming, would rather their children did not take up farming and preferred they choose routine jobs "outside." A 55-year-old farmer and father of young farmers recalled: "I wanted them to go to school and study. Start a business or get a job. If they couldn't then they could start working in the farm. I wanted them to get a job because there isn't much land left for farming."

Just as the fathers of today's young farmers wanted their children to finish schooling, with farming as merely a fallback option, the young farmers of today state that they would not give up land because it would be a fallback option for the children. Some of them want to buy land to ensure that this fallback option does not fail. One young farmer who is fortunate to have the means to do so told us that: "I managed, I am old now...but land is becoming smaller. So for my children, this would be helpful, so I bought 4.5 acres...when the land was still inexpensive" (35 years, ST, Chhindwara).

At the same time, several respondents see their own future in farming in a positive light. Several aspire to invest in a water source, diversify into different crops or ancillary activities, and expand the farm. Those who do not have such aspirations are from the second group of young farmers who are merely waiting to exit farming.

A farmer in his early twenties summarized the problems faced by youth in farming:

> The young do not focus on farming because their land is not suited for cultivation. They have inherited land and because their fathers were farmers, they identify themselves as farmers but this is just a name. If their land quality could be improved, land levelled, stones and boulders removed, if possible, water for irrigation provided, then they could produce more and truly be a farmer. This will reduce migration and farmers may start taking more interest in farming.

Another shed light on the predicament of young farmers by pointing to the larger challenges of farming: "There is no ideal farmer... because if someone has a little water, he does not have motor, if there is motor then knowledge about seeds is lacking. So something or the other is lacking. That is why there is no ideal farmer in this or any neighbouring village."

There are heartening stories as well of women's interest in farming. One young farmer told us that all of his children had stopped studying, but all of his sons wanted to work in jobs not related to farming. He added: "My daughter helps me more than my sons in farming… She doesn't like anything else. She likes farming."

Concluding Remarks

The experience of young farmers in Madhya Pradesh provides evidence that the perception that the youth want to leave agriculture is not entirely valid. We found several young farmers who, given a choice, would rather engage in agriculture as a full-time activity. Both young men and women farmers define good farmers as those who can live solely off of the farm, do not have debts, and are able to reap profits from agriculture. In this, the main barriers for young farmers are in the form of larger constraints such as water availability, the quality of land, and these farmers' limited ability to expand farm size through land purchase. To be sure, several young farmers feel stuck in agriculture and are waiting to find off-farm jobs. Even for this group of farmers, however, agriculture seems to provide a fallback option, in a context where these off-farm jobs are difficult to obtain and often are of poor quality. In the study areas within Madhya Pradesh, CSR initiatives and NGOs seem to play a key enabling role for youth to pursue agriculture, with a limited role for social media and greater reliance on traditional extension and peer networks. There is evidence that the state, despite problems, seems to offer some support via the MGNREGA or through other subsidies. As elsewhere, men and women have different experiences as young farmers; whereas today's young men appear to have a greater choice of whether to farm, it appears that young women have less say and see farming and domestic work as intrinsic parts of their married lives. As in Tamil Nadu, the distinction of becoming and being a farmer are quite blurred. People are already farming well before they have independent charge of their land, which often coincides often with the previous generation's passing or upon the young male farmer's marriage. Women rarely get independent charge—as daughters, daughters-in-law, or spouses. Our study suggests that many young farmers would rather continue farming if the larger constraints to agriculture, such as water, were addressed. This seems to challenge our popular perception that all young farmers overwhelmingly aspire for an urban lifestyle connected with work off-farm and desire to abandon farming.

References

Kumar, R. 2016. Rethinking Revolutions: Soyabean, Choupals, and the Changing Countryside in Central India (Delhi, 2016; online edn, Oxford Academic, 18 Aug. 2016), https://doi.org/10.1093/acprof:oso/9780199465330.001.0001.

Rao, N. 2017. 'Good Women do not Inherit Land': Politics of Land and Gender in India. (1st ed.). Routledge. https://doi.org/10.4324/9781315144610.

11

Youth and Agriculture in Indonesia

Aprilia Ambarwati, Charina Chazali, Isono Sadoko, and Ben White

General Background

This chapter reflects on the changing place of young men and women in Indonesian agriculture, based on available secondary sources and some preliminary local-level studies. Agriculture is important in Indonesia, not only to provide food for its 272 million population, but also as the country's single largest source of employment. Around 28 per cent of the total labour force (34.6 million people), and 48 per cent of the rural labour force, report their primary occupation as agriculture (BPS 2019). Despite widespread rural diversification and multiple-sector livelihoods, agriculture, and particularly the food crops sector, is still the main livelihood

A. Ambarwati • C. Chazali • I. Sadoko
AKATIGA Center for Social Analysis, Bandung, Indonesia
e-mail: apriliambarwati@akatiga.org; charina@akatiga.org; isadoko@akatiga.org

B. White (✉)
International Institute of Social Studies, The Hague, The Netherlands
e-mail: white@iss.nl

© The Author(s) 2024
S. Srinivasan (ed.), *Becoming A Young Farmer*, Rethinking Rural,
https://doi.org/10.1007/978-3-031-15233-7_11

activity of rural Indonesia. Contrary to general perceptions or expectations about youth, agriculture still employs a much higher proportion of young people than industry or any other sector,[1] and this proportion has been relatively stable in recent years.

To date there has been very little research on young people and agriculture, and most of this research has not gone much further than discovering that young rural men and women aspire to non-agricultural futures. To understand the position of rural youth and their (possible) futures in agriculture, more comprehensive research is needed.

In this chapter we first provide a general picture of agrarian structures in Indonesia. The next section then summarizes what we know about the changing position of young men and women within these structures, including: the age and gender of farmers, modes of intergenerational transfer of farm land and property, young people's apparent turn away from agriculture, patterns of rural youth labour mobility, agricultural education, and institutions representing rural youth interests. The main part of this chapter concludes with some reflections on policy. In the final part, we explain the selection of locations and the basic shared methodology for the three local case-study chapters that follow.

Agrarian Structure

Who Owns What?

Historically, post-colonial Indonesia did not inherit a class of large landlords who also dominated regional and/or national politics (in contrast, for example, with parts of the Philippines or India). It does, however, have a historical legacy of large-scale corporate plantations in such crops as rubber, tobacco, sugarcane, tea, coffee, and, more recently, oil palm. These are owned either by the state (many former Dutch and Belgian plantations nationalized under the Sukarno regime in the late 1950s) or by domestic conglomerates and domestic-foreign joint ventures. As seen

---

[1] The next largest sectors of rural employment for youth are trade (13.8 per cent) and manufacturing industry/handicrafts (11.3 per cent) (BPS 2019).

Table 11.1 Land area in major large-scale plantation crops, 2000–2020 ('000 ha)

Year	Rubber	Oil palm	Cocoa	Coffee	Tea	Sugarcane	Tobacco	Total
2000	549	2991	158	63	90	388	5	4246
2010	497	5162	92	48	66	437	3	6307
2015	545	6725	42	47	61	217	0.6	7368
2020	407	8560	18	24	60	174	0.3	9243

Source: BPS (2022a)

in Table 11.1, most large-scale plantation crops have remained stable or contracted in the last 20 years, but all are dwarfed by the rapid expansion and huge area of oil palm plantations.

There are also large areas of export and cash crops grown by smallholders, whether independently or on contract to agribusiness (Table 11.2). Here again we can see the rapid expansion of oil palm; the total area planted to plantation and smallholder-based oil palm will soon overtake the area planted to Indonesia's main staple food crop, rice.

Smallholder agriculture dominates staple food production and horticulture, with no significant plantation sector. Table 11.3 shows the area planted to the major food crops and their growth/decline over the previous four years. The area devoted to rice, maize, and soya has been expanding in recent years, while for cassava, groundnuts, mung beans, and sweet potato, it has been declining.

Indonesia's last (2013) Agricultural Census recorded 26 million smallholder farm households cultivating a total of about 22 million hectares (ha) of land (BPS 2013). Farm sizes in the smallholder sector tend to be very small: in 2013 three-quarters of all smallholder farms were under 1.0 ha and almost half were under 0.5 ha (Table 11.4).

AKATIGA's study of 20 rice-producing villages in Java, South Sulawesi, and Lampung found varying degrees of land concentration and landlessness. Large land ownership (in this type of village) does not lead to large farm sizes, but to increasing rates of tenancy (particularly share tenancy) as the larger owners parcel out their land to sharecroppers (Ambarwati et al. 2016). This appears to have been the pattern since the late colonial period, at least for Java (White 2018). Reviewing more than 30 local studies and reports on land distribution from different parts of Java in the 1930s, Ploegsma was adamant that where land concentration was found,

Table 11.2 Area planted to major smallholder cash and export crops, 2000–2020 ('000 ha)

Year	Rubber	Oil palm	Cocoa	Coffee	Tea	Sugarcane	Tobacco	Coconut	Cloves
2000	3046	1190	641	1322	67	n.a.	163	3602	n.a.
2010	2948	3387	1558	1163	57	278	213	3697	462
2020	3305	6004	1509	1221	51	229	230	3365	566

Source: BPS (2022b)

Table 11.3 Area planted to major food crops and recent trends

Crop	Area planted (million ha.) 2018	Change 2014–2018 (%)
Rice	16.0	+16
Maize	5.7	+49
Cassava	0.8	−21
Soya	0.7	+10
Groundnuts	0.4	−25
Mung beans	0.2	−5
Sweet potato	0.1	−29

Source: Deptan (2019)

Table 11.4 Smallholder farm sizes, 2013

Farm size (ha.)	Number (millions)	% of total
<0.1	4.3[a]	17
0.1–0.19	3.6	12
0.2–0.49	6.7	26
0.5–0.99	4.6	18
1.0–1.99	3.7	14
2.0–2.99	1.6	6
≥ 3.0	1.6	6
Total	26.1	100

Source: BPS (2013)

[a]The number of farms under 0.1 ha is widely believed to be under-enumerated in the 2013 Agricultural Census due to definition changes, resulting in a large apparent drop since the 2003 Agricultural Census in the total number of smallholders and especially those under 0.1 ha. In 2003 the corresponding number—with a different definition of "farm household"—was 9.4 million. The 2018 Intercensal Agricultural Survey (BPS 2018) arrived at a total of 27.7 million smallholder farmers, including an apparent jump in those under 0.5 ha from 14.6 to 16.2 million; it is unlikely the number would have declined sharply between 2003 and 2013 and risen again between 2013 and 2018

"it certainly does not lead to large-scale [farm] enterprise. The accumulated holdings will be sharecropped or rented out, and agro-economically speaking nothing changes, the small-farm enterprise persists" (Ploegsma 1936, 61).

Outside the densely populated regions of Java, Bali, and parts of some other islands where irrigated rice farming is practised, some two-thirds of Indonesia's total land area is claimed by the Ministry of Environment and Forestry as state-owned land under its jurisdiction. In these regions, peasant households occupy land under customary tenure, inherently insecure.

There are no formal barriers (and in most of Indonesia, no customary barriers) to women's ownership and inheritance of land. One exception is West Manggarai, Flores (see Chap. 12); another is the island of Bali, where Hindu customary law prevents daughters from inheriting ancestral lands (Saitya 2021). On the other hand, there are numerous "cultural" barriers (both in the bureaucracy and in rural communities) to women's discursive and material recognition as farmers. Nonetheless, 11 per cent of *petani utama* (the self-defined "primary farmer" or farm head in farm households) are female, as seen in Table 11.7; this number undoubtedly underestimates the reality due to the discursive cultural barriers just mentioned.

As in so many other parts of the world, land prices in Indonesia are rising rapidly, and not only in urban and peri-urban regions. Land is a safe investment and in many parts of Indonesia, speculative investment and absentee ownership are becoming more common, although absenteeism is technically illegal under Indonesia's Agrarian Law. Absentee-owned land is one of the sources of land for share rental. Buying land is becoming an increasingly unrealistic option except for those who are already rich. In the 12 rice-producing villages that AKATIGA studied in 2013–2015, the price of one ha of irrigated rice land varied between about IDR 100 million[2] (US$7143) and IDR 1500 million (US$107,143). Agricultural worker wages at that time were generally around IDR 50,000 (US$3.60) per day, and informal-sector earnings—for those with little

---

[2] US$1.00 is approximately 14,000 Indonesian Rupiah (IDR).

capital—were generally not much more than IDR.1 million (US$71.40) per month. Migrant worker wages in factories, or in oil palm plantations in Malaysia, were around IDR 2.5 million (US$179) per month. Therefore, even if a young migrant could save IDR 500,000 (US$35.70) per month out of those earnings, it would take him or her between seven years (in the cheapest location in South Sulawesi) and 100 years (in the most expensive in Central Java) to buy a rice farm of only 0.4 ha. This crude illustration underlines the fact that for landless rural youth, saving to buy any significant amount of land is no longer a realistic prospect unless they have access to a lucrative overseas migration opportunity.

Compared to Indonesia's "green revolution" period of the 1970s and early 1980s, smallholder farming in Indonesia receives little government support, and much of the available support does not reach small farmers. Government-sponsored cooperatives have generally failed, and small-holders face oligopolistic trading markets for both inputs and outputs. Subsidized smallholder credit schemes no longer exist, and crop insur-ance—increasingly important in the context of climate change and high-input agriculture—is in its infancy.[3]

Pluriactivity—household livelihoods composed of a combination of farm and non-farm activities—has been common for a long time, at least in densely populated regions, among both large and small-farm and landless-worker households. In general, larger farmers transfer surpluses into investments in relatively high-return, non-farm activities such as trading and shopkeeping, agro-processing and transport, while small farmers and landless farm workers transfer labour without capital into low-return activities—often providing less income per day than agricul-tural wages—such as petty trade and handicrafts (Ambarwati et al. 2016). Alexander et al. (1991) give some historical examples of this pattern from the late colonial period, White and Wiradi (1989) for Java in the "green revolution" period, and Ambarwati et al. (2016) for recent years.

---

[3] Lately, the Ministry of Agriculture, through the state-owned insurance company Jasindo, has initiated a crop insurance programme for landowners or sharecroppers of irrigated land. But on the ground, however, this scheme is still very limited.

Who Gets What in Indonesian Agriculture?

Looking at various agricultural commodities gives us an introductory idea on Bernstein's (2010) "who gets what?" question. In rice-producing areas of Java, Sumatra, and Sulawesi, large numbers of rural households—sometimes more than 50 per cent—are landless or have very small holdings and work as sharecroppers or pure wage labourers. The majority of rural households in these regions still need to buy rice for their own family for part of the year (i.e., they are net buyers). As already mentioned, their livelihoods are derived from various sources, both farm and non-farm activities. For landless and near-landless workers, wages in manual harvesting work (using the sickle) still provide the highest return to labour when compared to other work. In a few regions, the subsidized introduction of small combine harvesters has threatened harvesting opportunities.

Smallholders in areas of high-value vegetable production such as in West Java and North Sumatra are in a similar situation to rice farmers, but more dependent on middlemen collectors for marketing. The risks in commercial vegetable farming are higher than for staple food crops, but in a good season, the profits can be much better than rice. Urban young people and green groups who are interested in farming are often involved in these activities.

In export cash crops like coffee and tobacco—which, as can be seen in Tables 11.1 and 11.2, are mainly smallholder-grown—the main players are big agribusiness corporations. They operate in the upstream and downstream of farming rather accumulating land. Since the markets are relatively narrow, market channels are the key. The big players do not necessarily have land but dictate the prices, giving smallholders the inputs and training/dictating to them on how and when to plant. Small farmers obtain low returns while the big players capture the value-added in high-return processing.

Indonesia is the world's biggest producer of oil palm, as shown in Tables 11.2 and 11.3; plantations now cover more than 14 million ha, mainly in Kalimantan and Sumatra, with a government target of further expansion to 29 million ha. Big corporations have "grabbed" large amounts of land where the occupants do not have formal ownership certificates and the land falls under the jurisdiction of the Ministry of Forestry, as discussed above. Most of the oil palm is formally or informally under the control of big plantation actors, sometimes operated on classic plantation lines, sometimes combining this with smallholder contract-farming schemes. About 10 million people (2 million workers and their families) now live in the oil palm zones and depend on the plantations for income once the land frontier is closed. This level of employment (with only one worker per 5 ha) is very low, even compared to other plantation crops; in rubber, for example, the ratio is closer to 1:1 (Li 2018). Plantation expansion often leaves the original landholders in place, but confined in enclaves on which they may be able to continue some kind of farming on a reduced scale; the real squeeze begins a generation later when the remaining land in the enclave proves insufficient for the needs of young (would-be) farmers. As one elder in West Kalimantan explained to Tania Li: "'When the company came we thought our land was as big as the sea.' But more companies came. Now his children and grandchildren are landless. They are marooned in a sea of oil palms in which they have no share" (Li 2018, 59). These large-scale land deals have closed off the smallholder option, not only for today's farmers but also for members of the next generation who face permanent alienation from land on which they, or their children, might want to farm, and in the absence of livelihood opportunities elsewhere.

Young People and Agriculture

For rural young people in Indonesia, agriculture is the largest sector of employment (see Table 11.5). The next two largest sectors of rural youth employment are trade and manufacturing. In 2019, 38 per cent of the rural youth labour force (15–34 years) worked in agriculture; this increased to 40 per cent in the following year (the first year of the pandemic and related economic disruption). In 2020, agriculture still employed a much higher proportion of rural youth than trade (17 per cent) or manufacturing (12 per cent).

Table 11.5 Percentage of the rural youth labour force employed in three main sectors, 2019 and 2020

	Sector[a]		
Year	Agriculture	Trade	Manufacturing
2019	38	16	14
2020	40	17	12

Source: BPS (2019, 2020)
[a]This proportion only for rural youth labour force

To the best of our knowledge, there are almost no studies of young farmers available in Indonesia, besides an exploratory study on rural youth by AKATIGA in 12 rice-producing villages (Nugraha and Herawati 2015) and the study by the Indonesian Institute of Sciences (LIPI) on the "crisis of agricultural re-generation" in three villages of Central Java (2015). Both of these studies focused more on young people's aspirations and apparent turn away from farming rather than seeking out young people who wanted to (or had already) become farmers.

Age and Gender of "Primary Farmers"

Some data on the age and gender structure of Indonesia's farming population in 1983, 2013, and 2018 are shown in Tables 11.6 and 11.7. These data are drawn from the Agricultural Censuses of 1983 and 2013 (a complete enumeration) and the Inter-census Agricultural Survey 2018 (a sample survey). They show the age of those members of farming households who self-report themselves as the *petani utama* ("farm head").

Table 11.6 shows that the average age of farm heads has been rising significantly over the period 1983–2013, and if we add in the 2018 Sample Survey data, the trend has continued after 2013. In the space of one generation, the proportion of farm heads under the age of 35 has roughly halved, while those 55 years and older have roughly doubled.

Table 11.7 shows the gender of these self-reported farm heads in 2013 and 2018. These data suggest that (1) only 11 per cent of Indonesia's farm heads were female in 2013 (with a slightly higher proportion, 13 per cent, in the 2018 sample survey); (2) the female percentage among farm

Table 11.6 Changing age of smallholder farm heads[a], 1983–2018

	% of all farm heads		
Age group	1983	2013	2018
<25	3	1	1
25–34	22	12	10
35–44	31	26	24
45–54	25	28	28
≥55	18	33	36
Total	100	100	100

Sources: BPS (1983, 2013, 2018)

[a]Farm head (*petani utama*) in this table and Table 11.7 is defined as "the farm holder who represents the [farm] household. The farm holder selected was the highest income earner from agricultural undertaking amongst the farm holders within the household. If two farm holders had the same income, then the [one with] the largest activity in agriculture was selected" (BPS 2013, 78)

Table 11.7 Age and gender of farm heads in smallholder farming, 2013 and 2018

	2013				2018			
Age group	% of all farm heads	% male	% female	Total (millions)	% of all farm heads	% male	% female	Total (millions)
≤24	1	90	10	0.2	1	89	11	0.3
25–34	12	94	6	3.1	10	94	6	2.9
35–44	26	93	7	6.9	24	91	9	6.7
45–54	28	89	11	7.3	28	88	12	7.8
55–64	20	85	15	5.2	22	86	14	6.1
65+	13	79	21	3.3	14	79	21	3.8
Total	100	89	11	26.1	100	87	13	27.7

Sources: BPS (2013, 2018)

heads rises with the age of the farmer—possibly associated with widowhood and/or divorce; (3) the population of "farm heads" is still relatively youthful with 39 per cent of farm heads under 45 years of age (35 per cent in 2018) and only 33 per cent over 55 years (34 per cent in 2018); and (4) however, only 1 per cent of farm heads are under 25 years of age and a further 12 per cent between 25 and 34 (2013), or 10 per cent, in 2018. Table 11.6 shows that even in 1983, the proportion of farm heads under 25 years of age was very small (only 3 per cent). At that time, most boys in rural areas were leaving school at age 15, and girls often at age 12. Thus, in the past as in the present, there was a long gap between the age of leaving school and the time at which young people could take over management of a farm.

Looking at these statistics, we can ask: are farmers being forced to continue farming into their old age because of the lack of successors—this is the most commonly assumed explanation—or are they living and/or staying healthier longer and therefore not ready to hand over farms to their successors? Is the problem that the young are unwilling to start farming, or that they are unable to start because the old are unwilling (or unable) to stop? Or is there another, more complex dynamic at work, as Jonathan Rigg (2019) argues based on his research in Thailand, meaning that these are the wrong questions to ask and that we need to reconsider the ways that we think about ageing and occupational change, about what is a farmer and what is farming?

Modes of Intergenerational Transfer of Farm Land and Property

As stated earlier, in most parts of Indonesia, both male and female heirs can inherit land and other family property. Shares are sometimes equal, and sometimes daughters receive less than sons. In Kupang (E. Nusa Tenggara province) male children inherit more land than daughters. Daughters may keep the land they are cultivating after marriage, but when they die, the land reverts to their parents or male siblings or their descendants (Ruwiastuti et al. 1997, 30). In Western Lombok (West Nusa Tenggara province) inheritance rules follow the *sistim nina nyenyon mama melembah* (the woman carries one load on her head, the man two loads on a shoulder pole), that is, male heirs receive twice the share of female heirs. The same principle, *sepikul segendong*—comparing the two-basket *pikul* shoulder pole carried by men with the single basket which women carry on their backs—is often reported as customary norm in parts of Java, but not always followed in practice. In some cases where landholdings are too small to be further sub-divided, daughters do not receive a share, but depend on the male heir(s) to give them a share of the harvest (Ruwiastuti et al. 1997, 30). In Hindu-majority Bali where daughters are customarily barred from inheriting ancestral property, they may inherit property acquired during their parents' lifetime, but in practice sons still receive larger shares of non-ancestral property (Saitya 2021, 49).

Besides the Bali study just mentioned, we have not found any detailed ethnographic studies on the processes of intergenerational farm transmission. AKATIGA's study in 12 rice-producing villages in Java and South Sulawesi found that land could be transferred either when a son/daughter married, when the parents became sick or too weak to continue farming, or on the parents' death. Children waiting to inherit land may either stay in the village and help on the farm or—more frequently—migrate to work in various non-farm occupations. Cases where children had been able to become independent farmers (rather than farm helpers) while their parents were still living were rare. When grown-up children help on the parental farm, the parents may give them a share of the harvest (Nugraha and Herawati 2015). In some regions, such as our Kulon Progo research village discussed in Chap. 14, it is not uncommon for children to farm their parents' land on a share tenancy basis, under the same conditions as prevail between landowner households and their landless share tenants.

In many parts of the world, the transfer of farmland and assets and their division among (potential) heirs are sources of great tension between generations and/or between siblings, and sometimes a taboo subject that is almost impossible to discuss openly within the family (White 2020, Chapter 4). In Indonesia to date, there have been very few studies of these dynamics, which require ethnographic research. Our case studies in the following chapters go some way towards filling this gap.

Young People: Turning Away from Farming?

As in many other countries (White 2020: Chapter 4), available research and anecdotal evidence—in the absence of systematic survey research—suggest that many young rural Indonesians aspire or intend to work outside agriculture, and many of their parents have the same ambitions for their children. A LIPI report warns about the "regeneration crisis" in the agricultural sector (LIPI 2015).

A 2014 study by the Koalisi Rakyat untuk Kedaulatan Pangan (People's Coalition for Food Sovereignty) and Oxfam in various regions in Indonesia found that 63 per cent of rice farmers' children, and 54 per cent of horticulture farmers' children, did not want to become farmers. Moreover, 50 per cent of rice farmers and 73 per cent of horticulture farmers did not want their children to become farmers (Wiyono et al. 2015). This study, however, makes the classic logical jump of assuming that these children's preferences represent a future reality. The AKATIGA study also notes a strong expressed preference for non-farming futures, but also underlines the need to see this preference in the context of the agrarian structures, which mean that many (often most) young people have no prospect of inheriting land, and certainly no prospect of obtaining parental land while they are still young (see below). The same study also notes—although information on this is limited—that many of today's older farmers also previously chose to migrate—as their children do today—returning to the village and to farming only when land became available (Nugraha and Herawati 2015).

Young people's apparent aversion to the idea of farming futures is partially related to the image of farming as occupation and of rural life generally, but economic and structural issues are certainly also an important cause. The AKATIGA researchers have been studying these issues since 2013, in 12 rice-producing villages in West Java, Central Java, and South Sulawesi. We talked with young men and women between the ages of 13 and 30 from different backgrounds. Some were children of landowners, others from smallholder, tenant farmer, or landless families. When we look closely at these rural young people's views and hopes, the picture is quite complex, as is summarized briefly below.[4]

In most of these rice-producing villages, the landholding structure means that most young people have no realistic prospect of becoming independent farmers, or at least not while they are still young. Landlessness is widespread and less than half of farmers own the land they cultivate. The only people who have some chance of owning land while they are

---

[4] More details are given in Nugroho and Herawati (2015) and AKATIGA and White (2015)

still young are those who come from wealthy land-owning households. But they typically go to university and aim for a future in a secure, salaried job; their parents also have the resources to get them into these jobs. They may look forward to inheriting and owning land, but as a source of income through rent—they have no interest in farming it themselves.

Meanwhile, young people growing up in smallholder farming families may eventually inherit a piece of land, but their parents have too little land to hand over part of it to their children while they are still young. As a result, many young adults become share tenants on their parents' land. They may be in their 30s or 40s when they finally receive land from their parents. For those whose parents are landless, there is only the prospect of becoming a sharecropper or farm labourer, unless they can find another way to access land. Share tenancy conditions are quite burdensome, with the tenant providing all of the purchased inputs as well as their own labour, and delivering half of the crop to the landowner[5] (Wijaya and White 2019). For these young people, the only possible way to become an independent farmer is to first find work outside of agriculture (and often outside the village) and hope to save enough money to buy or rent land.

Due to either its image, its vulnerability, or its low incomes—even though the actual levels of income in available urban occupations may be no better—smallholder farming is not really an attractive prospect for many rural youth. On the other hand, the great interest of speculative finance and trading mafias in agriculture, and the growing markets for agricultural products, suggest that agriculture can potentially offer promising futures for smallholders, if given the necessary support. Current conditions and trends, however, are certainly not in favour of young farmers. It is hard for a young (would-be) farmer to become an independent farmer owning his or her own land unless they are first able to accumulate capital in other sectors or through other activities.

It is not surprising, then, that so many young rural men and women decide to migrate to work in various kinds of paid jobs or informal-sector work, often in other regions and sometimes as far away as Malaysia,

---

[5] This is in contravention of the Law on Share Tenancy, which stipulates that the crop should be divided after the deduction of input costs.

Taiwan, Hong Kong, or the Gulf states. But young people's decisions to farm or not to farm, and to stay in the village or to migrate, are not permanent decisions. As already noted, many of today's older farmers themselves migrated when they were young, returning home when they had saved money or when land became available.

Meanwhile, the large-scale plantation sector offers few attractive labour or career opportunities to young people. Wage levels and labour conditions in this sector are generally very poor. To date, there is only one study available focusing on young people's prospects in this sector. Li's (2018) study of oil palm plantations in West Kalimantan concludes that once land frontiers are closed, opportunities for plantation-related wage work are very limited, and the corporations make no provisions for either land or jobs for the next generation.

> …low wages, impoverishment and fragmented families are the future that lies ahead if Indonesian's oil palm plantations continue to expand. The prospect of 20–30 million hectares of oil palm, much of it in plantation mode, is dismal indeed. An intergenerational perspective helps clarify why many people who live in plantation zones are in despair, and the social devastation that will come unless there is a radical change of course. It also clarifies why 'sustainable development'…is fundamentally incompatible with expanded plantations. (Li 2018, 71)

Patterns of Rural Youth Labour Mobility

After graduating from secondary school or further education, poor (landless and near-landless) rural youth start to explore various options of nonfarm income opportunity. Young women may try to work in factories, in petty trade, as shop assistants in urban areas, or as domestic workers in Indonesia or abroad. Young men are less visible in factory and trading sectors, but more in the construction sector.

A study by AKATIGA found that both young men and women tend to change jobs often, trying to gain experience and access better opportunities (Djamal and Pithaloka 2017). They often use creative strategies and stop shifting jobs when they have found a good opportunity. The

AKATIGA study found, for example, one woman who had the opportunity to work as nanny in an expat home. She tried to learn English and widened her network among the expat helpers to get maps of opportunity since her boss would probably not stay in Indonesia for very long. Another young woman who found a job as a nurse tried to be more professional and to become part of a good nursing agency. A young man who started selling coloured textiles for *batik* and Muslim clothing made the effort to better understand the market and then adjusted his product accordingly. These are examples of young people who have found relatively well-paying occupations and are able to accumulate some savings. Their capital will often be invested back in the village, mainly to buy land and housing or livestock as a form of saving; the livestock will be sold before the big Muslim holidays when they need money and the price is high. Relatives who stay in the village (such as siblings, spouses, or parents) will take care of the land, house, and/or animals. When women marry, get older, or find that it's becoming harder to find good jobs, they return to the village and utilize their savings to become a farmer or to finance other activities (e.g., a small grocery store, trading clothes from the city, or other non-farm activities). Often, young men working in the city leave their children and/or wife in the village, and young women who go abroad may also leave their child and/or husband in the village.

Young people from larger land-owning or wealthy farm households tend to inherit land from their parents, but generally are not interested in becoming farmers themselves; instead, they become landlords, sharecropping out their land in small parcels. Although their original source of accumulation may be their farmland, as time progresses, their main source of accumulation is no longer from farming. In this case, farming or land-based income is additional income, as savings in the form of land or as a buffer for their other businesses. Their main income sources generally involve supplying various products and services in the village: farm inputs and equipment, building materials, capital goods rental (machinery rental), large grocery stores, transportation (buying a truck for transport of goods to other areas), and so on. They may also work in speculative businesses such as buying and selling land. The big landowners tend to be able to expand their landholdings at relatively low cost, as poorer landowners who need money (in a medical emergency or to finance a migration, for example) will sell or mortgage their land to these landowners at relatively low prices (Ambarwati et al. 2016).

Education and Pathways into/out of Farming

Education is often seen as a road to a better future for rural youth and also for better futures in farming. Agricultural education and training, however, generally do not produce a new generation of young farmers. In 2017, various online media reported that Indonesian President Joko Widodo criticized that "many graduates of the Institut Pertanian Bogor (the top agricultural university in Indonesia) find jobs in banking, so who wants to be a farmer?"(CNN Indonesia 2017) Moreover, one of the professors at Institut Pertanian Bogor admits that in 1985–1986, more than 50 per cent of his alma mater worked in banking (Suryowati 2017). More often they seek employment in the financial sector in big cities that offers better salaries than many other sectors. Those graduates who do become involved in the agricultural sector are more likely to be involved in post-harvest trading and processing in urban areas (Hidayat 2017).

There are 1837 Agricultural Vocational Secondary Schools (SMK Pertanian) in Indonesia (Directorate of Vocational Education 2022). The fees are relatively low compared to other vocational or general high schools. The Ministry of Agriculture, the Ministry of Education, Culture, Research, and Technology, and various private-sector donors have established scholarship schemes to attract more students (Ernis 2022; Ulum 2017). Most of the students are children of smallholder farmers. The SMK curriculum framework includes an obligatory internship in collaboration with farmers and businesses that are located nearby the SMK.

However, most of the graduates of agricultural vocational schools— both SMK and Islamic Boarding Schools—that we have visited in Java and Flores during the course of this research are not working as farmers, even though the school provides them with extensive field experience, internships, and real-life involvement in agriculture and agribusiness activities. Most of the students attend these schools as a stepping-stone to higher education opportunities or to find employment in factories or the service sector in semi-urban areas, mostly not directly related to agriculture.

There are also agriculture-focused polytechnics at the tertiary level, sometimes focusing on specific branches. Examples are the Sekolah Tinggi Pertanian (Agricultural Polytechnic) known as the "Oil Palm University" in Yogyakarta, and the Politeknik Kelapa Sawit Citra Widya Edukasi in Jakarta (Citra Education Widya Oil-Palm Polytechnic), which is also focused on the oil palm industry. Private-sector investors have established these schools to meet the need for lower-level technical staff in the rapidly growing oil palm industry. They cooperate with various palm oil companies to channel their graduates into positions within these companies. The Ministry of Agriculture supports Sekolah Tinggi Penyuluhan Pertanian (Polytechnics for Agricultural Extension, STPP) in several cities and provides scholarships for both private/public agricultural extension or vocational students in agriculture to continue studying at STTP. In 2016, the Ministry of Youth and Sport launched its Youth Farmer programme, which targets young people who have an interest in agriculture and can promote this interest to other young people. In this programme, enrolees are trained in land management. The research team, however, was unable to locate sources or documents that explain how the programme is implemented.

New Types and Styles of Farming

Though still limited, there are various emerging types of "new farming" differing from the traditional pattern, all of which (we think, based on scanty and anecdotal evidence) mainly involve young men and women. One is the cultivation of non-traditional, high-value seasonal crops such as watermelon on rice fields (either independently or on contract) and medium-scale poultry farming on contract (see, e.g., White and Wijaya 2021). Another is organic farming. Organic farming products are available in large supermarkets in big cities, but in general, the market share is still very small; organic Arabica coffee exports are one exception. In the main areas of intensive food crop production, pure organic farming is not easy to achieve since the groundwater is often contaminated with nitrogen or other chemicals. Formal certification costs are also prohibitive for most smallholders; more realistic is "trust-based" organic or near-organic production in which groups of producers build nested markets with networks of consumers.

Another new form of farming is urban farming. Following the global trend, urban farming is often discussed in social media and linked to the recycling movement. In some cases, a small area of urban land is used to introduce urban schoolchildren to green activities and agriculture. There are dozens of communities formed to promote urban farming with names such as 1000-yard Community, Jakarta Farming, and Green Bogor, but the total area of urban land cultivated is still negligible. Some authors have pointed to urban agriculture as an alternative anti-poverty option, providing resilience in times of economic crisis or when urban development policies such as the development of shopping malls displace residents (see, e.g., Purnomohadi 2000; Siregar 2001, 2006; Suryana 2006). In his study of four urban-fringe locations in East Jakarta, Semiarto Aji Purwanto (2010) argues that such regions deserve our attention for a better understanding of the complexities of urban-rural relations. Peri-urban farmers who have migrated from rural areas often return to their villages and maintain social ties there, making them not "full" migrants.

Institutions and Initiatives Channelling Rural Youth Interests

Besides government, some independent farmer organizations and non-governmental organizations (NGOs) have programmes that encourage young people to learn about, and engage in, farming. Unfortunately, no systematic information on such initiatives is available and we provide only a few illustrative examples here. One example is Serikat Petani Pasundan (SPP), which has established *sekolah pertanian* (farming schools) for local farmers and their children. SPP raises funds to provide scholarships for farmers' children, and several of the graduates are involved in regeneration of SPP activities. As an example of an NGO initiative, Plan Indonesia's Youth Economic Empowerment (YEE) programme aims to improve the capacity of vulnerable young people, especially girls (80 per cent) to secure decent work or build an independent and sustainable enterprise. In some rural areas, this programme has assisted targeted young people and their communities in developing horticultural farming (Djamal and Pithaloka 2017). On a smaller scale, the NGO Sunspirit in East Nusa Tenggara province has programmes to develop the potential of young people in various economic sectors, including agriculture. They provide training in farming techniques and promote seed banks in cooperation with the farming community.

Other initiatives, particularly those aimed at wide audiences or operating at a national level, emphasize fostering agribusiness entrepreneurship and "smart farming" with sophisticated technologies. One of these is the youth branch of the state-sponsored All-Indonesia Farmers' Harmony Association (HKTI), which we will discuss in a later section. In 2014, the Innovation Community Youth and Agriculture, which various external donor NGOs sponsored, launched a series of annual competitions to identify 10 Young Agripreneur Ambassadors. The purpose was "to promote agripreneurship among youth and make agriculture more attractive as source of jobs for youth"; the ten chosen ambassadors were expected to "campaign to show the young generation that agribusiness is cool and there are young agripreneurs who have been successful in developing their agribusiness" (Wulandaru 2018).[6] On a larger scale, the International Fund for Agricultural Development (IFAD) has included Indonesia in its Youth Entrepreneurship and Employment Support Services (YESS) programme. In a pilot project in four provinces, the programme aims to support poor and vulnerable youth in 320,000 households. Specific targets to be achieved in the six-year project include 33,500 young farmers/entrepreneurs reporting a profit and 32,500 young people finding agri-sector-based jobs (IFAD 2018).

Another important requirement to increase opportunities for young people in the farming sector is democratic and rooted institutions that young (would-be) farmers can use to articulate their interest in agriculture and increase their bargaining position. Unlike their counterparts in

---

[6] This programme appears to have folded in 2018 as donors shifted to other priorities; both of these sites have been inactive since 2018.

neighbouring countries like Thailand and the Philippines, Indonesia's tens of millions of peasants and agricultural workers—the country's largest single occupational group—have no strong national movement, organization, or political party representing their interests. Two generations ago, in contrast, the Indonesian Peasants' Front (BTI) and Plantation Workers' Union (SARBUPRI) together claimed almost eight million members. Following Soeharto's takeover of power in 1966, the government dissolved the BTI and all other independent peasant organizations and replaced them with a single, state-sponsored monolith organization, the Himpunan Kerukunan Tani Indonesia (Indonesian Farmers' Harmony Association, HKTI), which was officially mandated to promote the interests of small farmers and farm workers.

In the more than 50 years of its existence, HKTI has done little towards fulfilling its mandate; it has "functioned as a figurehead organization with no effective role in voicing the concerns and aspirations of Indonesia's tens of millions of villagers" (Bourchier 2015, 175). In recent years, for example, it has been silent in the face of the forced expulsion of local peasants from millions of hectares of land for corporate agriculture (especially oil palm), airports, dams, and other infrastructure projects. Currently, the HKTI serves mainly as a vehicle for political ambition, providing support for political parties or candidates for high political office at the national or regional level. For a decade since 2010, the HKTI was locked in a leadership struggle between military and non-military business and political elites, with two rival HKTIs, two rival chairmen, and two rival websites claiming legitimacy (Hasan 2010). Neither website showed any vision of agrarian renewal or offered any programmes or activities aimed at rural youth. Since 2020, the two factions have been reconciled under the chairmanship of former General Moeldoko, who is also Presidential Chief of Staff, a businessman, and chair of the Democratic Party. A youth branch, HKTI Pemuda Tani, is now active. Its chairperson is a PhD candidate and board member of Bank Raya Indonesia, an agricultural arm of the state-owned Indonesia People's Bank (BRI); the Secretary General is owner of the agri-export and import Pinus Nusantara Group. Its aims are "to attract young people's interest and capabilities in agriculture," with an emphasis on technological innovation and agribusiness.

Unfortunately, more than two decades after the collapse of the Suharto regime, the rural institutions or associations supposed to serve the needs of small farmers are basically top-down imposed institutions inherited from the Suharto era. Water user associations, farmers' cooperatives, farmers' groups, and other institutions are used to channel government programmes and subsidies. Farmers' cooperatives are mainly busy channelling subsidies for seeds, fertilizers, and machinery. In general, although there are a few exceptions, very few farmers receive support from farmer groups. The subsidies are captured by local rent seekers who use the names of the whole farmer group to capture subsidies and other government programme opportunities.

However, there are several local-level movements and activities that promote the needs and interests of small-scale farmers. Serikat Petani Indonesia (Indonesia Peasants' Union), for example, is active in some regions and has strong links to the global organization La Via Campesina; others are the Serikat Petani Pasundan (SPP) in West Java and the Alliance of Agrarian Reform Movement (AGRA) in South Sulawesi. Aliansi Petani Lembor (APEL) tries to build and develop farmers sovereignty through media and local government. Since the emergence of these movements, starting in the last years of the Suharto period, they have tended to suffer from chronic fragmentation, a problem common to Indonesia's civil society landscape. Their campaigning priorities are often disconnected from the concerns of the mass of Indonesia's rural people, especially the young generation (White et al. 2023). An important question, therefore, is whether the emergence of more autonomous, democratic (young) farmer's movements (and maybe a national federation of such movements), and their efforts to include young people in their activities and in their policy lobbying, can present young rural men and women with a vision of a farming future that is more attractive to them as well as the needed support to realize such vision.

Recent years have seen the emergence of a number of locality-based young farmer movements and networks, such as the Bali-based Petani Muda Keren (literally Trendy Young Farmers, www.petanimudakeren. com) or the Yogyakarta-based Petani Muda (Young Farmers, www.petanimuda.co.id). Again, there is no systematic source of information on these initiatives. At first glance, they seem to involve relatively well-educated young farmers, particularly in commercial horticulture, and to focus on technological innovations beyond the reach of the majority of rural youth. Petani Muda Keren, for example, describes itself as "a movement that aims to integrate farming from upstream to downstream, based on the concepts of small-scale farming integrated with digitalization and IoT [Internet of Things]…with smart farming we can manage our farms at a distance, for example: irrigating, fertilizing, monitoring PH levels and humidity, checking by CCTV, etc" (Petani Muda Keren (n.d.)).

Indonesia has for the last two decades had a huge programme for rural poverty reduction (PNPM) that at its peak covered all villages in Indonesia. PNPM created rural revolving fund institutions and village development implementation teams that were democratically elected. In villages where these institutions have survived and still manage a lot of revolving fund money, the institution has been found to serve (relatively) the majority and to involve many women and the relatively young who have confidence to manage the funds transparently in front of all their fellow villagers. At present, these institutions are not targeted specifically at the young or at farmers, but could be utilized by the rural youth to further their interest in becoming farmers.

Regarding young (would-be) farmers' problems in gaining access to land, Indonesian administrative regulations continue to move in the direction of greater autonomy for villages to manage their own affairs. There is increasing scope for local-level adjustments to current land tenure structures. One village that we have studied in Kebumen (Central Java) has helped its landless and near-landless villagers to gain access to village public land through more appropriate tenancy arrangements, and to limit absenteeism and excessive concentration of ownership. Part of the block grants that villages now receive under Law 6/2014 on Villages— amounting to more than IDR 1 billion per village per year—could be used to increase the stock of public land. The targeted allocation of use rights over that land could be a means to give poor people, women-led households, and the young a better chance of obtaining a piece of farmland. Similarly, village governments and farmer groups should be able to insist on better support for smallholder production and reject inappropriate technologies (such as combine harvesters) that the Ministry of

Agriculture has introduced in some areas; these benefit only a minority of the richest families and jeopardize harvesting employment opportunities (Wati and Chazali 2015).

Young People and Farming in Indonesia: Concluding Reflections

There are many reasons why leaving the village may seem attractive, and farming futures unattractive, to young people. Mass media often portray the rural world and farmers as backward and poor. But many dimensions of rural life are changing fast. In many villages connectivity is now as good as in the cities, motorbikes are cheap and common, and young people are busy with smartphones and social media accounts. Young people engage actively with global ideas and global youth lifestyles, which may make them look at rural life and farming differently to how their parents did.

If Indonesia's food needs are to be met in future largely by smallholder farmers, rather than by the large corporate industrial food estates that technocrats favour, rural life and farming have to be made more attractive to young people. We need to have a clear idea of the main barriers—both practical and cultural—to young people's entry into farming, either while still young or as a later lifetime option. When we look at young people's migration and their apparent decision not to become farmers, we need to take a longer-term, life course perspective.

The issue of young people and access to land needs to be taken seriously. In Indonesia, this generational issue has attracted little attention in land policy discourse. There is a need to look at possibilities to take land out of private property markets and to allocate it in use-right form to young people as well as to find ways to curb speculative investment in land. The latter is bad for the economy—it is an unproductive, parasitic form of investment—bad for social cohesion in rural areas, and as we have seen, bad for young people's prospects. While men and women formally have equal rights to own land, there are many practical gender distinctions and barriers to young women's access to land and farming opportunities.

As was also the case in the Indian, Chinese, and Canadian studies in this book (see the concluding chapter) our Indonesian farmer respondents, both young and old, did not raise issues of environmental degradation or climate breakdown. Looking back, we wish that we had done more to explore their awareness of and concerns about these looming problems. In this connection, a recent national survey on climate change that the polling agency Indikator conducted has found that the great majority of young rural Indonesians are indeed concerned about climate change. Seventy-nine per cent of young rural people (aged 17–35) were concerned about "environmental degradation," which came narrowly behind "corruption" as a top level of concern; they also prioritized protecting and preserving the environment over the current national obsession with economic growth (CERAH 2021). This is clearly an important area for further research. Issues of climate breakdown and campaigns for the creation of millions of "green jobs" in ecological regeneration is one area in which rural youth movements can forge alliances with their peers in urban and environmental movements.

Indonesia's young people are the most important potential source of innovation, energy, and creativity in developing new, environmentally responsible, and highly productive farming practices. Much can be done in general education, the media, and particularly on social media to correct the prevailing images of farming and rural life. Concrete examples of young men and women farmers, practising new, smart, and creative ways of production and making a decent living, can potentially have powerful impact. For most young rural people, it is not rural life or agriculture as such, but the lack of local jobs and the poor incomes from smallholder farming under present conditions that reinforce their decision to leave their homes and villages.

Methodology and Introduction to the Five Sample Villages[7]

The case studies that we present in the following three chapters are drawn from field research conducted in three study villages in Java (Central Java and Yogyakarta), and two study villages in West Manggarai (Map 11.1). In these villages, we interviewed 109 young farmers, including 49 young women farmers. We first introduce the five villages and then describe the field methodology, including the selection of the young farmer samples.

Sidosari, Pudak Mekar, and Kaliloro (Java)

Our three sample villages in Central Java and Yogyakarta[8] reflect the characteristics of the region: densely populated, with very small farm sizes, significant rates of landlessness, and long histories of pluriactivity and out-migration—not always permanent—of young people. In all of these villages, both sons and daughters inherit land. Sidosari is a village in Central Java's lowland rice-bowl region with good canal irrigation. *Pudak Mekar* is closer to the southern coastal area (Indian Ocean) where almost all of the area is *tegalan* (rainfed land). These villages are located in Kebumen District, around 165 kilometres south of the provincial capital of Semarang. Kaliloro is a rice-growing village with good canal irrigation located between the river Progo and the Menoreh foothills, some 35 kilometres northwest of Yogyakarta City.

Langkap and Nigara (Western Manggarai, Flores)

Langkap is an upland village directly adjacent to the Mbeliling forest. Being developed as an ecotourism village, Langkap has both natural and cultural tourism potentials. Nigara is also mainly an upland village, but one of its hamlets is far separated from the rest in the lowland part of the

---

[7] All names of people and villages are pseudonyms.

[8] Yogyakarta is a Special Region in southern central Java, geographically an enclave within the province of Central Java.

Map 11.1 Study villages in Central Java, Yogyakarta, and West Manggarai. (Source: edited from https://www.freepik.com/, accessed on 3 July 2023)

village, which has irrigated rice fields. This hamlet is part of an area that was developed as a rice-growing area since the Suharto era (1967–1998). These villages have a combination of rice fields and dry land farming as well as a system of customary tenure in which land can only be allocated to men.[9]

In both of the West Manggarai study villages, some "land grabs"—not spectacular, but no less important to those who experience them—have been observed in recent years. Almost half of all customary land within these two villages has already been sold to national or international private investors who plan to develop tourist resorts in these areas.

Sample Selection and Interview Methods

We interviewed 109 young farmers (60 men and 49 women). For the purpose of selection, we defined "young farmer" as all of those farmers (male and female) under 45 years of age, following the guideline agreed in the four-country Becoming a Young Farmer project. This limit, which may seem high to many readers, is appropriate in the Indonesian context for various reasons. In many villages, as already mentioned, young farmers are former migrants who turn to farming only in their late 20s or early

---

[9] Widows in Langkap also receive a piece of customary land from the customary head.

30s. If we had restricted our sample to the standard United Nations' definition of "youth" (ages 15–24), we would have missed many young farmers. Furthermore, as we wanted to explore the experiences of young farmers, we did not want to have a sample only of those who had just recently begun farming. All of those farmers aged 40 or over in our sample started farming while in their 30s, and many of them in their early 30s or late 20s. The oldest age at which a respondent had become an independent farmer, among our 109 young farmers, was 38; the average age of starting was 24 years; and the modal age was 27.

In all of the research locations, we selected the young men and women respondents by a combination of information from key informants and snowball techniques. More details are provided in the case-study chapters. Data collection was mainly qualitative, including semi-structured interviews, but also included a short household survey questionnaire. The semi-structured interviews were inspired by the life-history method, with a focus on key moments over the young respondents' life course in the process of becoming a farmer.

We also interviewed several older farmers, parents of our young farmer respondents, mainly to obtain information on intergenerational changes in farming practices and intergenerational transfers of resources and farming knowledge.

In all of the villages, we tried to identify and interview respondents from different geographical locations within the village, as location may be an important influence on farming and other economic activities (e.g., in relatively remote neighbourhoods compared to those close to the main road).

For the duration of the research period, the research teams stayed with villagers. This enabled us to complement the interview-based methods with participant observation by taking part in everyday activities. Staying in the village was also important for generating rapport that made it easier for the research team to discuss delicate topics such as intergenerational and inter-sibling relations and inheritance. We often engaged our young respondents in informal conversation while joining them in day-to-day activities in and around the house. In this way, they felt freer to tell their stories because they did not feel they were being "interviewed." These conversations often happened in the kitchen while preparing food, in the early evening when women like to sit together and chat, or while enjoying the evening meal together.

In all of the research locations, in the case of married young farmer couples (where both were active farmers), we tried to interview husband and wife. In such cases, to complete the structured questionnaire (on landholdings, family structure, and other basic household-level data), we tried to interview them together. For the subsequent in-depth interviews, we tried, where possible, to interview women separately, as they felt more comfortable telling their stories without their husbands' presence.

References

AKATIGA, and B. White. 2015. Would I like to be a farmer? *Inside Indonesia*, 120, April–June.

Alexander, P., P. Boomgaard, and B. White, eds. 1991. *In the shadow of agriculture: Nonfarm activities in the Javanese economy, past and present.* Amsterdam: Royal Tropical Institute Press.

Ambarwati, Aprilia, Ricky Ardian Harahap, Isono Sadoko, and Ben White. 2016. Land tenure and agrarian structure in regions of small-scale food production. In *Land and development in Indonesia: Searching for the people's sovereignty*, ed. J. McCarthy and K. Robinson, 265–294. Singapore: National University of Singapore Press.

Badan Pusat Statistik (BPS). 1983. *Sensus Pertanian 1983* [1983 Agricultural Census]. Jakarta: Central Bureau of Statistics.

———. 2013. *Laporan Hasil Sensus Pertanian 2013 (Pencacahan Lengkap)* [Report of the 2013 agricultural census, full enumeration]. Jakarta: Central Bureau of Statistics.

———. 2018. Hasil Survei Pertanian antar Sensus (SUTAS) [Result of the inter-censal agricultural survey]. Jakarta: Central Bureau of Statistics.

———. 2019. *Keadaan Angkatan Kerja di Indonesia, Agustus* [Labor force situation in Indonesia, August]. Jakarta: Central Bureau of Statistics.

———. 2020. *Keadaan Angkatan Kerja di Indonesia, Agustus* [Labor force situation in Indonesia, August]. Jakarta: Central Bureau of Statistics.

———. 2022a. Luas Tanaman Perkebunan Besar Menurut Jenis Tanaman [Area of large scale plantations by crop type]. Jakarta: Central Bureau of Statistics.

————. 2022b. Luas Areal Tanaman Perkebunan Rakyat Menurut Jenis Tanaman [Area of smallholder plantation crops by crop type]. Jakarta: Central Bureau of Statistics.

Bernstein, Henry. 2010. *Class analysis of agrarian change*. Halifax: Fernwood Press.

Bourchier, D. 2015. *Illiberal democracy in Indonesia: The ideology of the family state*. New York: Routledge.

CNN Indonesia. 2017. Jokowi ridicules IPB graduates who prefer to work in a bank than in the fields. Accessed 13 June 2022. https://www.cnnindonesia.com/nasional/20170906110335-20-239766/jokowi-sindir-lulusan-ipb-banyak-kerja-di-bank-daripada-sawah

Departemen Pertanian (Deptan). 2019. *Data Lima Tahun Terakhir* [Data on the past five years]. Jakarta: Ministry of Agriculture. www.pertanian.go.id.

Directorate of Vocational Education. 2022. *Data Pokok Sekolah Menengah Kejuruan* [Vocational high school basic data]. Jakarta: The Ministry of Education, Culture, Research, and Technology. http://datapokok.ditpsmk.net/dashboard.

Djamal, Fauzan, and Viesda D. Pithaloka. 2017. *Final report: Youth migration pattern in East Nusa Tenggara*. Bandung: AKATIGA.

Ernis, Devy. 2022. Tertarik Belajar Pertanian? Beasiswa S1 Bakti Tani 2022 Dibuka, Cek Syaratnya [Interested in studying agriculture? Bachelor scholarship for Bakti Tani 2022 opens, check the requirements]. *Tempo.co*, February 3. Accessed 10 June 2022. https://tekno.tempo.co/read/1556943/tertarik-belajar-pertanian-beasiswa-s1-bakti-tani-2022-dibuka-cek-syaratnya

Hasan, Rofiqi. 2010. Munas HKTI Kisruh, Muncul Desakan Forum Tandingan [Chaotic HKTI National conference, emerging pressure for counter-forum]. *Tempo.co*, July 13. Accessed 20 June 2022. https://nasional.tempo.co/read/263172/munas-hkti-kisruh-muncul-desakan-forumtandingan/full&view=ok.

Hidayat, Rafki. 2017. Menjawab Sindiran Jokowi: Mengapa Banyak Lulusan Pertanian Kerja di Bank? [Answering Jokowi's satire: Why do many agricultural graduates work in banks?]. *BBC.com*, September 7. Accessed 13 June 2022. https://www.bbc.com/indonesia/indonesia-41175869

Indonesian Institute of Sciences (LIPI). 2015. Crisis regenerasi petani: masalah serius di perdesaan [Farmers' regeneration crisis: A serious rural problem]. http://lipi.go.id/lipimedia/single/lipi:-krisis-regenerasi-petani-masalah-serius-di-perdesaan/10832.

International Fund for Agricultural Development (IFAD). 2018. *Design completion report: Youth Entrepreneurship and Employment Support Services (YESS)*. Jakarta: International Fund for Agricultural Development.

Li, Tania M. 2018. Intergenerational displacement in Indonesia's oil plantation zone. In *Gender and Generation in Southeast Asian Agrarian Transformations*, ed. C.M.Y. Park and B. White, 56–74. London: Routledge.

Nugraha, Yogaprasta, and Rina Herawati. 2015. Menguak realitas pemuda di sektor pertanian perdesaan [Unmasking the reality of youth in the rural agricultural sector]. *Journal Analisis Sosial* 19 (1): 27–40.

Petani Muda Keren. n.d. Petani Muda Keren! [Cool young farmers!]. *petanimudakeren.com*. Accessed 13 June 2022. https://www.petanimudakeren.com/

Ploegsma, N. D. 1936. Oorspronkelijkheid en economisch aspect van het dorp op Java en Madoera [Origins and economic aspects of the village in Java and Madura]. PhD dissertation, Leiden University.

Purnomohadi, N. 2000. Jakarta: "Urban agriculture as an alternative strategy to face the economic crisis". In *Growing cities, growing food: Urban agriculture on the policy agenda. A reader on urban agriculture*, ed. N. Bakker, M. Dubbeling, S. Gundel, U. Sabel-Koschella, and H. de Zeeuw, 453–466. Feldafing: German Foundation for International Development (DSE).

Purwanto, Semiarto Aji. 2010. Bertani di Kota, Berumah di Desa: Studi Kasus Pertanian Kota di Jakarta Timur [Farming in the city, living in the village: Case study of urban agriculture in East Jakarta]. PhD dissertation, University of Indonesia.

Rigg, Jonathan. 2019. *More than rural: Textures of Thailand Agrarian transformation*. Honolulu: University of Hawai'i Press.

Ruwiastuti, Maria, et al. 1997. *Penghancuran Hak Masyarakat Adat Atas Tanah* [The destruction of indigenous people's land rights]. Bandung: Konsorsium Pembaruan Agraria.

Saitya, Ida Ayu Grhamtika. 2021. Gender and the intergenerational transmission of Property in Rural Bali. MA research paper, International Institute of Social Studies.

Siregar, Masjidin. 2001. Petani pinggiran kota: Suatu alternatif atau masalah baru? [Periurban farmers: An alternative, or a new problem?]. *Bulletin Agro Ekonomi* I (4): 8–11.

———. 2006. Peri-urban vegetable farming in Jakarta. Proceedings of the International Workshop on Urban/Peri-Urban Agriculture in the Asian and Pacific Region. Tagaytay City, Philippines, May 22–26.

Suryana, Asep. 2006. *Menjadi Pinggiran Jakarta: Dinamika Sosial Petani Buah di Wilayah Pasar Minggu 1921–1966* [Becoming an urban fringe: Social dynamics of fruit farmers in Pasar Minggu, 1921–1966]. Research report, Indonesia Across Borders: The Reorganization of Indonesian Society 1930–1960 project. Jakarta: Indonesian Institute of Sciences and NIOD (Institute for War, Holocaust and Genocide Studies).

Suryowati, Estu. 2017. Guru besar jelaskan alasan lulusan IPB banyak kerja di bank [Professor explains why IPB graduates prefer to work in banks]. *Kompas.com*, September 7. Accessed 13 June 2022. https://nasional.kompas.com/read/2017/09/07/05490031/guru-besar-jelaskan-alasan-lulusan-ipb-banyak-kerja-di-bank?page=all

Ulum, Miftahul. 2017. JAPFA Foundation Buka Program Beasiswa SMK [JAPFA foundation opens scholarship for vocational school]. *Bisnis.com*, January 31. Accessed 10 June 2022. https://surabaya.bisnis.com/read/20170131/531/763679/japfa-foundation-buka-program-beasiswa-smk

Wati, Herlina, and C. Chazali. 2015. Sistem pertanian padi Indonesia dalam perspektif efisiensi sosial [Indonesian rice farming from a social efficiency perspective]. *Jurnal Analisis Sosial* 19 (1): 41–56.

White, Ben. 2018. Marx and Chayanov at the margins: Understanding agrarian change in Java. *The Journal of Peasant Studies* 45 (5–6): 1108–1126.

———. 2020. *Agriculture and the Generation Problem*, Agrarian Change and Peasant Studies Series. Rugby: Practical Action Publishing.

White, Ben, and Hanny Wijaya. 2021. What kind of labour regime is contract farming? Contracting and sharecropping in Java compared. *Journal of Agrarian Change* 22 (1): 19–35.

White, Ben, and Gunawan Wiradi. 1989. Agrarian and non-agrarian bases of inequality in nine Javanese villages. In *Agrarian Transformations: Local Processes and the State in Southeast Asia*, ed. G. Hart, A. Turton, and B. White, 266–302. Berkeley: University of California Press.

White, Ben, Colum Graham, and Laksmi Savitri. 2023. Agrarian movements and rural populism in Indonesia. *Journal of Agrarian Change* 23 (1): 68–84.

Wijaya, H., and B. White. 2019. The persistence, expansion and dynamics of sharecropping in a Javanese village. In *Presented at the ICAS Conference, Leiden*, July.

Wiyono, Suryo, Masbantar Sangadji, Muhammad Ulil Ahsan, and Said Abdullah. 2015. *Farmer regeneration in rice and horticulture farming families*. Jakarta: Koalisi Rakyat untuk Kedaulatan Pangan.

Wulandaru, Maula Paramitha. 2018. Duta Petani Muda [Young Agripreneurz Ambassadors]. *Agrifocus.com*, August. Accessed 13 June 2022. https://agriprofocus.com/duta-petani.

Yayasan Indonesia Cerah (CERAH). 2021. *Presentation on the national survey on climate change*. Jakarta: CERAH. www.cerah.or.id/id/detail-program/climate-change.

Open Access This chapter is licensed under the terms of the Creative Commons Attribution 4.0 International License (http://creativecommons.org/licenses/by/4.0/), which permits use, sharing, adaptation, distribution and reproduction in any medium or format, as long as you give appropriate credit to the original author(s) and the source, provide a link to the Creative Commons license and indicate if changes were made.

The images or other third party material in this chapter are included in the chapter's Creative Commons license, unless indicated otherwise in a credit line to the material. If material is not included in the chapter's Creative Commons license and your intended use is not permitted by statutory regulation or exceeds the permitted use, you will need to obtain permission directly from the copyright holder.

12

Young Farmers' Access to Land: Gendered Pathways into and Out of Farming in Nigara and Langkap (West Manggarai, Indonesia)

Charina Chazali, Aprilia Ambarwati, Roy Huijsmans, and Ben White

Introduction

This chapter describes rural young men and women's pathways out of and (back) into farming in two villages in West Manggarai district, Flores island, Eastern Indonesia. The chapter has seven sections. First, we describe the methodology and our sample in the two sites. The second section then provides the geographical and social context and describes livelihood patterns in the research villages. Next, we present illustrative cases of young people's pathways out of and (back) into farming, both young men and women, followed by an account of the tensions arising

C. Chazali (✉) • A. Ambarwati
AKATIGA Center for Social Analysis, Bandung, Indonesia
e-mail: charina@akatiga.org; apriliambarwati@akatiga.org

R. Huijsmans • B. White
International Institute of Social Studies, The Hague, The Netherlands
e-mail: huijsmans@iss.nl; white@iss.nl

© The Author(s) 2024
S. Srinivasan (ed.), *Becoming A Young Farmer*, Rethinking Rural,
https://doi.org/10.1007/978-3-031-15233-7_12

through the process of land transfer between generations in the fourth section. The fifth and sixth sections focus on young people's farming practices and how the government supports young farmers. In the final section, we reflect on how gender, generation, and class combine to shape the pathways of young farmers out of and (back) into farming and their access to agrarian resources.

In West Manggarai, social structures are organized patrilineally and inhabitants practise village exogamy; with very few exceptions, land is inherited only by sons or male relatives, and when women marry, they move to their husband's village. In this respect, the situation is more similar to the Indian case-study villages than to the other (Javanese) villages described in this book.

Methodology

This chapter is based on the authors' field research conducted during August and September 2017[1] in Nigara village (Lembor sub-district) and Langkap village (Mbeliling sub-district), West Manggarai (see Fig. 12.1).[2] Both are agricultural villages but differ in a number of important respects, making for a stimulating comparison. Langkap is the larger of the two locales. It is located relatively near the port of Labuan Bajo and due to its stunning views, Langkap has been targeted by tourism investment. Nigara village offers an interesting land settlement situation as it combines both lowland (rice producing) and upland cultivated areas (e.g., candlenuts).

Our data collection was based on qualitative techniques, and we interviewed 50 young farmers (32 young women and 18 young men). The life-history method inspired the semi-structured interviews, leading us to focus on key moments over the life course in the process of becoming a young farmer. For the duration of the fieldwork, the research team lived at the field sites. This enabled us to complement the interview-based

---

[1] Charina Chazali and Aprilia Ambarwati spent one month in the field sites and carried out all interviews, joined for shorter periods of time by Roy Huijsmans, Isono Sadoko, and Ben White. Hanny Wijaya provided valuable inputs to develop this chapter.

[2] The names and villages and respondents are pseudonyms.

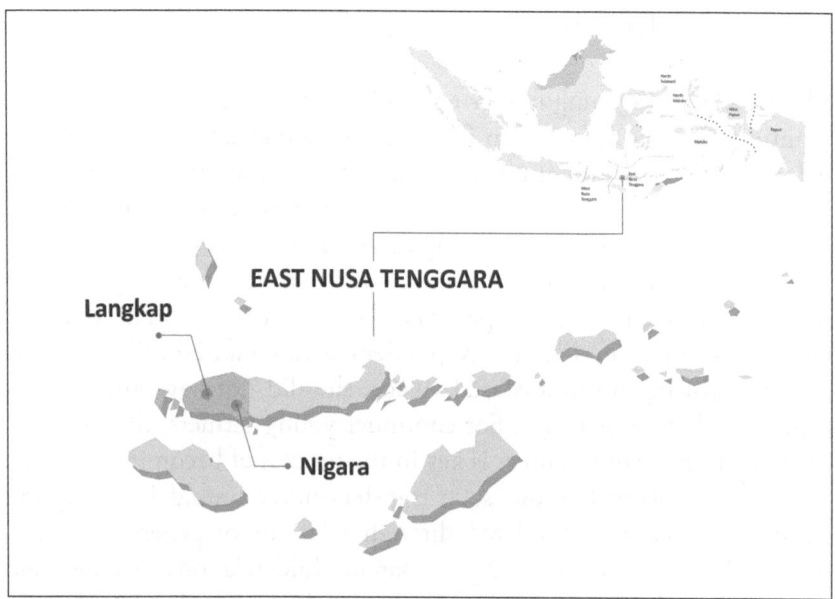

Fig. 12.1 The location of Nigara and Langkap (West Manggarai, Flores Island, East Nusa Tenggara)

method with participant observation by taking part in everyday activities; staying in the village also helped to generate rapport, which allowed us to discuss delicate topics such as land inheritance.

Across the two villages, we identified interviewees through snowballing. We were introduced to the first young farmer respondents with the help of a village official and an active member of the local farmers' group. These initial respondents then helped us to identify further respondents, and so on. We also interviewed several parents of our young farmer respondents to obtain information on intergenerational changes in farming practices and transfers of resources and farming knowledge.

In the case of married young farmer couples, we tried to interview both husband and wife. The depth of information obtained varies from case to case. While women work in almost all stages of farming, they do not own the land. Sometimes we encountered awkward situations when speaking to women about land inheritance as they were hesitant to talk about their husbands and in-laws.

Who Are the Young Farmers?

Almost all of our young farmer respondents, both men and women, are "continuers," who grew up in farming households and have experience helping with farm duties in their childhood (Monllor 2012; White 2018, 708). Some have a history of engaging in non-farm activities before returning to farming and inheriting agrarian resources while others have taken over the farming activities and resources from the older generation without any off-the-farm experience. Next to these young continuer farmers, we met one (female) respondent who could be called a "newcomer"; growing up in a coastal village, she did not have any farming experience before marriage. For continuer young farmers, the intergenerational transfer of resources is key in the process of becoming a farmer. As further explained below, such transfers never unfold in a uniform manner as they are negotiated through relations of generation (birth order relations between siblings or parent-child relations), gender, and class. Out of the 50 young farmers that we interviewed across the two villages, most are between the ages of 25 and 34 (see Table 12.1).

The majority of our respondents had only completed formal education until primary or lower secondary school level. However, the younger respondents had greater access to senior high schools due to government policy on education subsidies since 2016. The majority of the young farmers that we interviewed were married (see Table 12.2). Most of the unmarried respondents are in the 15–24 age group. However, some female respondents who are in this age category are already expecting to be married and some are promised as part of arranged marriages. Conversely, none of the men that we interviewed in this age category plan to marry in the near future.

Table 12.1 Age range of young farmer respondents

Sex	Age			
	15–24	25–34	35–40	Total
Female	7	18	7	32
Male	4	8	6	18

Source: Analysed by research team from primary data

Table 12.2 Marital status of young farmer respondents

Sex	Not married	Married	Total
Female	5	27	32
Male	5	13	18

Source: Analysed by research team from primary data

In terms of migration, only 10 of the 32 female respondents had a history of out-migration. They had worked in informal-sector jobs such as shop or food stall assistants or as domestic workers in other locations on the island (Labuan Bajo, Ruteng, and Ende) or on other islands (Makassar in Sulawesi, Surabaya in Java, Bali, and Kalimantan). In addition, very few of them had migrated to pursue an education; this is a privilege of wealthy families' children. The other 22 respondents had never moved away from the village—for school or for work. The reasons for not migrating, they mentioned were: (1) family responsibilities to dependent household members; (2) marriage—a woman will live in her husband's village; and (3) for some respondents from more remote upland areas of Nigara, the poor roads and transport facilities. In their spare time, most of our women respondents from this area still weave cloth to be sold in the nearest market or to middlemen.

Among our 18 male respondents, 7 had a migration history. Only one had left the village to pursue an education (university) in Makassar; the others were labour migrants. The latter travelled to other islands (Makassar and Pare-Pare in Sulawesi, West Papua, and Kalimantan) and worked in informal-sector jobs in restaurants, shops, as plantation labour, or in small-scale mining. Those who did not migrate generally worked as farm labourers after finishing school, work that they combined with non-farm income such as *ojek* (motorcycle taxi drivers) or in security.

Profile of the Fieldwork Sites

West Manggarai District is known nationally as an important agricultural region of eastern Indonesia, both for staple food crops (irrigated and rain-fed paddy as well as rain-fed maize and soya), horticultural crops, and tree crops. It is one of the more important rice-producing regions in this part of the country, and Lembor sub-district (where Nigara is located) is the biggest rice producer in the district.

Geographically, Nigara village is one of the largest rice producers in West Manggarai (on both irrigated and dry land) and also produces coffee, candlenut, and fruits from its dry land. Nigara's population is 1874 with 434 households (BPS 2018a). The distance between Nigara and the district capital of Labuan Bajo[3] is 79 kilometres and can be accessed by car or motorcycle.

Nigara has a unique settlement pattern (between upland and lowland neighbourhoods), which is related to gender and generation. Parents and the elderly generally live in the upland settlements with the younger generation in lowland settlements. Until recently, many children attended school up to the completion of elementary level (six years) in the upland locations where the Nigara State Primary School is located. Here, they live with their grandparents while their own (young) parents work the paddy land in the lowland part of the village. The national government has invested in maintaining and developing irrigation channels in the village since the 1960s. Some of the upland families migrated to the lowland, especially after irrigation massively improved yields between the late 1960s and the 1980s. Families who live in the hills for most of the year move down to the lowland areas for the seasons of peak activity in irrigated rice cultivation (usually for several days or a month). They then return to the hills for the dry land harvest season.

Nowadays, due in part to the construction of a new elementary school in the lowland area, upland families—especially those with school-age children—are starting to spend more time in the lowland part of the village.

In contrast, Mbeliling sub-district (where Langkap is located) is the district's most important producer of tree crops. Langkap inhabitants rely mainly on tree crops such as coconut, coffee, cacao, and candlenuts, with paddy, maize, and fruits also cultivated. With its spectacular upland scenery, it is also a destination for domestic and foreign tourists. Langkap has a population of 1105 (BPS 2018a). Langkap is less than two hours by car

---

[3] Labuan Bajo is the base for tourists visiting the famous Komodo monitors (Komodo dragons) on nearby Komodo island.

from Labuan Bajo and is famous as a tourist destination due to its upland scenery and traditional cultural performances. Some of our respondents have part-time jobs as *caci*[4] dance performers for tourists.

Table 12.3 shows the farm sizes among the sample. We define farm size as the total land area that the young farmer manages (whether inherited, gifted by parents, rented in, or share cropped). From the sample, the average and median land that the sample households in West Manggarai manage is slightly bigger than the government's official definition of a *gurem* (marginal) farmer, which refers to a landholder with less than 0.5 hectares (ha) (BPS 2018a).

The majority of sample households in the West Manggarai site own land (whether through inheritance of privately owned land or rights to customary land) as shown in Table 12.4. Among the 50 respondents, 80 per cent are pure "owner" operators, farming land that is either inherited from or given in trust by their parents or held as use-right on customary land. Meanwhile, 12 per cent of respondents manage their own land while also cultivating other people's land as sharecroppers. The sharecrop system means that two parts of the harvest are for the tenant and one part is for the landowner; the tenant's net share is two-thirds, but all of the

Table 12.3 Farm size range in the sample (hectare)

Site	Smallest	Largest	Average	Median
West Manggarai	0.06	2.40	0.70	0.56

Source: Analysed by research team from primary data

Table 12.4 Land access in the sample (percentage)

Site	Pure owner operator	Pure share tenant	Part own and share tenant/ rent	Pure rent	Combination rent/share tenant
West Manggarai	80	8	12	0	0

Source: Analysed by research team from primary data

---

[4] Sacred traditional arts with the concept of war dance. Then, it has developed into a special show for important guests as well as tourists to Flores traditional villages.

production costs, excluding harvest labour, are the tenant's responsibility. Usually, in cases of crop failure, the landowner and tenant will negotiate how to share the remaining yield, but the tenant is still responsible for the production costs. The remaining 8 per cent of young farmers are pure share tenants. It does not mean that they come from landless families, but at this stage in their life course, they have not yet obtained land from their parents, and therefore they have to manage other people's land. There is also a system of cash rent—paid in advance, unlike the share tenancy system—that only applies to rice fields. However, this system is rare, and our young farmer respondents cannot rent the land as they do not have the needed capital or cannot bear the risk of crop failure.

In the last decade, many families have acquired motorcycles as their everyday mode of transportation for travelling between upland and lowland areas of the village, taking children to school, and other activities. Those who do not have a motorcycle usually walk to their farms or use *angkot* (public transportation). Old and young women usually sell the family's agricultural products, including fruits, candlenuts, cacao, and vegetables, in traditional markets near the city.

In addition, we still find *julu* or *dodo* (exchange labour) systems, which are mostly practised by women. This exchange labour is mostly seen in rice planting, weeding on dry and wet land, and harvesting. It is done in small groups of three to five people composed of family members and close neighbours. Women of all ages also engage in wage labour for planting, weeding, and harvesting. Men are usually the ones who hoe the soil, plough with buffaloes or tractors, and spray pesticides or herbicides. To reduce labour costs, many landowners have shifted from manual weeding to herbicides.

Gendered Pathways into and Out of Farming

Becoming an independent farmer who has access to land and makes decisions at every stage of the farming cycle is a long process. We asked all our young respondents about their initial farming experiences. As continuers, they have experience in helping on their parents' land since childhood. At this stage, both boys and girls are taught to plant, pull the weeds, use the hoe, and even plough the land with a buffalo. Damar (male, age 38, Nigara) remembers the time that his older sister taught him how to plough the land, first guiding him on how to hold and tie a buffalo and then how to direct the animal. Apart from farming, all respondents were also assigned household chores such as fetching water, washing, cooking (for girls), collecting firewood, taking animals to graze or bathe, and cleaning the house. Meanwhile, the eldest daughters are also expected to take care of younger siblings.

When helping on the land as children, parents sometimes offered pocket money to increase motivation. Sesil (female, age 56, Nigara) reminisced: "In my childhood, I used to eat from our garden like cassava, sweet potatoes, vegetables, and corn. No rice like nowadays. But my son (now 34 years old) had started eating rice since childhood. If my son helped me on the land, I gave him rice mixed with cassava and he was happy."

Farming at a young age is a fun memory if done together with peers. Sika (female, age 21, Langkap) started to help her parents in the field when she was eight years old. At school, she had a group of eight friends, both boys and girls, who reminded each other to help on their parents' land. "When school was over, we would rush to the rice field. When we were tired, we would eat and rest. If I hurt my hand, we would take some leaves, chew them, and put it on my wound. When our parents came home in the afternoon, we looked for firewood in the forest. We would sing loudly, maybe the villagers could hear it. But I don't see children doing this nowadays."

Children's work changed when schooling began to take more of their time. Girls in the two research villages are expected to focus on farm work rather than spend additional time on their education. For many of the young women farmers that we interviewed, their involvement with farming started at an early age. Our interview materials show that during childhood and youth, girls and young women were more actively involved in farming than boys and young men. One of the reasons for this is that in many families, sons are prioritized over daughters when the decision is made to invest in a child's education. This was Jeni's (female, age 33, Langkap) experience; she is the youngest of five siblings with two elder brothers and two elder sisters. Jeni began to help hand-pound paddy, weeding, and harvest using a sickle when she was nine years old. Meanwhile, her elder brothers were busy with school. After finishing junior secondary school, Jeni's father forbade her from continuing her education at the upper-secondary level because "as a daughter, you will be married out of the family," implying that her education was not a practical investment for the family.

Jeni continued: "We daughters cried when we had to stop school. My elder sister only completed primary school. At that time, my father sold a buffalo to pay for our brothers to continue to college, but my brothers refused to study further. And my father was disappointed [but still did not allow the girls to continue their education in place of their brothers]."

Parents expect their daughters to stay at home and help with the crops and the housework, rather than continuing their education, until they marry and move away from home. Sometimes they are also expected to support the education of male siblings through farm work or labour outside of agriculture. Migration for young people, both male and female, can be a means of gaining some freedom and experience and a way of gaining autonomy for themselves, for example, earning their own money until they return to the village.

Viska (female, age 34, Nigara) was unemployed after graduating from high school. She did not want to help her parents on the land and followed her cousin to Makassar to work as a house maid. Her father supported her because he needed money to support her two older brothers' college fees. She remembers that her older sisters also worked on her parents' land to support her brothers. Her salary in Makassar was IDR 500,000 (USD 35) per month. At first, she sent money to her parents, after that she kept all of her salary for her own needs; she says that at least she was no longer a burden on her parents. She was very happy to work and earn her own money. Three years after she left the village, she met her husband and married. They then moved to her husband's village where he would obtain land to farm.

In the process of becoming a farmer, young people gradually shift from only helping their parents, to working on other people's land, until they can finally manage and take decisions on their own farm. Nonetheless, this process does not occur uniformly or in a simple and linear progression following their increasing age. We found that some young people, still teenagers, were already making farming decisions on their parents' land, including choosing which crops to plant. They shared the crop with their families and used their parents' capital. Belen is an example of a young teenager who earns wages as a farm labourer on par with adults and makes decisions about crop choice in the family's home garden.

Belen (female, age 14, Langkap) is in the second grade of junior high school. She lives with her mother and an older cousin who is a builder. Her late father bequeathed a small piece of dry land to Belen's mother. Her mother is always busy working as a wage labourer because they do not have a rice field and the candlenuts on the dry land do not provide enough money and rice for the family to survive. Belen often helps her mother as a wage labourer, planting rice and picking candlenuts. Belen is paid the same amount as adult women—IDR 30,000 (USD 2) for half day. She uses some of the money to buy books and snacks before giving the rest to her mother.

In her mother's yard, Belen plants vegetables such as tomatoes, egg-plant, and shallots. She says: "This was mom's idea to make it easy for me to cook before going to school, but mom was too busy working, so I immediately tried planting myself. I asked for seeds from my aunt, I gave them fertilizer and watered them. Mom teaches me, but I do it all by myself." Impressed by her initiative, her mother asked her to start plant-ing maize in the yard. Belen instead chose *daun ubi* (yam leaves) because they can be eaten as a side dish and she can feed them to the pigs. Despite her success with her home garden and her work as a labourer, Belen wants to continue her education until high school, but she does not want to work far from her home. "I don't know what it will be like in the future. I want to have my own money, but if I have to work far away from home, I prefer to work in the village. But if I'm only working in our yard, maybe I will look for jobs in town."

Belen was able to decide what crops she wanted to plant and be respon-sible for cultivating the home garden, but she still has the desire to work outside the village. Meanwhile, young men have different options, espe-cially those from relatively prosperous families. They have more flexibility to experiment on parental land and using parental money, such as decid-ing on a new crop that requires high maintenance.

Adi (male, age 30, Langkap) has been cultivating the land with a plough and buffalo, and with a hoe, carrying candlenuts, and planting rice since he was in the fourth grade. He did not continue his education after graduating from junior high school at the age of 16. As the only son who will inherit the land—both rice fields and dry land—he wanted to start farming more seriously. He tried to plant cacao because he knew of many middlemen offering a better price for this product at that time. He planted cacao in two locations on his father's land—60 cacao trees in the first plot, and a year later, 100 trees in the second plot. He asked his neighbour for cacao seeds and he bought equipment from neighbours and the store. "I made the *koker* (polybags for the seedlings) just by looking at other farmers." At that time, he got the money by selling some of his parents' rice (before this, he used to sell some of his parents' rice to buy cigarettes and for his own needs). Many of his neighbours do not own rice fields, so it is not difficult to find buyers. After several years, pests have damaged many of the cacao trees, but Adi is still able to harvest from the remaining trees for his own needs.

Marriage is a key moment for young women as it determines where, how, and with whom they will live and farm. In this virilocal marriage and patrilineal inheritance system,[5] as we have explained above, married women work on land that their husbands own/rent/sharecrop. Vroni (female, age 21, Langkap) received a marriage proposal from a man in another village and they will marry in a few months. As a woman, Vroni knows that she will not inherit land; her own family's land will go to her brothers. She accepted her fiancé's proposal because he stands to inherit some mixed-garden land. He is 25 years old and works as a construction labourer. After she marries, Vroni knows that she will be expected to help her parents-in-law on the farm, although she still dreams of finding a job that pays better in Labuan Bajo. "If I'm not yet married, I want to work and earn money in Labuan Bajo. But if I'm married, what can I do? I will have to follow my husband. And certainly, if he's often working in construction, I will be the one who has to help on the in-law's farm."

---

[5] Social practice of newly married couples taking up residence with or near the husband's family, and an inheritance practice running through the male line.

After marrying, moving to the husband's village, working and managing the husband's or in-laws' land, young women are usually immediately entrusted to manage their husband's land; the men will normally be occupied earning non-farm income as construction workers or ojek driver. Jeni feels that she gained confidence and more space to manage the land after proving to her husband that farming is not an easy job.

Jeni married when she was 20 years old and moved in with her husband's family. It was not until one year into their marriage that Molana, her husband, was given some of the parental rice fields to farm. The young couple moved into their own house, which Molana's parents helped them to build. Molana now owns two *bujur* (one bujur is 25 × 25 metres = 625 m^2) of rain-fed rice fields and a little mixed-garden land planted with coffee trees, banana trees, pineapple plants, and a few candlenut trees. He sometimes helps on the farm but only when there is no work for him as a construction worker. Jeni recalls that in the beginning of their marriage, Molana underestimated her and was a little bossy, even telling her that he would teach her to farm. When they planted the clove seeds together, her husband said that growing cloves was easy, and Jeni could just wait until the cloves could be harvested. Jeni did just that and deliberately did not take care of the clove trees that her husband had planted. "I saw it as a competition, I do not want to take care of my husband's trees, I do not pull out the grass around them, I do not cut the rotten branches. I only took care of my trees. Now my clove trees are tall and can be harvested, while my husband's trees are short. I just want my husband not to talk carelessly and understand that we need to take care of our crops." Jeni feels that she increasingly has autonomy to make farming decisions such as planting, choosing the crop varieties, marketing, and even managing income from the harvest, because she has proved to her husband that farming is not an easy job.

In contrast to respondents who grew up in farming families and could earn money as wage labourers since childhood, the one newcomer farmer in our case-study villages struggled for years to obtain trust from other people to be hired as a wage labourer. Leti (female, age 40, Nigara) was born in a coastal area in Maumere, East Nusa Tenggara, and had no farming experience prior to her arrival in the village. Since she was five years old, she fished with her father at sea. After graduating from high school, her sister invited her to move to Surabaya, Java to work in a shoe factory. Leti met her husband in Surabaya and married at the age of 27. After her marriage, she continued to work in Surabaya until her father-in-law fell sick and asked her husband to return to the village to work the land. They moved to the village, but her husband soon returned to Surabaya to work in a furniture factory, and at age 33, Leti is now responsible for her husband's rice fields. She felt overwhelmed, but her husband's aunt taught Leti about rice and vegetable cultivation, and she followed the older woman's example. She recalls:

> At first, my husband's family laughed a lot because I often pulled out weeds very slowly, even slower than the children here. It took three years for me working the land before finally someone asked me to work as a wage labourer. At that time, I got IDR 30,000 (USD 2), just the same as the other adult women here. When I work on other people's land, I have never been scolded publicly, but I was told at home by my sister-in-law that I need to work faster on other people's land.

Access to Land

Lack of access to arable land is the main problem that young people face in becoming farmers, even if they come from families that own land and even if they expect to inherit land in the future. This access is not only an issue of land availability, but also influenced by their position as young people who the older generation do not yet trust to fully manage the land. For male respondents, becoming an independent farmer offers the prospect of earning some money, but when still working on their parents' land, their involvement in farming impedes their earning money needed to finance their aspirations. As the case of Fian and his mother Theresia illustrates, this easily leads to tensions and frustrations.

Fian (male, single, age 19, Nigara) earns money as a casual labourer. His daily activities are casual wage work on others' rice farms, helping on his mother's farm, and occasional construction work. His mother, Theresia (female, age 48, Nigara), owns half a hectare of land inherited from her deceased husband, which is irrigated rice land laid out in two blocks. She works this one one-quarter hectare block with help from Fian and his elder brother Tomi (male, age 27, Nigara), while the other block is rented to a relative for IDR 500,000 (USD 35) per cropping season. Fian says: "it's only been rented out for one season; I don't know if it will be rented out next season or not. It's my mother who will decide and let me and my brother know." Fian feels that if he could farm his mother's land, it would bring in more than the rent it currently yields. The other block usually yields seven to eight *gabah* (sacks of grain) each harvest. Fian cannot explain how the money is used because his mother makes all of those decisions. He and his brother participate in all of the stages of work, without pay, as the harvest is in his mother's hands.

> My mom said there was no one to help her, she said Tomi was too lazy and I was always busy earning wages on other farms. So that's why she rented the land out to a relative. It's true, I prefer to work outside as I get IDR 60,000 (USD 4) and a pack of cigarettes for a day's work hoeing on other's land. And I also get paid to help applying pesticides or fertilizers. I can always get work, especially in the rice planting season. I would also like to work on our own rice farm. If we did that, we could earn much more than what we get from the rent. But Tomi and I are just helping our mother. If we need to buy fertilizer or pesticides, I'm the one who goes to the market because I know what to buy. But the one who takes paddy to the rice mills or goes to the rice miller for a loan, is always mother. If it's about money, *Mamak* (mom) doesn't trust us (laughing).

Fian dreams of becoming a share tenant on another's land but share tenancies on rice land are hard to find, and most landowners prefer to rent out their land for cash. "If I have to rent the land, how can I get the money to pay in advance? As a share tenant [paying the rent as a share of the harvest], I think I could manage. But for the moment it's best to work for a daily wage. If I'm sick or sleepy or I want to have fun, I can rest. But as a share tenant, I would have to work even harder."

Lack of land access makes young people aspire to work and earn a living outside the village and hope that one day they can save enough money. Young people tend to not have space to raise their concerns regarding access to land, either in their family or at the village level. Both

Nigara and Langkap have extensive *tanah adat* or *tanah ulayat* (customary, community-owned lands), including irrigated rice fields, dry land, and residential land. Apart from the inheritance of privately owned land, customary land could potentially be allocated to young people who aspire to be farmers.

In contrast to our research in Java where village-owned land (*tanah kas desa*) is offered on a temporary use basis, in our West Manggarai villages, customary land is assigned to individuals as a form of ownership right. Women cannot hold these rights, with the exception of widows who, in some cases, can retain land allocated to their husbands.

In Nigara, there are 25 hectares of customary land in the upland hamlet that are in the process of being sold to a company. This dry land has never been used for any purpose and is supposed to be distributed to the local people. The customary leader and village elders made all of the decisions regarding its sale. Dalis (male, single, age 25, Nigara) is a returned migrant who previously worked on an oil palm plantation in East Kalimantan. In the village, he helps his parents on their land, works as a wage labourer, and sometimes as an ojek driver. He and his friends did not want the customary land to be sold, but on the other hand, people said that the land would be too small to be distributed to everyone in that hamlet. When the sale was finalized, Dalis received money as compensation—IDR 500,000 was given to each young adult and adult male villager.[6] He used the money for clothing and gasoline for his motorcycle. His father owns 2 hectares of dry land, and as a son, Dalis is quite sure that he will inherit land. However, he has three male siblings, and he wonders? how his parents' small area of dry land will be divided among four boys. Therefore, he decided to migrate again to East Kalimantan to reengage as a palm oil worker. He says: "With work like this now, (I) will not be able to buy land. Working in the village is like being *setengah mati* (half dead). You can work for a day, (and then) you cannot find work in two weeks."

In both villages, besides being ineligible for customary land allocations, women generally cannot inherit land from their parents, even

---

[6] The distribution of money is only given to men, both married and unmarried, adult and young adult.

though they are much more engaged in farm work than men. As we have described, the established practice is that upon marriage, women leave their natal household and move to the husband's village where the wife manages the husband or father-in-law's land. Women who are married to men with land can manage the farm, but do not have a say in how the land will be distributed to the younger generation.

Even though Jeni manages the farm on her husband's land, she cannot decide how the land will be distributed to her children in the future. She has two daughters, but Jeni and her husband want to have another child. "If I have a son in the future, the land must be given to my boy. Later my daughters can have land from their husbands, just like me (while laughing). But actually, this is not my land, it's up to my husband because it is his land."

A daughter can inherit land if her father and brothers agree to the allocation. Usually, the land given is dry land or a piece of land for houses—not a productive rice farm. If a father gives part of his land to his daughter, her brothers or her nephews can ask for the land to be returned when the father dies. One way to secure the land is through the use of a legal agreement, which the father, the brother(s), and the village head must agree to and sign. The daughter also needs to finance a small traditional ceremony to commemorate the land handover. This strategy is rare in our case-study villages. We found, for example, only one female respondent who could obtain this entitlement because she was born into an elite village family—her father is the hamlet's customary leader—and he has plenty of land to distribute to his children.

Meanwhile, for young people, especially those from landless families, it is impossible to buy land with their income as wage labourers; land prices have been increasing rapidly in the area due to tourism. In 2011, a plot of less than 1 hectare of dry land with 30 candlenut trees was sold for IDR 500,000,000 (more than USD 40,000). Prices have continued to rise.

Sita (female, age 25, Nigara) and her husband Beni (male, age 25, Nigara) are a landless couple. Sita does not own any land because she did not receive an inheritance, and her brother did not ask her to manage any land. Meanwhile, Beni did not inherit any land because his father sold all of his land due to chronic debt. Sita only completed elementary school and then migrated to the city to work as a shopkeeper and housemaid. The couple married when Sita was 21 and the couple returned to Beni's village to work as casual daily labourers.

After their return, the young couple won the trust of a landowner and became share tenants of a 1250 m² rice farm. From the 15 sacks of harvested paddy (15 kilogrammes per sack), three sacks are given to the harvest labourers, four sacks are paid to the landowner, and eight sacks remain for Sita and Beni. Meanwhile, the couple bear all of the costs for managing the land. From their wages, they purchased a pig and intend to breed pigs as another source of income. Sita also grows vegetables in their yard to reduce food costs.

Unfortunately, after three seasons (around 18 months), the owner ended their agreement as he wanted to manage the land himself. Sita says: "if (the land) is taken back by the owner, we cannot do anything, we have to return it." She hopes that one day she can become a tenant again. The couple was once offered the opportunity to rent a piece of land but did not have the cash to pay the rent in advance. Since this upheaval, Beni has been spending more and more time working as a farm labourer and construction worker when he can find opportunities. Sita must limit her daily work because she is pregnant with their second child.

Lack of Government Support

Poktan (kelompok tani), the government-sponsored farmer groups in the research villages, were created to help solve farmers' problems and functioning to channels the young (or older) farmers' aspirations to solve common agricultural problems. The poktan is a government institution and therefore focuses on offering government-proffered guidance. On paper, every farmer (who owns farmland or manages a farm) is a member of the farmer group. In practice, the poktan group's activities are limited to coordinating the provision of seeds and agricultural tools, especially in cases where the group leader is active and close to the government. For instance, even with the most basic problems like the provision of subsidized fertilizers—poktan cannot function as they were intended. These supplies are supposed to be distributed to each region based on a regularly updated database of farmer groups and farm sizes but the database is unreliable due to the manipulation of government data regarding subsidized fertilizer demand.

Umang (male, age 29, Langkap) is member of a poktan but is unaware of its activities. "Even though there is a farmer group, I still have to go to the city (Labuan Bajo) to buy fertilizer. It is difficult to find fertilizer here (in the village). Meanwhile, if I buy fertilizer from a different area, I have to provide my identity card, so I used my relative's identity card who lives in the area."

Likewise, the majority of young female respondents have never participated in poktan activities other than receiving seed assistance for crops like vegetables, fruits, and cacao. They only became aware of this organization when seeds arrive in the village and the poktan leader invites the women to his house or the village office to collect them.

At poktan meetings, the role of young female farmers is also limited to providing refreshments. Olin (female, age 25, Nigara) recalls being suddenly summoned by the poktan leader one afternoon. While the men (mostly old men) discussed the poktan plan and activities, Olin was asked to join several other young women in the kitchen and take care of the food and drinks for the men. In poktan meetings, the women never joined in the conversation or had the chance to share information about their farming problems.

Although they are much more involved and invested in farming activities than men, young women farmers' work and commitment has not received the poktan leaders or the government's attention. One poktan in Nigara received government assistance in the form of onion seeds, fertilizers, and pesticides. The district agricultural officer asked the poktan members to plant the onions and report on their progress. One member offered a part of his land as a *lahan percontohan* (demonstration plot). Young women, though, were the ones who planted, watered, and sprayed the test plot. Olin and Sita are among those who received daily wages for its care but neither know about the yield or how the income from the plot will be divided.

In addition, *petugas penyuluh lapangan* (government-employed agricultural extension worker) is often too busy with administrative matters to try to understand farmers' daily problems. Farmers need to discuss and share farming techniques, including pest eradication options. Unfortunately, middlemen and input providers who profit from the sale of pesticides have subsumed this role. As a result, farming practices (especially for rice crops) are becoming increasingly unfriendly to the environment.

Hedi (male, age 38, Langkap) always asks middlemen if he needs advice on pest eradication or plant diseases:

> I have never received farming assistance from the government. I heard that there are field extension workers, but I have never met them. So, I trust the middleman. If he recommends some brands for particular pests, I just buy them. I can also borrow money from him, but there will be interest. If I pay in three months, the interest is only 5 per cent for all. Later, the interest will increase if you borrow the money longer... we work (as a farmer) like a fighter.

Farming Practices and Dependence on Debt

As explained earlier, West Manggarai is one of the biggest rice producers in eastern Indonesia. For decades, there has been a substantial increase in the use of industrial fertilizers and other agricultural inputs to increase productivity. Chronic pest infestation is an acute problem and contributes to make farming unattractive to the younger generation. Coupled with the lack of support from petugas penyuluh lapangan (field extension workers) and an increasing reliance on inputs suppliers, farmers are facing increasing pressure to spend more on costly industrial inputs and entering chronic debt in much greater numbers than the previous generation.

The changes in farming practices include the use of herbicides that replace exchange labour, massive fertilizer usage, and the excessive use of pesticides that are no longer effective in eradicating pests. All respondents, except those from prosperous families, take loans from rice millers to finance the production process, and sometimes for their daily needs.

Damar (male, age 37, Nigara) started farming after his marriage at age 18. At that time, his father already used pesticides, especially for *wereng batang coklat* (brown planthopper, *Nilaparvata lugens*). However, since the 2000s, the pest infestations were no longer responding to eradication attempts. He says: "Pests are getting worse and ruining my land. I use pesticides more often than before... I have to spray pesticides eight times per season (one season on a rice farm is equal to four months)."

Currently, he borrows his working capital from the rice mill owner. He purchases his farming inputs and equipment—fertilizer, pesticides, sacks, and farming tools—from the same source. Damar will sell harvested paddy to the rice miller, taking only one sack of rice home; he will borrow rice from the rice miller if supplies at home run out. The following season's harvest will pay off that season's debts.

Those who can borrow from middlemen or rice millers are farmers who own or manage land. For Sita, who is a landless young woman, she only dares to borrow a small amount of money and rice from her neighbours. To pay the debt, she will cook if a neighbour has an event or celebration.

In our research, we also discovered an intervillage farmer alliance that a national environmental non-governmental organization (NGO) supports. This alliance aims to bring back locally grown foods and reduce chemical inputs. The farmer alliance encourages planting different crops to reduce the villagers' dependence on rice as a staple food. The alliance's members are largely men, young and old, who are active in the village government programme. Suitable for dry land and with relatively low maintenance needs when compared to rice, sorghum is a source of carbohydrates that the older generation grew and consumed before the government introduced rice massively in the late 1960s and the 1980s. Since 2013, farmer alliance members have planted sorghum in their yards and on dry land. It is consumed by its members, sold to other farmers, and the NGO also helps the men sell their product to urban consumers.

At the end of 2017, this alliance planted sorghum on 10 hectares of dry land offered by older male farmers who own relatively large amounts of land. Young male farmers tend to be spokespersons, communicating with NGOs and the district government throughout the programme and sometimes taking care of the crops. Both young and old women farmers in the village are involved as wage labourers to take care of the sorghum. In 2018, this farmer alliance invited local government representatives to an event to promote the successful yield and the initiative to support locally grown food. The attendees consumed sorghum together and a large part of the harvest was sold to urban dwellers with the NGO's support.

Concluding Reflections

Through our in-depth interviews with 50 young women and men farmers, we have explored the pathways by which young people move (back) into and out of farming in two villages in West Manggarai. Land is the most important resource for farming, and we have shown how gender, class, and generation affect access to land. Even though young women farmers are engaged in almost all stages of farming, customary law denies them the opportunity to inherit land, except in exceptional circumstances. Therefore, young women farmers can only farm their husband or male relatives' land. The role of the father and the brother and the expectation for the daughter to be kind and take care of her brothers determines each family's land distribution decisions.

This study has also revealed two directions of change in young people's farming practices. On rice farms, their farming practices are far from environmentally friendly, even worse than the previous generation. The positive initiatives include an NGO-supported farmer alliance that promotes locally grown food and aims to reduce locals' dependence on rice as a staple food.

Our research found no government-initiated activities aimed at engaging young people interested in farming futures in this region. Parallel to this, there are no active youth organizations that could exert some pressure on the local government to provide support for young people who aspire to be farmer. Older men and landowners still dominate these groups and their decision-making, while young women have limited participation. This condition raises questions as to why there is little evidence of young people exercising collective agency.

References

Badan Pusat Statistik (BPS). 2018a. Lembor dalam Angka 2018 [Lembor in numbers 2018]. Manggarai Barat. https://manggaraibaratkab.bps.go.id/publication/2018/09/26/2a7c5d23d62e191747ba8b88/kecamatan-lembor-dalam-angka2018.html.

Monllor, N. 2012. *Farm entry: A comparative analysis of young farmers, their pathways, attitudes, and practices in Ontario (Canada) and Catalunya (Spain).* http://www.accesstoland.eu/IMG/pdf/monllor_farm_entry_report.

White, B. 2018. Marx and Chayanov at the margins: Understanding agrarian change in Java. *Journal of Peasant Studies* 45 (5–6): 1108–1126.

13

The Long Road to Becoming a Farmer in Kebumen, Central Java, Indonesia

Aprilia Ambarwati and Charina Chazali

Introduction

This chapter presents the trajectories of young men and women into farming and is based on the authors' research in two villages in Kebumen Regency, Central Java. In our study, we have prioritized the perspectives of young people themselves in respect of their own experiences and pathways as farmers. The chapter consists of four sections. First, we describe the background of the selected location as well as the methodology and techniques of field data collection. The second section provides a picture of the economy, society, and agrarian structure in the two research locations. The third section describes various dimensions of young people's trajectories into farming and how they respond to the various obstacles and challenges that they face, using the respondents' own words where possible. The last section offers some conclusions and reflections.

A. Ambarwati (✉) • C. Chazali
AKATIGA Center for Social Analysis, Bandung, Indonesia
e-mail: apriliambarwati@akatiga.org; charina@akatiga.org

© The Author(s) 2024   **361**
S. Srinivasan (ed.), *Becoming A Young Farmer*, Rethinking Rural,
https://doi.org/10.1007/978-3-031-15233-7_13

Research Context and Methodology

As explained above, this chapter is based on our July 2017 field research in two villages in Kebumen Regency, which we call Sidosari and Pudak Mekar (see Fig. 13.1).[1] In selecting the two research locations, we aimed to ensure contrasts in patterns of access to land, types of farming, diversity of livelihood sources, and topology.

Sidosari village was selected as a lowland village, located near the Kebumen Regency capital (approximately 10 kilometres away). The villagers' main crop is irrigated rice. Technical irrigation channels water the fields, enabling two rice crops per year; in the third (dry) season, rice is planted with green beans with minimum treatment—the seeds are sown and waiting for harvest.

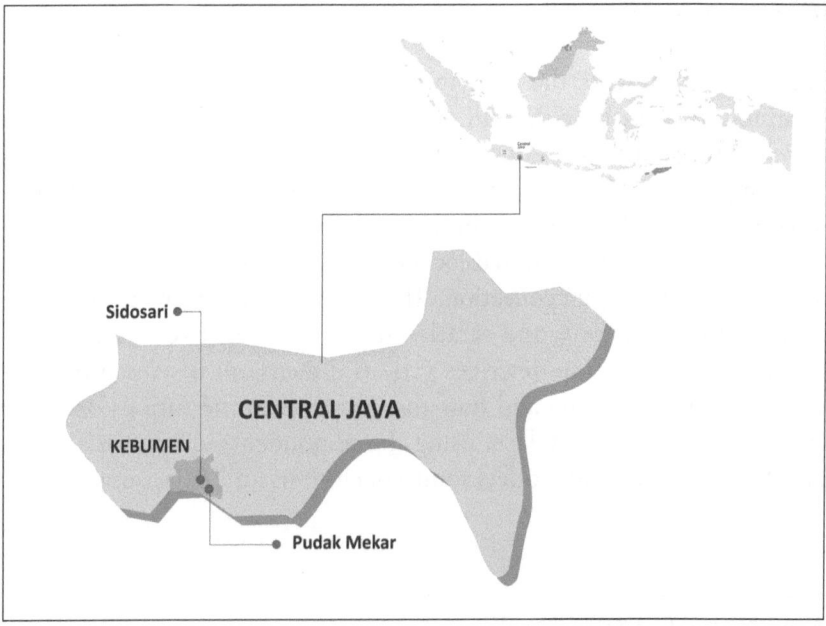

Fig. 13.1 The location of Sidosari and Pudak Mekar (Kebumen Regency, Central Java)

---

[1] The names of villages and respondents are pseudonyms.

Pudak Mekar village, in contrast, is a coastal village further from the city centre (approximately 20 kilometres). Most of the villagers are dryland farmers; others are fishermen. The farmland is dry, with sandy soil, and difficult to irrigate. Rainfed rice can be planted only once a year in the rainy season. Other common crops are vegetables and fruits.

The study employed mainly qualitative methods, including in-depth, life-history interviews with 29 young farmers—11 females and 28 males aged 17–43 years—with an average age of 32 years. Where possible, we also interviewed the spouses of our male and female respondents.

We identified our respondents in both research villages in two stages. First, a staff member of a local civil society organization (CSO) and village officers provided us with a list of potential young farmer respondents. This effort produced names of mostly male respondents. We then used the snowballing technique to identify young female farmers and modify the list, replacing several of the listed males with female respondents. Aside from the interviews with young farmers, we spoke to various key informants, including the parents of young farmers, village officials, and staff of the Kebumen Regency Department of Agriculture.

During our field research, we stayed in each village for 10 days. This allowed us to observe everyday activities and interactions and to build rapport and help the local people to feel comfortable in our presence. In 2014, other AKATIGA researchers had conducted research in Sidosari for a food sovereignty project. In 2019, one of this chapter's authors returned to Sidosari for a different research project and was able to update some of the information on young farmers in the villages.

Local Context and Agrarian Structure

Kebumen is regarded as a "food-basket" regency with approximately 83,000 hectares (ha) of rice fields and producing surplus rice for other districts in Central Java (BPS 2019). In Sidosari, according to the Sidosari village profile for 2017, the area of irrigated rice fields is 98 ha with 44 ha of dry fields. In the coastal village of Pudak Mekar, in contrast, there are no irrigated rice fields and the area of dry fields is about 66 ha.

There exist two types of land ownership in the villages: individual and village-owned land (commonly called village treasury land). The village-owned land generally has two functions: "prosperity land" and "*bengkok* land." Bengkok land is allocated to village officials in lieu of salary and as a retirement benefit after they have left office. The area of bengkok land allocated to each village official varies depending on the area of land that the village owns as well as each village's policy on its allocation. On average, the village head is allocated around 4–5 hectares of bengkok land, the village secretary gets 1.5–3 hectares, and other officials get less than 1 hectare.

Village treasury land can also be designated as "prosperity land." Each village has its own policies for its management. This land is usually rented to the villagers and the village officials manage the rental income as *PA Desa* or "village own source revenue"; it can be used for operational and overhead expenses needed for village administration. Pudak Mekar village does not possess prosperity land; all of the village land has been designated as bengkok land due to the limited area of village-owned land—only 9.2 ha compared to 19.7 ha in Sidosari (see Table 13.1).

In Sidosari, 7 ha of the total 19.7 ha of village-owned land is rented to villagers on a rotating basis, divided into 77 land plots with an area of about 38–77 *ubins*[2] or approximately 100–500 m² for each plot. For the last decade, the village government has rented out this land at a lower price than the current market rate. In 2019, the rent price of prosperity land for two growing seasons ranged from US$120 to US$180 per 100 ubin, depending on location and the quality of available irrigation.

Table 13.1 Village treasury land in Sidosari and Pudak Mekar

Village Treasury Land	Sidosari (hectares)	Pudak Mekar (hectares)
Bengkok land	12.7	9.2
Prosperity land	7	–
Total	19.7	9.2

Source: Sidosari and Pudak Mekar village profiles, 2017

[2] 1 ubin is equal to 14 m².

With this rotating rent system, every poor/marginalized household has a fair chance to rent the land. Registration for land rental is opened annually, and those who have not had rented land in the past are prioritized in the next rental period. Only poor households may register. This policy has helped to provide agricultural land access for the landless (Ambarwati et al. 2016). Landless young couples appreciate this opportunity as an affordable way to access agricultural land.

Among the 77 prosperity land plots, there is one plot called the "youth farm," which is 77 ubin or around 1000 m². The Sidosari village government uses the rental income from this plot (about US$90 annually) to fund Karang Taruna, the local youth organization. Previously, the youth community never cultivated this land. In early 2019, the village government allocated a new smaller farm plot of 56 ubin (almost 800 m²) with number one quality land—closest to irrigation—for the youth community to manage. This land is farmed collectively by a number of young couples. They manage the revenue earned from the harvest as collective petty cash. As a collective farming activity on a small plot of land, this activity does not provide significant extra income to youth community members. However, this young people's farming initiative has gained the trust of village officials, and the local government recently—two years— provided a new, more fertile plot for the youth community to manage.

Election candidates for village head in Sidosari and Pudak Mekar use the bengkok land as a political tool; the newly elected village head always gives the use rights of some bengkok land to his supporters. In Sidosari, the village head promised that if he were elected, he would give some of the bengkok lands (use rights) to every hamlet head and *musholla* (little mosque) in order to make revenue available for operational activities. In Pudak Mekar, the village head used his bengkok land as a form of reward for his campaign team and local supporters. These use rights for bengkok land are doled out for the village head's six-year term of office.

In both research villages, land control (access and ownership) among our young farmer sample is less than 2800 m² on average (median of 2100 m²). The young farmers accessed their land through a variety of channels: inheritance, a gift from living parents, purchase, cash rental, mortgage, and share tenancy. As shown in Table 13.2, the great majority of the 29 respondents farmed a combination of owned and rented or

Table 13.2 Land access and land ownership of 29 respondents

Land access and land ownership	Percentage
Pure owner operator	21
Pure share-tenant	0
Part own and share-tenant/rent	66
Pure rent	10
Combination rent/share-tenant	3
Total	100

Source: Becoming a Young Farmer survey, 2017

Table 13.3 Education and marital status of young farmer respondents

	Education level (per cent)			Marital status (per cent)	
Respondents	Primary school	Junior high	Senior high	Married	Single
Female	27	55	18	100	–
Male	33	61	6	56	44

Source: Becoming a Young Farmer survey, 2017

sharecropped land; only 21 per cent were pure owner—owned all of the land they worked—10 per cent were pure tenants who paid rent in cash, with the remaining 3 per cent pure tenants combining cash rental and share tenancy. Among those who owned some or all of their land, the most common road to ownership in both villages was a *hibah* (gift) of land from still-living parents to their children.

In our young farmer sample, there were no significant inequalities in farm sizes or socio-economic conditions. As shown in Table 13.3, the majority of male and female respondents are junior high school graduates, while only 18 per cent of young female farmers and 6 per cent of young male farmers are senior high school graduates.

Table 13.3 also shows that all of our female respondents are married, while many (44 per cent) of the male respondents are single. Large numbers of both young women and young men migrate for work after leaving school. After marriage, most women remain in the village and involve themselves in farming while many married men continue to migrate.

Young People's Pathways into Farming

The majority of young farmers in both villages, as in the rest of Java, are "successor" farmers or, as Monllor (2012) calls them, "continuers." This refers to those who take over the family farm. Many of these farmers, however, are "late continuers"—those who first leave the parental farm to engage in other work (whether inside or beyond the village) and return to farming later in life as land becomes available (White 2019: 22). Most young farmers have previous migration experience, working in non-agricultural jobs before returning home to engage in small-scale farming combined with other income sources (White and Wijaya, in this book). For both younger and older generations, farming has never been their only source of income. Apart from farming, they are still engaged in other sources of livelihoods in or outside the village, sometimes including temporary or seasonal work in large cities.

Access to agricultural resources significantly affects the agricultural practices of young farmers. Agriculture resources, particularly land, are mostly accessed through a hibah of land from still-living parents. Those whose parents own more than a small plot of land are more likely to receive land from their parents and also find it easier to negotiate in order to obtain access to the land sooner than young people whose parents only have a small plot. This land transfer process impacts the timing of their return to farming. During the waiting period, as indicated above, young people migrate and engage in non-agriculture work. Their involvement with farming before marriage is largely to assist their parents. After marriage, those who have acquired land from their parents will have more independence in their farming practices. Others who are still waiting for land become their parents' helpers or enter into a sharecropping agreement with them.

The young farmers who we interviewed have no formal agricultural training, but since childhood, they have been observing and participating in their parents' agriculture activities. As continuer farmers, they do not always practise the same modes of farming that they learned as children. Their migration experience—the opportunity to meet different types of people—and access to information technologies often encourage them to be more creative than their parents. These young farmers tend to be "risk takers" in crop choice, seeking out investment to start farming or experimenting with different farming techniques. While their parents prefer to grow staple food crops such as rice or maize, the younger generation's farming methods are adjusted based on their own experience and the context of technological development in their environment. Their innovations are commonly paired with creativity,[3] not in terms of the introduction of sophisticated equipment or technology, but rather creative experimentation with new crops and new ideas that aim to improve productivity.

Girls are generally less in contact with farming than boys while still attending school. As they age, leave school, and often leave for work outside the village, marriage becomes their way to return to the village—both their village or husband's village—and farm. Women are involved in most farming processes, but ironically, they receive less recognition and support as female farmers.

Farming Experience in Childhood

All of our respondents told us that they have experience helping their parents on the farm or in the rice field when they were children. In both villages, children were, on average, 11 years old or in the fifth grade of primary school when they first start helping their parents on the farm (median age of nine years or third grade). This involvement, however, usually halts with a child's transition to secondary school, with its longer school hours and homework requirements. Besides providing help to their parents, the children played around the fields as well; they learn by observing how their parents work, preparing the ground, planting, weeding, and harvesting. Such childhood experiences still linger in their memories. Mahdi (male, age 34, Pudak Mekar) recalls: "I was first asked to go to the farm by my parents when I was about eight. At that time, I was

---

[3] We understand creativity as involving the generation of ideas, while innovation is about implementing these ideas in practice (Rietzschel et al. 2016)

ngasak [gleaning][4] the rice, sweet potato, or other vegetables. The yield from ngasak was sold for our daily needs. Previously, I was asked to collect grass to feed our family's livestock."

In Pudak Mekar, boys and girls gather grass. Farmers in Pudak Mekar own more livestock than those in Sidosari; the majority of households in the former own one or more goats or cows. Mila (female, age 34, Sidosari) remembers: "I used to go to the farm when I was in sixth grade, I just went along with my parents. I did ngarit too (gathering grass to feed the livestock) together with other friends."

As they are now adults and many have become parents of their own children, they do not encourage their children in the same way as their parents did—playing around and helping out on their parents' farm. The long school hours, including travel time and homework, are the main reasons that they do not ask their children to help on the farm. Secondary school is a full-time activity from early morning to late afternoon so they don't want to burden their children by asking them to help out; instead, they encourage them to stay home and rest.

Andi (male, age 45, Sidosari) was helping his parents' *derep* (harvesting another's farm) as a teenager. Derep can be done individually or in groups. Derep in a group, in Sidosari and in other neighbouring villages, is largely composed of male farmers, unlike some other areas of Java. Women are assumed to be slower in rice harvesting work.

Unlike him, Andi's oldest daughter (Ina, age 19, Sidosari) has never helped him in the fields. She only helps serving meals to the farmers. His wife Susi (age 38, Sidosari) adds: "Girls nowadays don't want to help on the farm at all, they are afraid to get tanned. No girls now want to do *tandur* (rice transplanting), they just let their mothers do the work instead. It's different compared to the previous generations; we began to be hard workers in our early life."

Just as experienced by Andi and Susi, Imah (female, age 32, Sidosari) spent much of her childhood—since the third grade—helping her parents to farm. She felt no shame or laziness in helping out her parents at that time.

---

[4] *Ngasak* (gleaning): picking up the scattered grains or stalks that the harvesters leave behind that then belong to those who collect them.

I can clearly remember when I was little, I used to run errands helping out on the farm while playing around at the same time. It was playing yet working. In the dry season, I helped out planting peanuts on another farmer's land and earned US$0.04. I was very excited since it was my first time earning money myself. I didn't give the money to my mum; I spent the money myself.

Sam (male, age 38, Pudak Mekar) gives pocket money to his children so they are willing to help on the farm. For example, he told us that he gives his son Rasif, a sixth grader, pocket money for his help watering the chillies and cutting grass to feed the livestock.

Access to Farm Land

Farming requires land capital. "Land is important, the most important,…" says Rajif (male, age 35, Pudak Mekar). "As long as we own land, even though we don't have money for production, we can go to the middleman—just tell him that you want to farm and he will give you some money as your working capital. He won't lend you any money if you don't have land to plough."

Land is one of the most basic needs for young people in both villages to start farming. As noted above, access to land can be gained through inheritance, a grant from still-living parents, purchase, cash rent, mortgage, and share tenancy. As seen in Table 13.2, most young farmers in our sample own part or all of the land that they cultivate. Many of their parents, especially in Pudak Mekar, have granted their land to their children as use rights. When the parents die, ownership will transition to the children. Of our respondents, 48 per cent started farming on land that their parents had granted to them, while another 28 per cent had inherited land. One reason that parents, even when owning only a small amount of land, feel able to grant land to their children is that their livelihoods are based on multiple income sources, so that giving some land to their children does not disturb their household income too seriously.

Parents generally give land to married children. They will pass on the land with the hope that their children will learn how to farm independently. In both villages, it is assumed that after marriage, young people begin in earnest to learn how to farm. It is easier for children from rich farmer families to ask their parents for land. Tari (female, age 43, Sidosari) asked her parents to purchase for her a small farm when she turned 25 and married. Her parents own a roof tile factory in the village. She knew that she would receive an inheritance from her parents, so she dared to express her wish to have her own farm. Her elder brother received 35 ubin of land from his parents after his marriage. Tari shares: "I dared to tell my parents that the land owned by my younger brother-in-law was for sale, I think it's better if I buy it rather than it's sold to other people. I asked my father to buy the farm and I offered myself to manage it. My parents agreed on that and bought the farm right away."

In another case, prosperous parents in Pudak Mekar gave 100 ubin of land to their son Dandy, (male, age 28) even though he was unmarried. Dandy's father runs a sawmill business and Dandy cultivates horticultural crops on the land. "I asked my parents for some land. I used to harvest papayas on another farm, I had been wanting to cultivate by myself, then my father allowed me to do it."

For farmers who do not own large farms and/or do not have significant non-farm income, intergenerational land transfers are more difficult. There are many young farmers who, although they are already married, are still helping on their parents' farm without any revenue-sharing agreement and are still dependent on their parents for their daily needs. Ano (age 27, Sidosari) and his wife Eni (age 25, Sidosari) still live in Ano's parents' house and help cultivate their land. Even though his father (Kamsi, age 57, Sidosari) thinks that Ano seems to be ready to farm independently, he does not entrust the farm to Ano and Eni yet because he thinks that there is not enough land to share among all of his children. Ano's father cultivates 43 ubin and says that he will continue managing it himself. Another rice field of 40 ubin that Ano and Eni manage is rented from his father for a period of two years. His father needed cash at that time and asked Ano to rent his rice field. The rental price that the young couple pay is regular market price. Kamsi says: "I don't know yet about inheritance. Even though I am already old and don't have much energy left, I don't think of passing on the land to my children. I will offer them share tenancies instead, so that I still can enjoy part of the harvest."

Renting land is also common in Kebumen. It can be accessed from other farmers (or a parent) or from village treasury land, as explained above. Especially in Sidosari, the village government has a particular way of managing prosperity land that supports poor villagers. The charged rental price is lower than the regular rental price. Moreover, unlike private land rentals where the rental must be paid in advance, the rental payment can be paid one month after the bidders have been granted a tenancy on the prosperity land. For young (married) people with low and uncertain incomes, access to prosperity land is a good opportunity for them to start farming because they can afford to pay the rent. "I gladly support the new auction[5] system, which no longer uses the higher price to get a plot of land," says Imah (female, age 32, Sidosari). "If we rent the land from auction, expenses for meals can be saved for additional investment for the next renting period."

The situation is the same for Eni: "So far I have farmed on village treasury land three times. The village treasury land auction is very much in demand because the price is much cheaper than the general rental price. In the past, auctions were not like now; in the past those who could rent were villagers who had money because it was rented to the highest bidder."

Intergenerational land transfers work through the inheritance system. In contrast to the children of rich family farmers, Sugi (male, age 38, Sidosari), the son of a poor farmer who owns less than 60 ubin (840 m²) of land, explained that he received access to family's land only after his parents passed away, six years after his marriage and when he already had three children. "I didn't dare to ask my parents for land, I was too young (32 years old at that time). Even after I got married, I was still dependent on my parents. As a son, I just accepted all things. Alhamdulillah, I got the land (448 m² of rice field) after my father died."

Land inheritance practices in our two research villages are not solely based on Islamic inheritance law where sons get a two-thirds share and daughters receive one-third. Some parents divide the lands equally among their children and others do not. Both male and female children who are considered more responsible may inherit a larger portion than those

---

[5] Although still called an "auction" (*lelang*), the land is no longer auctioned off to the higher bidder, as it was in the past, but allocated among applicants via lottery.

considered less responsible. Dimas (male, age 26, Sidosari) remembers: "I started to farm independently [meaning: all decisions and harvest time are done by himself] in 2010 after my parents passed away. I was 19 at that time, when I managed the 100 ubin (1400 m²) of land that my parents owned. Now, some of the land is already mine."

Dimas inherited a relatively larger share than his siblings. The youngest child, he stayed in the village and took care of his ailing parents until they died because his two brothers had migrated to Sumatra. He himself had migrated to Bandung (West Java) but had to return home to take care of his parents. His parents asked him to help on the farm because they were elderly and sick. They offered to give him a plot of land if he returned home permanently. Parents commonly make inheritance decisions; for the children, raising questions about inheritance matters is considered impolite or taboo because it suggests that they hope that their parents will die soon.

Rajif says that all of the decisions about the division of land among his siblings will be made by his father. "Maybe shares will be more or less equal. But honestly, it hasn't happened yet, so I don't know how it will be. There has been no discussion yet, informally or formally, with the family members on how the lands will be divided in the future."

Migration and Pluriactivity

For many young people, out-migration is an option during the "waiting" period until they can access land. In Sidosari, earlier migration patterns influence migration destinations; for instance, young men tend to migrate to Riau and Palembang, Sumatra to work on oil palm plantations or to Jakarta where they work as freight container labourers in Tanjung Priok Harbour. Young women mostly prefer to migrate to Bandung or Jakarta where they find work as shop assistants or housemaids. In Pudak Mekar and Sidosari, some young migrants have found work abroad in Malaysia, Saudi Arabia, and Hong Kong. In both villages, the majority of young women and almost all of the young men in our sample have migration experience, as shown in Table 13.4.

Table 13.4 Migration experience of young farmer sample

	Female (n: 11) (%)	Male (n: 28) (%)
Ever migrated	64	94
Never migrated	36	6

Source: Becoming a Young Farmer survey, 2017

Youth migration is not a new phenomenon. When the parents of our young farmer sample were younger, they generally experienced migration as well. One of the reasons for migration is to save money for farming activities. Kamsi, for example, migrated and worked in the city while saving money to rent land in his home village; he thought that he could not be a farmer without owning land after he returned to the village. His parents were landless tenant farmers, so he had no prospect of inheriting land. "I got married when I was 25; I migrated while I was still single. We needed money to rent a farm (land) at that time, so I went away to earn and save some money. Migrating is best in the dry season when we can't earn anything by working as tenant farmers."

Ano also regularly migrated to Jakarta to work on road construction projects for two months at a time, returning to the village for the harvest season and migrating again to a different destination depending on the available work opportunities. Ano stopped migrating after his first child was born. Since then, he and his wife have focused on farming while also earning money weaving *caping* (a traditional conical-shaped farmer's hat made of woven bamboo).

Nearly all of the villagers in Sidosari—both young and old, male and female—are engaged in *caping* making. Previously, there were several *caping* distributors, but since 2016, only one distributor remains and the piece price for the finished product has declined. The income earned for 20 *caping* ranges between US$1 and US$2 with a production duration of 1.5 days. Earlier, the payment for *caping* from middlemen was often delayed, so the crafts(wo)men had to wait a few days despite having delivered the product. In Pudak Mekar, young women engage in another low-return handicraft activity—making floor mats from coconut fibres. They first wind the fibres into long ropes. It takes three hours to make a five-metre rope, which sells for US$0.15. One floor mat sells for only IDR 3000 (US$0.20).

Options for non-agricultural incomes in Pudak Mekar are more varied than in Sidosari. Apart from farming, most farmers keep cows or goats. Even though they only have a few animals, they say that keeping livestock is a way to save money for expensive needs like house building or repairs, hospital bills, and wedding parties. Their crops are used for their daily needs. Many also plant *rumput gajah* (elephant grass, *Pennisetum purpureum*) on part of their land as livestock fodder.

Some young men in Pudak Mekar are fishermen. They take daily fishing trips during four months of the year. Every trip requires a substantial investment. Besides the boat and equipment, one needs at least US$7143 for logistics, fuel, and food supplies for one boat with three to five fishermen on board. Even though they rarely go fishing, some of the interviewed respondents, including Na'im (male, age 31, Pudak Mekar), say that fisherman is the formal occupation that is written on their identity card:

> My occupation written on my ID Card is fisherman. I usually go sailing in Sura month (Muharram month in Islamic calendar), it could be every day in that month, and I temporarily leave my farming job. I pay for tenants to cultivate my farm. If I get lucky and the sea is not too rough, I can earn millions. But if I'm not lucky, the money is only sufficient to cover the fuel costs.

Another non-farm income source is sand digging in the Luk Ulo River, which is located on the edge of Pudak Mekar. Only young men do this work because it is physically demanding. Sand digging starts in the evening and ends the next morning. The diggers stand in the river all night, digging the sand from the river floor and carrying it to a pickup truck. For one night's work (about 8–10 hours), they can earn up to US$11. Sand digging is mostly done during the growing season because there is less work to do than during the harvest season. The money that wage workers earn from sand mining is often used to finance horticultural crop planting, which requires a fairly high investment when compared to rice, cassava, or maize. Most young farmers in Pudak Mekar choose to plant horticultural crops because the schedules of farming and non-farm work can be flexibly combined. As Rajif explains:

> Most young men earn money by working as daily labourers, as sand diggers or sawmill labourers. The sawmill workers go briefly to the fields from 6 to 7 in the morning, then work at the sawmill from 8 in the morning till noon. From 5 p.m. till dusk, they go back to the fields. If they need more money for farming, they can work as sand diggers in the evening. Farm incomes are very uncertain; sometimes the harvest is good, but the price is low. Sometimes the price is high, but weevils ruin the crops. Sawmill or sand digging work provides stable earnings, but it doesn't make people smart because they work monotonously. In the sawmill the work is in silence, just cutting the wood. And in sand digging, we don't have much time to chat with other because we have to reach the target. When we're farming, we use our brain, thinking about how to eradicate the weevils, sharing with other farmers, and we can enrich our knowledge.

Innovation: Different Ways of Farming, New Sources of Information

Compared to their parents, young farmers in both villages are more adventurous in trying new farming practices. This includes taking risks, choosing methods of farming that require more investment, new seed varieties, new farming techniques, and more intensive practices. In Pudak Mekar, young farmers mostly grow horticultural crops while their parents prefer to plant rice, cassava, or maize, which have a relatively longer growing season, but require less maintenance and lower production costs.

Most young male farmers, including Dandy (male, age 28, Pudak Mekar), think that horticulture is more interesting and the revenue is relatively fast to earn: "It's different now and then. People in the past grew peanuts, rice, maize, or cassava, just all of them, without variation. Now we grow bitter gourds, chillies, or long beans. The planting process as well as the treatment is different." Rajik confirms that young people tend to prefer horticulture. "It's different with maize, where when we have finished planting, we just have to wait till harvest time."

Muhtar (male, age 28, Pudak Mekar) says that the move to farming was his own initiative after he left his job in a garment factory in nearby Kebumen city. "When I went out I saw a garden planted with bitter gourds. They looked good, so I wondered what if I try it, and I immediately tried planting them by myself." At first, he worked together with his brother. They cultivate their parents' 7,1 ubin (100 m²) of land. According to him, farming together with his sibling is quite easy. "It is less of a burden if we farm together, one can pull up the weeds and another one can water the plants." For Muhtar and his brother, even though their parents are farmers, planting horticultural crops is new for them; their parents prefer to plant elephant grass to feed their livestock.

Muhtar uses the internet to locate information about horticulture. He often searches for the best treatment for weevils or other pests and diseases that could potentially damage bitter gourds. He also looks to his farmer friends who have experience in planting horticultural crops. Toni (male, single, age 37, Pudak Mekar) also often uses his smartphone to search on the internet and to expand his knowledge about weevils and other pests, such as the silver-leaf whitefly (*Bemisia tabaci*), aphids (*Aphidoidea*), and thrips (*Thrips tabaci lindeman*) that often damage his vegetable plants: oyong (*Luffa acutangula*), chillies, cucumbers, tomatoes, and long beans. He also learns about the active ingredients in the various recommended pesticides. The pesticide brands recommended on the internet are often not common in Pudak Mekar or in nearby farm stores, so he looks for other brands that have the same active ingredients. He prefers to search for information online because it can be accessed anytime and is more practical instead of having to ask other farmers or the fertilizer merchant.

Uji (male, age 21, Sidosari) also uses the internet on his smartphone to learn how to make anti-plant-hopper pesticide (*Fulgoromorpha*) using natural ingredients. Based on this information, he makes a liquid pesticide from boiled soursop leaves and sweet flag (*Acorus calamus*, a kind of rhizome with a specific smell). It is applied by spraying the liquid on the lower branches of the plants. He also uses a different rice planting method from his parents. The *legowo* system, which involves reducing the space between plants in each row, but leaving an empty row after every fourth row, ensures better photosynthesis and ease of plant protection. He acquired this knowledge at a briefing meeting that the local Department of Agriculture hosted in the village office. He went to the meeting in his parents' place because only older farmers were invited. According to Uji, the legowo system is better because it requires less seeds and produces better yields. He also applies natural fertilizers that he makes from livestock manure.

The young farmers frequently experiment with new practices to improve crop productivity. Rajif uses plastic mulch on the raised beds of his chilli plants to prevent weed growth, a technique that he has been using for the last five years. For him, plastic mulch is very pricy; it costs US$35–50 a roll with a width of 1–1.5 metres and a length of 250–500 metres. As a cheaper substitute, he sometimes purchases used plastic from shrimp farm owners, paying only US$14–21 for the equivalent of a double large size roll of mulch. The farmers contact the shrimp farmer by phone or SMS to order the used plastic. Mahdi has tried using MSG (food flavouring, Monosodium Glutamate) as an alternative growth stimulator for his chilli plants, combining it with milk powder in water. He learned this tip from successful farmers in the neighbouring village.

Young farmers are also more willing to borrow capital from other parties when compared to their parents. There are several options available, including loans from middlemen, banks, or relatives. Dandy borrows money from a local vegetable trader in Pudak Mekar; according to him, borrowing money from a local middleman is easier than borrowing from the bank because the middleman will buy the harvested vegetables right after they are harvested. The middleman's purchase price, Dandy says, is not much different from the prices that other brokers pay. The loan then provides Dandy with his working capital and a guaranteed buyer for his produce. Dandy told us that he once borrowed IDR 15 million (around US$1000) for his investment in planting bitter gourds. The loan enabled him to earn a net profit of US$145 in the short growing season (40 days). In contrast with Dandy, Toni and Mahdi prefer to take bank loans. They are registered for KUR (small-scale enterprise credit) at the BRI Bank. Toni has accessed KUR several times for IDR 5–6 million (US$350–420) loans, using his BPKB (motorcycle ownership certificate) as a guarantee. The loan is repaid after each harvest or after six months (two harvests) for his small tomato and cucumber farm. For Toni, "to become a farmer, capital and land are the important things. The first thing is land: if there is no land, we cannot farm, but we can start by renting from others, so capital is the most important."

Lack of Support for Young Farmers

Young farmers in both villages do not receive the same recognition as farmers when compared to older farmers, and this can affect their access to resources. As in the other Indonesian research sites (Flores and Kulon Progo), older farmers or larger landowners always dominate the village-level, state-sponsored *Poktan* (Farmers' Group). Almost all of the villages in Java have Poktan groups, and many of them are members of the local *Gapoktan* (Association of Farmers' Groups). The Poktan is the official conduit for government training, counselling, and subsidies, commonly in the form of fertilizers, seeds, or hand tractors. In reality, almost all farmers' groups do not engage in group activities and tend to be passive recipients of subsidies.

Na'im and several other men in Pudak Mekar prefer to state their occupation as fisherman for their identity cards, even if most of their working time is spent farming. One of the main reasons for this choice is that those registered as fishermen receive more support from the government when compared to support for small farmers. Nai'im received a fishing boat, complete with nets, as aid from the government. But in his work as a farmer, he has never received any support from the village, regency, or central government. Muhtar concurs:

> The Farmers' Group in this village is passive; it does have some activities, but they don't run well. The Poktan here doesn't even have a *mantri tani* (a government-employed agricultural extension worker also called a PPL); we only have visiting veterinarian (for livestock) from the Kebumen Regency Department of Agriculture. Village officials have no programmes that support young people in this village in their farming activities. The village should have a programme to help young people to do farming. I would recommend the village officials to support them with agricultural equipment.

For young farmers like Dimas in Sidosari, although owning land is their right, they do not have access to Poktan membership. "I have never been invited to join the meeting of the farmer group. There was once an announcement that a representative of the Department of Agriculture was coming to the village office to provide training to the farmers, yet I wasn't invited." He has never met a government agricultural official.

Our female young farmer respondents, including Imah, claim that they did not even know that there is Poktan in their village:

> I have never known, let alone joined, the meetings related to agriculture, either in our neighbourhood or in the village. I don't know anything about Poktan either; have never been there, I'm just doing this kind of farming (as a landless tenant farmer). I have never met the PPL in the fields or in the village office. If there is a problem with my crops, I go to my brother for advice—he is an experienced farmer. I and my husband have never tried to look for information on how to farm or how to exterminate pests or disease through the internet. As long as I have been farming, I have never received fertilizer aid from the government. It says that there is sub-sidized fertilizer, but I know nothing about that. In fact, fertilizer is hard to get and more expensive; we buy it from the nearest merchant. My farm that I plant is not large, so I buy the fertilizer in retail. TS (super phosphate) fertilizer costs US$1.1 per five kilogrammes, while urea is US$0.8.

Santo (age 55) is Secretary of the Kebumen Regency Department of Agriculture. He told us that agricultural mechanization is one way to attract young people to farming. Yet, they—Department of Agriculture officers—realize that agricultural support or aid is always distributed through the Poktan, whose members are mostly older farmers.

> Institutionally, young farmers don't yet have a (formal) organization. We (Department of Agriculture) don't dare to establish young farmer groups due to concerns about violating rules or legal sanctions. The farmers' groups that have been approved by the law at village level are Poktan, Gapoktan, Farmers Association, and the Food Security Board. These are the official groups that have the right to access aid from the local and central governments. All supports/aids from the Department of Agriculture are distributed through these groups.

Conclusion

Becoming a young farmer in Sidosari and Pudak Mekar is a long process. Even though their parents are also farmers and they grow up around the farm, young men and women wishing to start farming face various challenges, including access to land, access to capital, and the lack of adequate recognition and support for young farmers. Their experiences of migration and work in various non-agricultural activities have coloured their pathways to becoming a farmer during their "waiting period."

Young farmers admit that they must be more willing to take risks than their parents. They want a more profitable farming model, different from the agriculture that their parents practice. Access to information, either through direct contact with the fertilizer merchant or via the internet, has become their preferred channel for learning and training opportunities related to agriculture.

There are many government programmes on paper that feature the jargon of millennial farmer programmes. But in many cases, such as in our two research villages in Kebumen, young farmers have not received any significant support. The government's subsidy and training programmes are still biased towards, and dominated by, older generation male farmers. As we have seen, in responding to the challenges of being a farmer, young farmers in Kebumen have largely sought their own solutions.

References

Ambarwati, Aprilia, Ricky Ardian Harahap, Isono Sadoko, and Ben White. 2016. Land tenure and agrarian structure in regions of small-scale food production. In *Land and development in Indonesia: Searching for the people's sovereignty*, ed. J. McCarthy and K. Robinson, 265–294. Singapore: National University of Singapore Press.

Badan Pusat Statistik (BPS). 2019. Kabupaten Kebumen Dalam Angka 2019 [Kebumen Regency in numbers 2019], Kebumen: BPS.

Monllor, N. 2012. *Farm entry: A comparative analysis of young farmers, their pathways, attitudes, and practices in Ontario (Canada) and Catalunya (Spain).* http://www.accesstoland.eu/IMG/pdf/monllor_farm_entry_report.

Pudak Mekar Village Government. 2017. Profil Desa Pudak Mekar [Pudak Mekar village profile]. Kebumen: Pemerintah Desa Pudak Mekar.

Rietzschel, E.F., H. Zacher, and W. Stroebe. 2016. A lifespan perspective on creativity and innovation at work. *Work, Ageing and Retirement* 2 (2): 105–129.

Sidosari Village Government. 2017. Profil Desa Sidosari [Sidosari village profile]. Kebumen: Pemerintah Desa Sidosari.

White, B. 2019. *Rural youth, today and tomorrow.* 48 IFAD research series. IFAD.

14

Pluriactive and Plurilocal: Young People's Pathways Out of and into Farming in Kulon Progo, Yogyakarta, Indonesia

Ben White and Hanny Wijaya

In this chapter we explore young people's spatial and sectoral mobility, specifically their trajectories out of and into farming, in the Javanese village of Kaliloro, focusing on young men and women from small-farm and landless families who make up the majority of the population. The study is based on field research in Kaliloro in 1972–1973, 1999–2000, and 2016–2018 and thus provides the opportunity for an analysis with some historical depth.

Our main data sources are as follows:

1. Household surveys covering all households in 5 of the village's 26 neighbourhoods in 1973 (411 households), 2000 (473 households), and 2017 (519 households).

B. White
International Institute of Social Studies, The Hague, The Netherlands
e-mail: white@iss.nl

H. Wijaya (✉)
Sekar, Yogyakarta, Indonesia

© The Author(s) 2024
S. Srinivasan (ed.), *Becoming A Young Farmer*, Rethinking Rural,
https://doi.org/10.1007/978-3-031-15233-7_14

383

2. Sample surveys of about 50 small-farmer and farm-worker households in 1973, 2000, and 2018.
3. Detailed time-budget studies from 20 small-farmer and landless households covering a one-year period and all children and adults from age four and up (1973 and 2000). A detailed study of time allocation is highly labour intensive (involving the recording of several thousand person-days of time use, even in a small sample of 20 households) and also highly intrusive on the private lives of those we study. For both of these reasons, time allocation research was not repeated in our most recent field study.
4. Qualitative interviews with 35 "young" farmers and smaller numbers of older farmers in 2017–2018, focusing specifically on the trajectories out of and into farming that are the main focus of the second half of this chapter.[1]

The chapter is organized as follows. In the next section, we present a snapshot of Kaliloro in 1973—in the early years of the Suharto period and of Java's "Green Revolution"—focusing on agrarian structure and livelihoods. We then summarize the main changes in infrastructure, education, and livelihoods in the four decades since 1973. We then describe the general (and quite dramatic) changes in the lives of young people in the same period, before focusing specifically on young men and women's contemporary trajectories out of and into farming.

Kaliloro in the Early 1970s[2]

The village of Kaliloro[3] lies about 30 kilometres to the northwest of the city of Yogyakarta in southern Central Java. It lies on a thin plain of rice terraces and settlements some two kilometres wide between the foothills of the Menoreh mountain range to the west and the Progo River to the east (Fig. 14.1).

---

[1] Thanks to Aprilia Ambarwati and Charina Chazali who joined us in conducting these interviews.
[2] Sources for this section are White (1976a, b) and Stoler (1977).
[3] The name is a pseudonym, as are all names of persons mentioned.

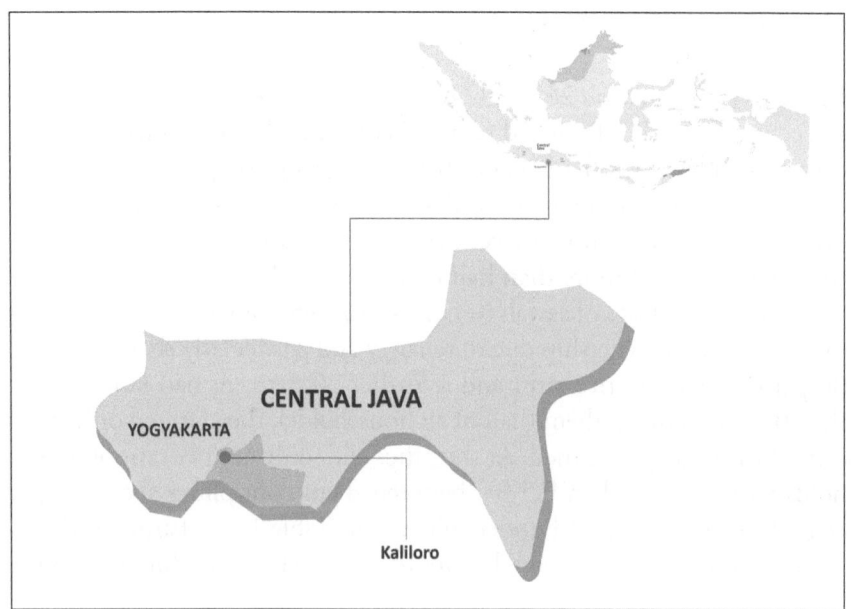

Fig. 14.1   Central Java and Yogyakarta, showing the location of Kaliloro

It is a large village, the result of fusion of five smaller villages in 1946, with about 2800 households and 10,500 people in 2017. It shares the basic features of many densely populated rice-growing villages in the Yogyakarta and Central Java region: widespread landlessness, high tenancy rates (mainly share tenancy), extremely small average farm sizes, intensive cultivation practices, a high degree of pluriactivity (multiple income sources) in both rich and land-poor households, relatively high levels of education in the current generation (with most boys and girls now completing secondary school, even in poor households), and high out-migration rates of these relatively well-educated young men and women. We further explore these features below.

Since 1968, year-round irrigation became available from the large-scale Kalibawang canal, which snakes through the village along the lower edge of the Menoreh foothills. Despite quite frequent breakdowns in the early years, during 1968–1973 most farmers were able to plant a second (dry season) rice crop.

Land Ownership and Access

In the five of Kaliloro's 26 neighbourhoods surveyed in 1972, almost 40 per cent of households owned no rice fields (*sawah*), and a further 23 per cent owned less than 0.1 hectares (ha); this group (62 per cent of households) between them owned less than 10 per cent of all the land. At the other extreme, the top 6 per cent of households with holdings of more than 0.5 ha owned more than half of all the sawah (see Table 14.1).

Operated holdings of sawah (= farm sizes) were somewhat more equally distributed than ownership due to tenancy and particularly sharecropping: 30 per cent had no rice farm and a further 20 per cent had farms of less than 0.1 ha. Between them (half of all households), they farmed only 9 per cent of the total area farmed. At the other end, only 4.4 per cent of households farmed more than 0.5 ha, between them controlling about 24 per cent of the total farm land under cultivation (Table 14.2). Farm size distribution was thus also unequal, but more equal than ownership, due to the

Table 14.1 Ownership[a] of *sawah*, 1972 and 2017

Year	1972		2017	
Area owned (m²)	% of all households	% of all sawah	% of all households	% of all sawah
0 (none)	38.6	0.0	50.1	0.0
1–1000	23.3	8.6	28.3	23.2
1001–2000	21.1	18.6	13.7	28.7
2001–3000	4.9	7.0	3.3	11.1
3001–5000	5.8	13.6	2.1	10.9
>5000	6.3	52.3	2.5	26.0
Total	100	100	100	
All households	(411)		(519)	
Owner households	(253)		(259)	
Average area owned	0.17 ha		0.08 ha	
(all households, owners only)	0.27 ha		0.15 ha	

Source: Authors' own household surveys, 1972 and 2017

[a]The sample for Tables 14.1 and 14.2 comprises all households in 5 of Kaliloro's 26 neighbourhoods (*dusun*). For the purposes of this table, "ownership" includes both owned land, *tanah bengkok* (village-owned salary lands allocated to village and neighbourhood government officials for the duration of their term of office) and *pengarem-arem* (allocated as pension after completion of their term of office)

Table 14.2 Area of *sawah* cultivated (operated farm size), 1972 and 2017

Area operated (m²)	Number		% of all households	
Year	1972	2017	1972	2017
0 (none)	122	252	29.7	48.6
1–1000	82	127	20.0	24.5
1001–2000	111	81	27.0	15.6
2001–3000	38	28	9.2	5.4
3001–5000	40	22	9.7	4.2
5001–7500	18	9	4.4	1.9
Total	411	519	100	100
Total rice farmers	289	267		
Average farm size (ha)	0.21	0.17		

Source: Authors' own household surveys, 1972 and 2017

Table 14.3 Tenure status of rice farmers, 1972 and 2017

Status	1972 %	2017 %
Pure owner-operator	54.3	40.3
(Part) share tenant	15.6	22.4
(Pure) share tenant	21.5	31.0
Rent/mortgage	4.8	4.9
Combination share tenant and rent	3.8	1.5
Total	100	100

Source: Authors' own household surveys, 1972 and 2017

prevalence of tenancy: 46 per cent of farmer households were cultivating land that they did not own, or only partly owned, nearly all under share tenancy agreements (Table 14.3). Average sawah ownership was 0.17 ha (among all households) and 0.27 ha among owner households. The average size of sawah farms (counting the farm households only) was 0.21 ha.

With such tiny farm sizes, it is not surprising that pluriactivity—diversification of occupations and income sources at the household and often also at the individual level—was already quite striking in 1973. There were very few households, rich or poor, for whom income from rice cultivation, or the time devoted to it, constituted a major part of the household's total productive activity. Households in all land-owning strata engaged in non-farm activities, but for different reasons. A detailed, year-long study of incomes and work in 20 small-farm and landless households found that only 27 per cent of their incomes were derived from sawah cultivation and 19 per cent from home gardens (*pekarangan*), with the remaining 54 per cent deriving from off-farm work (agricultural wages) and non-farm work. One characteristic feature of such non-farm work was that it provided, on the whole, lower returns per hour of work than own-farm production or agricultural wage labour (White 1976a, b). Some non-farm activities have declined or disappeared in previous decades. *Batik*-making on a putting-out basis for Yogyakarta merchants had completely disappeared since the 1930s recession; home-based handloom weaving still employed some 40 people but was on the decline. Other activities, however, had increased, notably bamboo and pandanus-mat weaving as well as petty trade. More than one-quarter of all adult women were involved in some form of trade, and another 10 per cent in production of food for sale.

More women than men were involved in agricultural wage labour, but as a secondary activity for the majority of those involved; this reflects the highly seasonal nature of agricultural wage employment for both men and women and the consequent necessity of other sources of income besides farm labour. The hand-pounding of rice as a source of wage income for women had recently disappeared after the introduction of rice hulling machines in the late 1960s. Rice harvesting, in contrast, still used the traditional finger knife (*ani-ani*) and was a major source of income for women in small-farm and landless households (Stoler 1977).

During the 1960s and the 1970s, population growth in the village was not significant, not because of low birth rates but because of high out-migration rates. Out-migration, nearly exclusively of young adults, has been a constant feature of village life for several generations. In the late colonial period, many young men and women left for working in plantations in North Sumatra or to pioneer settlement regions of Lampung and South Sumatra.[4] In 1972, among the children of Kaliloro residents who had already left the parental household, 55 per cent had left the village and the great majority of these had moved outside the district, with 103 (22 per cent) having moved to destinations outside the

---

[4] Kaliloro has a long history of migration to Sumatra since the contract labour and colonization schemes of the 1920s and 1930s. One resident of Kaliloro was a professional labour recruiter for Sumatran plantations during that period. After independence, the presence of established kin in Sumatra made it easy for young people to move there without government assistance (White 1976a, 356).

Table 14.4 Indicators of out-migration, 1972, 2000, and 2017. Location of all sons and daughters of current residents who have left the parental household[a]

Location	1972 %	2000 %	2017 %	Note
Kaliloro	45	27	32	
Outside Kaliloro, Yogyakarta region	7*	12	19	*1972: Kulon Progo district only
Other region in Java	26	41	37	
Outside Java	22	20	14*	*includes 0.5% (3 persons) overseas (Malaysia, Taiwan)
Total	100	100	100	

Source: Authors' own household surveys, 1972, 2000, and 2017
[a]This table shows, for all resident adult women, the current location of any of their children who now reside outside the parental household. Only children of resident women are counted to avoid possible double-counting in cases of men who are divorced and remarried

island of Java (Table 14.4). These movements all represent the period before the rapid expansion of labour-intensive, export-oriented manufacturing in the late 1970s that drew many teenagers and young adults, particularly women, into the textile, garment, and footwear industries of West, Central, and East Java (Mather 1983; Wolf 1992).

Changing Village Economy and Livelihoods, 1973–2017

Returning to Kaliloro in 1999–2000, and again in 2016–2018, some of the most obvious changes are the following.

The population has grown modestly (by only 23 per cent in 46 years), with the number of households growing faster (by 40 per cent) and average household sizes falling from 4.3 to 3.7.[5] The city has come closer to the village, in many ways. In the 1980s, a new bridge across the Progo River, a few kilometres to the south of Kaliloro, reduced the distance and

---

[5] These numbers derive from village-level statistics. In the five neighbourhoods covered in our own household surveys, the number of households grew by 26 per cent (from 411 to 519) between 1972 and 2017.

travel time to Yogyakarta. In 2015, a new bridge was opened in Kaliloro itself, cutting the distance and travel time again by about 40 per cent. Daily commuting to Yogyakarta, though still rare, is now a possibility. The improvement and widening of the asphalt road and various bridges along it, and a big increase in the frequency of public transport, have made Kaliloro's main north-south road a busy thoroughfare. The quality of smaller roads and concrete pathways entering residential areas has also greatly improved. In 1973, there were only a few motorcycles and one four-wheeled motor vehicle in Kaliloro. The main mode of local transport for people and goods was by foot and bicycle, and several people kept small packhorses for the transport of goods. By 2000, the horses had disappeared, replaced by 34 private cars and minibuses, 45 trucks (most of the latter owned by one person), and almost 300 registered motorcycles. In 2019, there are so many motorcycles that the officials no longer keep a register.

Around Kaliloro market and in other parts of the village are many new shops, kiosks, and food stalls with a wider variety of goods for sale. At the village's main crossroads and near the marketplace, there are now about 130 offices, shops, and small businesses, including several banks and credit providers, six photocopy shops, two motorcycle dealers, a laundry, a notary's office, and a pharmacy as well as more than 50 shops of various kinds and more than 30 small food stalls (*warung*) offering a variety of foods.

The quality of housing is also much improved. Houses with wooden frames and woven bamboo walls (*gedek*) are now quite rare and most landless or near-landless households have been able to build brick houses with the support of reciprocal labour, in combination with some hired craftsmen (Abdullah and White 2006).

Kaliloro was connected to the State Electricity Company grid in the mid-1980s, and by 1999, 90 per cent of households surveyed were connected, officially or not. Besides the replacement of oil lamps with electric light, electrification has made possible various other innovations such as the two busy photocopy shops near Kaliloro market, the desktop computers and laptops in the village office and in some private households, commercial laundries using washing machines, "play-station" booths along the main road as conduit for the pocket money of school children and unemployed youth, and of course the enormous increase in the number of television sets. Since the first village-owned set was installed in front of the village office in 1974, more than half of the households that we surveyed in 2000 had televisions at home and, by 2017, virtually all households had one. In the early 1970s, there was no telephone of any kind in the village; by 2000 there were a couple of public telephones, and by 2017, the great majority of households had at least one mobile phone.[6]

Small-scale piped water (PAM) projects have brought running water to many houses on the eastern side of the river and to some on the western side; 20 per cent of our surveyed households had running water. Besides reducing the time spent in fetching water (from their own or nearby wells or from streams), running water has also made possible the irrigation of home gardens and the construction of year-round fishponds in many hamlets.

Since the early problems of the irrigation channel were overcome in the mid-1970s, regular double-cropping of rice has been assured. A rigid regime of water supply and crop calendars has made a tightly scheduled crop cycle of paddy-paddy-*polowijo*[7] universal. Improved irrigation, relatively high levels of inorganic fertilizer application (around 250 kilogrammes per hectare), improved pest control, and some improved practices[8] have brought paddy yields to about twice their earlier levels, that is, between 5 and 6 tonnes of barn-dry paddy per ha (or about 3.5 tonnes of milled rice). In pre-Green Revolution times, a rice farm of 0.2 ha was needed to provide an average-sized household with enough rice to eat in a normal year. In 2019, a small plot of 0.1 ha can provide about 0.7 tonnes of milled rice per year, more than the food requirement of a family of four or five persons.

Paddy tractors have replaced buffalo-drawn ploughing or hand hoeing on most of the *sawah*, thus reducing opportunities for male wage labour.[9]

---

[6] The exceptions are a few elderly individuals or couples who do not have their own electricity supply.

[7] *Polowijo* crops are the rainfed cash crops of soya, groundnuts, mung beans, and so on. Some farmers leave their fields fallow in the polowijo season.

[8] Straight-row (*sipatan*) planting, which a few farmers were using in 1973, is now universal. Some farmers now use urea fertilizer tablets in place of loose powder.

[9] Exceptions are terraces too small or too steep to allow tractor access.

The tight cropping calendars, with all farmers planting within a few weeks, have made the peaks of labour demand in transplanting and harvesting higher but also shorter in duration. Agricultural wage employment is therefore both more female and more seasonal than previously. In harvesting, sickles have replaced the *ani-ani* and husband-wife couples are often seen harvesting together in what was previously exclusively women's and girls' work. A more important change, however, is that much of the paddy produced in Kaliloro is now sold as a standing crop to *penebas*[10] who bring in their own teams of harvesters from outside the village. This is true for a majority of farmers in the first (rainy season) harvest, and smaller but still significant numbers in the dry-season harvest. Penebas pay their harvesting teams a wage reportedly of between 1/10 or 1/12 of the amount harvested. On the remaining plots where "normal" *bawon*[11] harvesting is practised, the stratified system of bawon payments that Stoler described in 1977 endures: neighbours are often paid one-sixth (and with sickles, can now harvest up to 150 kilogrammes per day, thus earning some 30 kilogrammes of paddy), and while most farmers told us that they pay not less than one-eighth, harvesters from hamlets in the western part of the village told us quite definitely that they receive only one-tenth when they harvest for farmers from the eastern part (Abdullah and White 2006).

By 2000, a few farmers had begun cultivating watermelons on sawah, an intensive crop grown on plastic sheeting, demanding much greater inputs of capital than polowijo and requiring daily attention, and providing greatly increased profits if it is a successful harvest. Small numbers of farmers had shifted to high-value vegetables (chillies, tomatoes, etc.) on part of their sawah; these are mainly younger farmers, as we will see later.

In spite of all of these changes, the overall pattern of distribution of (sawah) landholdings changed little between 1973 and 2000—ownership had become slightly more unequal and the landless and near-landless (less than 0.1 ha) groups had grown faster than others. Changes in landholdings seemed to have accelerated, however, during the post-Suharto

---

[10] *Tebasan* is the sale of standing rice crops in the field, negotiated shortly before harvest time. The buyer (*penebas*) pays in cash and brings his/her own team of harvesters.

[11] *Bawon*: the harvester's wage, paid in kind as a proportion of the crop she/he has harvested.

years (2000–2017). Table 14.2 suggests that both centrifugal (differentiating) and centripetal (levelling) tendencies are at work: there are fewer owners in the largest size categories, but at the other end, greater numbers with no land or less than 0.1 ha. Less than half of all households now own sawah land (Table 14.1) and only 52 per cent operate a rice farm (Table 14.2). Half of all rice farms are now less than 0.1 ha in area and there are no farms larger than 1.0 ha (Table 14.2). Tenure statuses have also shifted, with the numbers of rice farming households owning none or only part of the land they cultivate rising from 46 to 57 per cent. Moreover, over half of all land is now cultivated by a tenant rather than its owner and nearly all of this land is operated on a share tenancy basis (Table 14.3). There is thus an increasing tendency for land owners with a large amount of land not to become farmers themselves, but to parcel out their land in small sections to sharecroppers; the small (or micro-)farm pattern remains, as in other regions of Java (White 2018). The average size of rice farms has declined from 0.21 to 0.17 ha. This should not, however, be seen as indication of immiseration, as a farm of 0.17 ha now produces much more than a farm of 0.2 ha previously. The village now produces a surplus of rice above its own requirements. However, due to unequal access to land, only one-third of households were self-sufficient in rice in 1999–2000 and one-third had to buy rice for more than half of the year.

Occupations and Pluriactivity

Changes have also occurred in non-farm activities, which continue to provide a substantial portion of livelihoods for both richer and poorer families. Brick making was formerly a common activity, beginning after the rice-planting season, and involving men, women, and children; the bricks were either sold or used to build or expand a family's own home. This household-based brick making is no longer found in Kaliloro; it has been replaced by small breeze-block (*batako*) industries, using simple machinery and employing three or four local male workers.

Two other formerly common non-farm activities in poorer households—the tapping of coconut trees and making of palm sugar (*gula*

Jawa) and weaving pandanus mats (*tikar pandan*)—have not completely disappeared but are found in only a few households and involve elderly people in their 70s or 80s. A few households have shifted to making dried reed (*mendong*) mats with materials that have to be purchased at the village market, but mat weaving is still (as it was in the 1970s) one of the activities with the lowest incomes per hour of work. Many women in their 30s and 40s have recently begun weaving laundry baskets for export, or a putting-out basis, collecting the raw materials from the (male) exporter in the neighbouring district. They are expected to meet a target of ten completed baskets in two days. They receive IDR 4000 (US$0.30)[12] per completed basket, which provides an income of about IDR 2000 (US$0.14) per hour, or less than one-fifth of the hourly wage for agricultural (transplanting) work.

In the 1970s, 1980s, and 1990s, many young women in their 20s and 30s still worked as farm labourers (transplanting, weeding, and harvesting). In recent years it is now mainly older women (age 40 and above) who are found in this work (Wijaya 2016). Younger women prefer to work in the various factories that have appeared within easy commuting reach of the village (producing, for example, women's underwear, bags, wigs, and handkerchiefs) and where they can earn between IDR 700,000 and IDR 1,200,000 (US$50 and US$86) per month. Others find work in the growing number of local shops and food stalls or as nannies or housemaids in the homes of their wealthier neighbours. For men, there have been fewer shifts in the kinds of work available and the majority work in construction or in small-scale animal husbandry.

Meanwhile, the richer households in the village have profited from developments in village infrastructure. We found four wealthy people active in the contracting business. They established contracting companies to tender for infrastructure projects including construction/repair of roads and public buildings and so on. They also rent out construction equipment and supply construction materials and workers. While there are some "new rich," most of the village's current economic elite are descendants of the old village elite, particularly the descendants of Kaliloro's five pre-1946 village heads, who own extensive residential and

---

[12] US$1 is approximately 14,000 Indonesian Rupiah (IDR).

home-garden (*pekarangan*) land in strategic locations near the main road. Today's elite occupy positions in village government as civil servants or are businessmen/women. Others have built rows of "kiosks" along the main road for rental or opened shops that hired workers staff. Others are active in agribusiness, including some medium-sized poultry farms with hundreds or thousands of birds or sales of farm inputs and animal feed. Another avenue of accumulation is the purchase of pekarangan in their own neighbourhoods, for rental or resale as the value rapidly rises.[13] The sons and daughters of these wealthy families generally complete tertiary education and hope to join the civil service, armed forces, or police. When parental land becomes available, they will generally not become farm managers themselves but parcel out the land to tenants.

Changing Lives of Young People

> Look, when we were children [in the 1920s] we used to run around naked—now the children all wear clothes and go to school and are able to do household chores by the time they are eight years old. (An old man reminiscing, 1972)
>
> Nowadays in the afternoon or evening after school the children rarely help their parents, they spend their time watching TV. (Primary school teacher, 1999)

The Prolongation of Childhood and the Emergence of Youth[14]

The prolongation of childhood (or "postponement of adulthood") in rural Java is a largely post-colonial phenomenon, resulting from more general and longer schooling, increasing age at marriage, and postponed

---

[13] The rising price of both farm land and residential land is an Indonesia-wide (and indeed, world-wide) trend, further accelerated in this region by the construction of a new international airport in the southern part of Kulon Progo district, and increasingly busy traffic on Kaliloro's main road, which is now part of a designated agro-tourism (*agro-wisata*) route.

[14] This aspect is analysed in greater detail, based on the results of time-allocation studies carried out in 1972–1973 and in 2000, in White (2012).

entry into labour markets. In the half-century since independence in 1945, each successive generation in Kaliloro has reached progressively higher education levels. While almost one-third of boys and four-fifths of all girls born in the 1920s and 1930s had no education at all, the numbers of children not receiving any education among those born in the 1950s were small for girls and insignificant for boys, and by the 1970s— despite widespread poverty—virtually all children attended primary school and many continued to lower secondary school (SMP). More remarkably, among those born in the 1970s (the early years of the Suharto regime), nearly all continued on to lower secondary school and around four-fifths (slightly more for boys, slightly less for girls) entered upper secondary school, thus continuing their education beyond the age of 15 at which compulsory education in Indonesia now ends. By 2017, nearly all the children born around the millennium—today's teenagers—are attending upper secondary school (SMA or SMK).

This has had important implications for the involvement of young people in work. A comparison of teenagers' time-use shows important changes between the early 1970s and the next generation in the early 2000s (White 1976b, 2012; White and Margiyatin 2016). In the early 1970s, both work and school were considered a normal and proper part of growing up. Virtually all children attended six years of primary school until completion, while more boys than girls attended junior secondary school (often stopping at age 15). Formal education, however, had not yet become disruptive of children's work involvement. Boys and girls of primary school age (6–12) and secondary school age (13–18) made significant contributions in both directly productive and domestic work. Boys and girls aged 6–12 worked for an average of around 30 hours per week; when hours in school are added, they were busy with work and school for around 50 hours each week. For teenage boys and girls, the gender differences were pronounced: boys spent 39 hours per week in work and a total (work and school) of 58 hours per week, while girls worked for 73 hours per week (as much as their mothers) and were generally no longer in school. Among landless and small-farm households, children contributed more than half of all working hours. Only a minority of these working hours were in agriculture; the importance of children's work lay mainly in their contributions in domestic work, firewood collection, animal care and feeding, and (for girls) handicrafts—all necessary tasks in which their contributions freed the labour of adult men and women to engage in agriculture (own farm and wage work), trade, and other activities directly productive of income. Most children in landless and land-poor households had experience of wage work and/or home-based work such as handicrafts, which generated cash income, which the children often used to provide for their own needs, such as clothes, school fees, snacks, and tobacco.

By 2000—a generation later when these children were adults and had their own children—education up to age 15 had become virtually universal and the majority of both boys and girls aged 16–18 attended upper secondary school; their education (attendance, travel, and homework) occupied increasingly more of children's time. Hours of "real" work had correspondingly declined, and this sometimes became a source of tension between parents and children, although the common parental complaint that children don't help their parents any more is an exaggeration. In the 13–18 age group, boys were still contributing 18 hours per week and girls 27 hours per week in various kinds of work, although both boys and girls were spending an average of close to 40 hours per week—which in the West would be regarded as a full-time activity—in school attendance and homework.

While children's work on the farm was not essential, both boys and girls still went to the rice fields at busy periods, particularly at harvest time. During our latest restudy in Kaliloro (2017–2018), however, we have found that today's teenagers are the first generation who, in many cases, have literally never set foot in their parents' rice fields; the process of deskilling and alienation from farming is well advanced.

The potential rupture in the regeneration of farmers and farm workers has various causes, including the increasing length of time spent in school and changes in young people's ideas and lifestyles as they engage increasingly with the outside world through the internet and social media. As already mentioned, nearly all children, including those of poor farmer and landless households, now complete secondary school. Many opt for vocational school in the hope of quickly finding work once they graduate. Lower and upper secondary school students don't get home until 3 or 4 pm and still have homework to complete. This affects not only the time they can spend helping at home or in the fields, but also the way that they think about themselves and their surroundings.

Nearly all secondary students in Kaliloro have access to a smartphone and social media accounts. This allows children and youth to bypass the old adult and adult-provided filters through which young people formed their ideas about the outside world (parents, school, religious teachers, the radio, and occasional newspapers). Nearly all of the young people that we have met during our study are active on social media (mainly Facebook and Instagram) and spend their free time in activities that have nothing to do with farming—enacting modern youth lifestyles. Once they have finished school, they hope to attain jobs in other sectors. Their parents seem aware of the changes that the new generation of teenagers are experiencing.

> Formerly, there wasn't much for children to do to amuse themselves. Now there are lessons, internet, gaming, all kinds of things to keep them busy. So they can no longer help their parents in the sawah, as I did in my child-hood. (Kukuh, 38, who farms land belonging to his father and an uncle)
> Ya what can be done? Kids today don't want to go to the sawah. Times have changed… maybe the generation of the 70s would go to the sawah, but today's kids run off to the factories. (Santoso, a 44-year-old share tenant)

Even those children who opt for the agriculture stream in vocational school generally hope to secure a factory job, as some local SMK teachers told us; less than 5 per cent of SMK agriculture graduates, they said, go into small-scale farming. But—and this is important in view of argu-ments that we will make later—one reason for this, they say, is that these young men and women don't have any prospect of access to a plot of land to cultivate themselves: "most of the children in this school come from sharecropper or landless worker families." The objective of an SMK edu-cation in agriculture, they say, is to produce graduates ready to take their place in the (agri-related) labour market. This is why the school has devel-oped links with various oil palm plantations in Kalimantan and Sumatra as well as with an agricultural machinery company in Yogyakarta. But most SMK agricultural graduates, one teacher explained, find jobs in fac-tories or shops. There is also another reason for the choice of agricultural vocational school, as a small farmer explained: "my daughter chose agriculture because it's cheaper than computer technology."

One small farmer's daughter attending the local agricultural SMK explained why she had no plans to become a farmer: as a farmer, it would be very difficult to make her family better off than it is at present, given all of the problems of small-farm size, risks of harvest failure, and low prices of farm produce. A steady job in a factory or some other business would be better, with a fixed monthly salary.

Most of today's teenagers do not see their future in farming, wanting to work outside of agriculture and outside of the village. What is less often understood is that this was also the case of many of their parents (the current generation of older farmers) and grandparents; aspirations are not a reliable guide to actual futures. As shown in previous sections, for at least three generations young people have "voted with their feet," moving to far-away destinations in search of employment. Out-migration (which, as seen above, was already common in the early 1970s) was even more prominent by 2017. As shown in Table 14.4, of all the children of current residents who had left the parental household, only one-third are still resident in Kaliloro. While some (19 per cent) remain close by in other villages in Kulon Progo district, the majority (51 per cent) have moved to other parts of Java or have left Java. What we do not know from these survey data is how many of this stream of migrants later return to the village and to farming; certainly, there are many migrants who have returned while still in their young adulthood (mid-20s and 30s) to take over land from their parents when it becomes available. This pattern continues today, as we will explore in more detail in the next section.

Pathways Out of and into Farming: Becoming a Young Farmer

Among the 35 young farmers—male and female—whom we interviewed,[15] we found both similarities and differences in their migration histories, the reasons they decided to become farmers, mechanisms of intergenera-

---

[15] Aprilia Ambarwati, Charina Chazali, and Hanny Wijaya conducted these interviews.

tional land transfer, crops planted, off-farm activities, and their involvement in farmers' groups. In this section, besides some general observations, we will present seven contrasting individual cases to illustrate the variety of pathways out of and into farming.

Almost all of these respondents came from landless or small-farmer families. Only one woman, who controls 1.1 ha of land, is a village official from an elite background. The others cultivate farms between 500 and 6000 m² (0.05–0.6 ha), mostly as tenants renting or sharecropping the land, whether from their own parents or other owners. One other exception is a young man who owns only 400 m² himself but sharecrops an additional 1.0 ha belonging to an owner who lives outside the village.

The majority of our respondents, both male and female, have a history of migration to the Jakarta region (Tangerang, Cikarang, Bekasi), Sumatra (Riau, Padang, Jambi, Batam), Kalimantan, or Malaysia after completing secondary school. Most of those migrating to Jakarta or Batam found jobs in factories (textile, footwear, dolls, automobile parts, machine assembly). Most of those who went to Sumatra or Kalimantan worked on oil palm plantations, while those who went to Malaysia worked in construction, except one who worked on a large-scale watermelon farm. Most of them reported using their wages for daily needs and savings for the future; only a few reported sending money home to their families in the village. Many of them used the services of a broker, both for travel and for finding work, while others were helped by relatives, friends, or neighbours who had gone before them. They also mentioned a variety of reasons for their decision to return to the village: expiry of their labour contract, not feeling at home in the city, to get married, or to care for ailing parents.

We only found three young farmers among the 35 who had no previous migration history, and only two of these had turned to farming as soon as they left school (the third remained in the village but doing non-farm work before turning to farming). The kinds of jobs available locally to school leavers who don't migrate include working in shops or restaurants, various kinds of casual labour, motorcycle taxi (*ojek*) driving, and in recent years, employment in the various manufacturing industries that are now within commuting distance from the village.

One of these "early continuers" is Yanto, who was 21 years old at the time of the interview.

Yanto, a young bachelor, lives with his father, grandfather, and two younger siblings. Since completing lower secondary school at about age 15, he has helped his father in the sawah and with collecting stones from the river bed. His mother died when he was still in lower secondary school and he felt an obligation to help with father in his rice fields. Presently, Yanto has 500 m² (0.05 ha) of sawah, inherited from his mother and registered in his name. His mother bequeathed all of her sawah to Yanto; his two younger sisters are still in (junior secondary and primary) school. Having such a small holding, his mother's main reason for leaving him all of her land was in consideration of who was most capable of using it productively and keeping the land in the family. During his childhood and adolescence, Yanto regularly helped his parents with farm work, and his mother saw that her son was more likely to become a farmer than his sisters.

He also helps his father with hoeing, weeding, and harvesting on the 1200 m² of sawah that he sharecrops from a relative. On his own land, Yanto uses a different crop mix than his father. Besides the usual rice and polowijo crops, he often plants chillies and vegetables. He explained that these are crops that bring in money, as their own family's rice harvest is not sold to a penebas but harvested for the family's consumption. For cash, he relies more on non-farm work; he looks after two cows, earns wages by transporting timber, and delivers manure to an organic fertilizer business. He uses the earnings to buy food for the family and to pay for his two younger siblings' school fees.

He says that he is happy to farm and care for animals. One reason is that he owns his own small plot of sawah, and another is that as a farmer, you are freer and more relaxed than a factory worker. "If you work in a factory and get sick or exhausted, you have to keep on working. But as farmers, we're free to manage our own time. If we feel sick or tired, we can just stop working for a while."

One main challenge that farmers face, says Yanto, is the low prices that they receive for their produce. Chillies and vegetable prices, for example, are low and unpredictable, which, he says, dissuades most young people from taking up farming.

A second case of a younger person who started farming as soon as he left school is Budiman, who is 31 years old and still single.

Budiman began farming when he was 18. His father owns 2000 m² (0.2 ha) of sawah in two separate plots. Part of it is planted in the traditional paddy-paddy-polowijo rotation, while the other is for horticulture (including chillies and vegetables). It was Budiman who suggested planting chillies and his father agreed; the father decided to plant the other vegetables, based on his experience as a transmigrant settler in Kalimantan in the 1990s. He decided to join the transmigration programme because they were promised 2.0 ha of rainfed (*tegalan*) land. The family stayed for only 10 years in Kalimantan, returning to Kaliloro because the father was often sick (some thought he was the victim of black magic). Since they returned, Budiman has helped his father on the farm; while they were away, a relative farmed the land. He is the only child living at home. His elder sister lives with her family in Sumatra and his younger brother died when still in secondary school. Budiman helps with all stages of farm work. A portion of their rice harvest is sold to a penebas, part is kept for home consumption, and all of their other crops are sold. Budiman gives all of the money from these sales to his mother. He says he has absolutely no problem with that, as the money is used for the family's needs, including his ailing father's medical care, food, electricity bills, and working capital for the next growing season. His mother also puts some money away as savings for future needs (including Budiman's marriage). When he needs money for himself, Budiman works for wages for other farmers on their land.

Budiman has managed the farm by himself for the past five years as his father is elderly and ailing. Budiman's father, although still the formal owner, allows Budiman to take all important decisions about its management. This is in contrast to many other young farmers, whose parents still retain control of both the land and decisions about its management.

Like Yanto, his decision to remain in farming was because he "feels he owns" the land that he cultivates, although it's still registered in his father's name. Budiman's future in farming is assured—he will inherit all of the land. His father (who was present at the interview) told us: "His sister is far away and won't be returning to the village. And she and her husband have their own land there." Budiman, however, quickly denied this, saying that it's possible that his sister will get half of the sawah; only they would have to make an agreement about its use and the sharing of the proceeds of the harvest.

Most of our young-farmer informants said that when they cultivate their parents' land, they still have to give the parents a share of the crop (and if their parents are share tenants, the harvest has to be divided not two but three ways—owner, parents, young farmer). This makes it even more important to have non-farm income. One such case is Darmi, a 45-year-old female farmer who is a share tenant on her parents-in-law's land.

Darmi is originally from Wonogiri (in the next district). She left school after junior secondary, then migrated to Solo and later Jakarta, working as housemaid or in restaurants; she also worked for a time in a textile factory in Karanganyar. She moved to Kaliloro in 2000 when she married a local man whom she met when he was working in construction in Wonogiri. His parents own 500 m² of sawah. Although Darmi's parents also own a small amount of sawah, she said she had never helped them with farm work; her first experience of this was helping her parents-in-law to plant and weed. Her mother-in-law also encouraged Darmi to join her transplanting rice for wages.

About five years ago, Darmi began to cultivate her parents-in-law's land. Although she and her husband now do all of the work—the husband does the hoeing, Darmi does all of the rest—they still have to give half of the harvest (half of 350–400 kilogrammes) to the parents. Darmi does not sell her share but uses it for home consumption. For extra income, she does agricultural wage work and weaves bamboo mats (*tikar*) at home. Her biggest source of income, she says, is agricultural wage work. At harvest time, she can earn up to 100 kilogrammes of unhusked rice. Other income comes from her husband's wages as a construction worker. He also makes furniture (tables, chairs) to order and works in a neighbour's small furniture business. They also raise catfish (*ikan lele*) in a fishpond near the house.

Regarding the future of her parents-in-law's sawah, Darmi is rather worried. Her father-in-law, she told us, plans to give it to his daughter who lives in Jambi (Sumatra), as all of the four other children have already been given some pekarangan land. She also has doubts about this plan, as her sister-in-law is unlikely to want to return home and farm the land. If the sister-in-law is indeed given the land, Darmi hopes to rent it from her. "What's for sure is that 500 m² of land can't be divided among six heirs—how much would each get?"

The uncertain prospect of land ownership makes Darmi push her two children towards other occupations. Her first daughter is working in an underwear factory in Sragen, a district in the eastern part of Central Java, while the second is still at vocational secondary school. "I hope my children won't become farmers and have the same difficult life as their mother. I'll be happy if they get jobs in the city, or even better in Korea with a good salary."

Gianto, age 36, also has a share tenancy relationship with his parents. Gianto and his father cultivate 4100 m² of sawah in four separate plots, only one of which they own. Gianto's father is the tenant of the other three share-tenanted plots. Gianto has no experience of migration, although he says that he would really like to migrate. "I was jealous when my younger brother went off to Cikarang as soon as he finished secondary school. He works in a factory. I really wanted to go to Jakarta, but then my father got sick with asthma, so I have to look after him and help with the [farming] work." At first, his father encouraged him to join him in collecting bamboo. He earned extra money as a motorcycle taxi driver (*tukang ojek*). Only a few years later did his father ask him to take over the work in the sawah. This was easy for him, as he had helped his parents in the sawah from an early age. His mother has a chronic health problem and has for many years been unable to help on the farm.

Although Gianto now contributes more of the farm work than his father, it is still the latter who takes all important decisions, including the timing of planting, choice of seed, fertilizer, and pesticides as well as decisions during the harvest period. Gianto's mother organizes the harvest, including how much paddy (or cash) will be given to Gianto after the owner has received his share. His father reserves part of the harvest earnings for the next season's cultivation costs.

For his own cash needs, Gianto buys stands of growing bamboo in the village, which he sells to a trader. His wife also runs a small food stall (*warung*) in front of their house. She also often helps with planting, harvesting, weeding, and sometimes fertilizing. Gianto also cares for two cows: one owned jointly with his father, the other in a *gaduh*[16] agreement with a neighbour. Parental dominance is clearly seen in this family. Although Gianto is active in the farmers' group (*Poktan*), it's his father who is registered as member.

Regarding his prospect of inheriting land, Gianto said the sawah will be shared between him and his younger sister, a factory worker in Cikampek. "My father told me that it will be shared, but that my share will be greater."

The intergenerational transfer of agrarian resources is indeed a sensitive issue. Most of our informants felt uncomfortable when we raised the question of land inheritance. They told us that making an issue of land inheritance before the parents have died is taboo and not right. This, however, makes it difficult for the younger generation to request land from their parents. In many cases, although they do the bulk of the work on their parents' land, the parents still have the right to the harvest so long as they live and will determine how much paddy or cash is given to the young farmer.

Regardless, many young people decide to enter into such a relationship, working on their parents' land. In many cases, the parents are sick or elderly and need their children to work the farm. Many also imagine that helping on the farm will ensure that they will inherit the land when the parents pass away; cultivating the land while the parents are alive is proof that they are capable of being good farmers. Another reason for choosing to work on parental land is that they are freed from (part of) the production costs, compared to when they become share tenants on another's land. Generally, the parent will arrange for the next season's cultivation costs to be reserved from the last harvest.

Most young farmers, then, will only gain full ownership and control of the land when the parent who provides the land dies; while parents are still living, they are only "helping on my parents' farm" (if the parent is still involved in farm work) or "working my parents' land" (if the parent no longer contributes work but still controls the product).

---

[16] *Gaduh* (agisting) is a form of livestock sharing where A raises an animal belonging to B and is given half of the sale price when it is sold, or half of the offspring.

There is no strict rule about the division of inherited land between sons and daughters—both normally inherit. There are cases where both inherit equally, and others where the sons' share was greater. The most important factor determining who actually has the opportunity to farm the land is which child has remained in the village and has farming experience. The siblings will come to an agreement on the division of the harvest, including a greater share for the one who cultivates the land. This often happens when the sawah in question is very small; a plot of 500 m², for example, will normally not be further divided but worked by one of the heirs with a crop-sharing agreement. Young (would-be) farmers with no prospect of inheritance, with only a tiny plot to inherit, or facing a long wait before taking over the land, will opt to enter a rental or share tenancy agreement with a landowner. An illustration of this is Santoso and his wife Watinah, a couple now in their early 40s.

Watinah and Santoso are from small-farmer families. They both completed secondary school and frequently helped their parents in the fields. Watinah's parents owned 600 m² of sawah, while Santoso's parents were landless share tenants. While still young, they both migrated for work. Watinah worked in a shoe factory in Tangerang as soon as she left school in 1993 and stayed until 2001 when she returned to the village to marry. Santoso first stayed in the village, helping his father on the farm and working as a casual labourer before becoming a travelling salesman in Jakarta from 1999 to 2001. He said that he had migrated to look for new experiences. In Jakarta, he met Watinah, also from Kaliloro. Before marrying, he bought a small plot of sawah (300 m²) with the help of a bank loan. Returning to the village, they say, was the natural decision to make when they wanted to form a family. "Life in the city is very expensive; our salaries are hardly enough to live on, and certainly not if we have children." They now have one daughter in junior secondary school.

Santoso's father had succeeded in obtaining a large (1.0 ha) share tenancy from an absentee owner. As soon as the couple returned to Kaliloro, they began to farm their tiny plot of sawah and helped Santoso's father on his tenanted farm. At first, the work was evenly divided between father and son, and the father often divided the harvest (paddy and/or cash) with him, after delivering the landlord's 50 per cent share. Since 2010, however, Santoso has completely taken over the cultivation, as his father is too old to work. Watinah helps with planting, weeding, and harvesting. They still give Santoso's parents a share of the harvest, in kind or cash. Meanwhile, Watinah's parents have given her 300 m^2 of sawah. The couple are over 75 years old and have divided their land among their children. Her elder brother received a plot of the same size; Watinah says her plot is better as her brother's plot is subject to flooding. Watinah then rented this land out, receiving IDR 2 million (US\$143) for a four-year lease; she gave all of this money to her mother.

With a relatively large farm to cultivate, Santoso and his wife are one of the few couples with no other (non-farm) source of income, besides their one cow and three goats. Santoso explained that he uses the proceeds from the sale of the animals to pay school fees, while the proceeds from the farm are for daily needs. They both said that they intend to continue farming since they now have some land of their own and also a sizable, tenanted area. They don't expect their daughter to become a farmer, as she has never helped in the fields and knows nothing about farming; a steady job in the city, they say, would be better for her.

Karso, age 41, is another illustration of a young farmer who has no prospect of land to inherit.

As soon as he had completed junior secondary school (at about age 15 or 16), Karso migrated to Tangerang to work in a textile factory. "At that time, 10 of us, all the same age, went to Tangerang together. Some of them are still there; others have returned to the village like me and become farmers." He stayed only two and a half years in Tangerang; city life, he said, is no good as wages are low and living costs high. Returning to the village, he worked for some time as a casual labourer until he was offered work on the village secretary's watermelon farm, looking after the plants from sowing to harvest. In 1999, he migrated again, this time to Cilegon to work in a relative's furniture business. He left after only one year as the wages were very low. In 2001, he married a Kaliloro girl and found work in Purworedjo (the next district) on a friend's watermelon farm. After a year, he decided to stop when his wife gave birth to their first child and has stayed in Kaliloro since then. He helps his father, a share tenant, on 1500 m² of sawah. As this is not nearly enough to support his family, he works part-time as a tractor operator and sometimes in construction or gathering stones from the riverbed; currently he earns wages both as tractor driver and works in a local batako.

According to Karso:

> the main barrier to becoming a young farmer is land. If you want to become a farmer, you must have land. Farming like I do, I'm really just a labourer, as I have to cultivate someone else's land.
>
> I became a rice farmer because I had no other choice; in fact I didn't want to, because there's no profit in it. As a share tenant, my share of the harvest is only IDR 700,000 (US$50), while the costs can be up to IDR 500,000 (US$36).

In fact, he would really like to shift to horticulture and grow chillies or watermelon, for example. But this requires a lot of capital. "I can't afford it," he told us.

The young women farmers in our sample experience the intersection of gender, generation, and class.

Yaya is 24 years old and married with a 4-year-old son. Orphaned when she was 5 years old, she was working at the age of 12, but her employer supported her education until she completed (vocational) secondary school with a qualification in secretarial work. After leaving school, she also left the village to work as a shop assistant and then in a food stall. When she was 20, Yaya married Jarwo and returned to the village. She is completely dependent on Jarwo's father for access to land. He owns only 700 m² of land used for rice cultivation, but as the neighbourhood head, he gets 0.6 hectares of village-owned irrigated rice fields in place of a salary. After two years of working for other farmers and helping her father-in-law, Jaya took over management of some of the land and now farms 2400 m², somewhat more than the average farm size in the village, as her father-in-law's share tenant. She gives him one half of the crop. Yaya has been the "main" farmer from the beginning. Jarwo does other work that brings in money more regularly than farming. "I decide almost everything," says Jaya, "and do almost all the work, choosing the seed variety, making the seed bed, germinating the seeds, levelling the field, making the lines for the planting, recruiting and paying the planters, weeding, fertilizing, spraying and checking the crop every day." Despite being the main farm manager, Yaya does not attend the meetings of the local farmers' group since it is assumed that the members are men.

Yaya and her husband are busy earning wages in a range of activities. Jarwo works as a tractor operator and for a coconut oil enterprise. Yaya earns wages both as a farm labourer (planting, weeding, and harvesting) and in handicrafts, making woven laundry baskets for export on a putting-out basis. Yaya and Jarwo have also organized a group of four *tebasan* harvesters, working with another young couple for a middleman in the next village and using a small portable thresher.

Thirty-six-year-old Partini, like Yaya, became involved in farming only after marriage. During her childhood, she never worked in the fields. As soon as she finished secondary school in 1997, she moved first to Riau island to work on an oil palm plantation, then to West Java to work in a shoe factory. A year later, she relocated to Batam island to work in a CD-ROM factory and after three years, moved back to West Java where she found a job in a toy factory. When she moved back to the village to live with her parents, she married Sarwidi. For the first nine years of their marriage, they had no land of their own. To make ends meet, Sarwidi worked in construction while Partini worked for wages planting and harvesting, and they have continued working for wages to the present, in addition to cultivating her parents' and parents-in-law's land as share tenants. Partini now farms 1800 m² (almost half an acre) of land in three different plots. About 1200 m² is two rent-free plots, owned by Partini's parents-in-law respectively, while she sharecrops the third plot of 600 m², which belongs to Partini's mother and aunt. Unlike Yaya, Partini plants both rice and vegetables: rice on the sharecropped land, and chillies, cucumber, and some other vegetables on most of her own land. This combination guarantees the supply of both rice and cash. Partini is involved in all stages of rice cultivation, and is the decision maker in most of them, including choice of seed variety and fertilizers as well as deciding when to plant, weed, apply fertilizer, and harvest. In the last planting season before the interview, when she decided to try fertilizer in tablet form, "[Sarwidi] just went along with it, leaving it to me as I am the one who applies the fertilizer." Partini, unlike Yaya, has no significant non-farm activities as she is busy every day looking after the chillies, while Sarwidi works in construction and tends their goats. Partini estimates that their non-farm income provides about 60 per cent of total income and farming 40 per cent.

Yaya and Partini's experiences clearly show that they are both real farmers (not just farm helpers) with knowledge and direct involvement in farm management. But there is no farmer organization or group that recognizes them as farmers. Neither Yaya nor Partini is registered as members of the local farmers' group—nearly all registered members of this state-sponsored farmer group are men. We have only come across one registered woman member, an interesting case because of her different position in the village class structure. She provides our final illustration.

Menik, now aged 39, manages her farm in a quite different way from Yaya and Partini—she uses wage workers. Menik comes from a wealthy family. Her grandfather was a village head and a large landowner, her father was a teacher and civil servant, and her mother was a housewife. Menik herself is a graduate (in agriculture) of Muhammadiyah University, which is located in the nearby city of Yogyakarta. After a short period of employment in Kalimantan, she obtained a position in Kaliloro's Village Finance Institute (LKM), and in 2009, she became an assistant village official. In this position, she receives 1.0 ha of village-owned *pelungguh* land in lieu of salary. She parcels out 0.7 ha of this land to various share tenants, but has decided to manage 0.3 ha herself, using wage workers. Since her parents were landowners but not farmers, is actually a "newcomer" farmer, although not in the sense in which the term is normally used (see the explanations of "continuer" and "newcomer" farmers in Chap. 1). She and her brothers also inherited land from their father; she and her two sisters received the same amount (0.15 ha), while her elder brother received a double portion of 0.3 ha. Neither her brother nor sisters farm their land themselves. Menik is the only one. She is also the first in the family to manage a part of her land as a commercial farmer. Her biggest farm profits come from chillies. Each harvest can earn her IDR 30 million (US$214), sometimes even more, after deducting all of the costs including hired labour. To this, she can add half of the proceeds of the rice harvest of her 0.7 of share-tenanted land; she always sells the standing crop so the income is in cash. She has invested part of the surplus earned from agriculture in non-farm enterprises, including a laundry, a poultry and livestock feed store, and a catering and wedding service that her husband runs. She has also established a commercial poultry farm that a neighbour manages on a profit-sharing basis. In turn, she has invested part of the proceeds of this non-farm income back into land, buying 0.25 ha of residential/garden land. Unlike Yaya and Partini, Menik is an active member of the local farmers' group, despite the general assumption that group members will be men; she attends meetings and voices her opinions. Owning land in her own right as well as her salary land, she has no concerns about her continued existence as a farmer.

Access to land, thus, has helped Menik to consolidate her position among the village bourgeoisie. Together with a mixed portfolio of income sources, these have allowed her to further accumulate land.

Collective Farming for Youth: The *Karang Taruna* Project[17]

In all Indonesian villages, there are state-sponsored youth groups called *Karang Taruna*. These groups are expected to be active in organizing sports, preparing for the national Independence Day festivities, and so on. In 2017, in one corner of the village, the leader of the local Karang Taruna group, himself a share tenant and former migrant now in his 30s, encouraged the younger members to apply to rent a plot of rice land from the village government, and experiment with collective farming. He wanted to find a way for these teenagers to learn the basics of farming, to ready them for the time when they may also decide to return from migrant work and become farmers. "With this collective farming project,

---

[17] This project has been analysed in greater detail in Ambarwati et al. (2017).

these teenagers who have never worked in the fields will know how to plant and do all the other tasks… If they don't make a success of life in the city, they'll certainly come home, and then what work is there for them except to become a farmer?" Despite initial opposition from the village government, the group lobbied until they got their way. They came in large groups to plant the rice, to weed it, and to harvest it. They were proud that despite their lack of experience, their harvest was no smaller than that of the neighbouring farmers. By 2022, they were into their ninth planting season and had developed various other income-earning activities as well as organizing training sessions on making organic fertilizer.[18]

Concluding Reflections

The main conclusion from this study is that nearly all of today's young farmers in Kaliloro have returned to farming after an initial period of out-migration. This confirms the importance of a life-course approach to the social reproduction of smallholder farming.

Thus, the typical "young farmer" in Kaliloro began farming in his (or her) mid-20s or even 30s and has a history of prior non-farm employment (usually involving a period of migration) before turning to farming. Many of them have no significant experience of helping on their parents' farm. Smaller numbers have stayed in the village to help their parents before taking over (part of) the parental farm land. There are a few from landless households who take over tenanted land that their parents formerly cultivated. Young farmers' livelihoods—like those of their parents—are built through pluriactivity: living from a small holding plus other sources (animals, wage work, petty trade, services, etc.). Young farmers also tend to be more innovative than their parents, though in modest ways, like growing vegetables on (part of) their rice fields.

While some of today's young farmers, who were teenagers at the time of our 2000 time-budget study, were used to helping their parents with

---

[18] Meanwhile, two other Karang Taruna groups in Kaliloro followed their example and started collective farming projects, but these appear to have been short-lived.

farm work during their adolescence, the current generation of teenagers is the first generation who have no or hardly any experience of farm work.

The Karang Taruna collective farming project gives some reason for optimism that despite their deskilling and relative alienation from farming, it is not farming as such that these young people are allergic to. They do not want to spend their young adulthood helping their parents in a position of dependency, and maybe in future, they do not want to farm in the same ways that their parents farm. But they—or at least some of them—are willing to consider other styles of farming for the future.

References

Abdullah, Irwan, and Ben White. 2006. Harvesting and house building: Decline and persistence of reciprocal labour in a Javanese village, 1973–2000. In *Ropewalking and safety nets: Local ways of managing insecurities in Indonesia*, ed. J. Koning and F. Hüsken, 55–78. Leiden and Boston: Brill.

Ambarwati, Aprilia, Charina Chazali, Hanny Wijaya, and Ben White. 2017. Rural youth and Karang Taruna: Collective farming in Kulon Progo [Pemuda desa dan Karang Taruna: bertani kolektif di Kulon Progo]. *Jurnal Analisis Sosial* [Journal of Social Analysis] 20 (1–2): 124–145.

Mather, Celia. 1983. Industrialization in the Tangerang Regency of West Java: Women workers and the Islamic patriarchy. *Bulletin of Concerned Asian Scholars* 15 (2): 2–17.

Stoler, Ann. 1977. Class structure and female autonomy in rural Java. *Signs* 3 (1): 74–92.

White, Ben. 1976a. Production and reproduction in a Javanese village. PhD diss. Columbia University.

———. 1976b. Population, involution and employment in rural Java. *Development and Change* 7: 267–290.

———. 2012. Changing childhoods: Javanese village children in three generations. *Journal of Agrarian Change* 12 (1): 81–97.

———. 2018. Marx and Chayanov at the margins: Understanding agrarian change in Java. *The Journal of Peasant Studies* 45 (5–6): 1108–1126.

White, Ben, and Ugik Margiyatin. 2016. Teenage experiences of school, work and life in a Javanese village. In *Youth identities and social transformations in modern Indonesia*, ed. K. Robinson, 50–68. Leiden: Brill.

Wijaya, Hanny. 2016. Women workers in the process of reproduction of rural households: Case study of Kaliloro, Kulonprogo [Perempuan buruh dalam proses reproduksi keluarga petani pedesaan. Kasus desa Kaliloro, Kulonprogo]. MA thesis. Gadjah Mada University.

Wolf, Diana. 1992. *Factory daughters: Gender, household dynamics and rural industrialization in Java.* Berkeley: University of California Press.

15

Conclusion: Youth Aspirations, Trajectories, and Farming Futures

A. Haroon Akram-Lodhi and Roy Huijsmans

Farming, the Demographic Challenge, and Justifying Case Selection

This book commenced with a question of global importance: in a world in which farming populations are ageing, who is going to provide the planet's peoples with the "sufficient, safe and nutritious food" that is needed to meet the "dietary needs and food preferences for an active and healthy life" (FAO 2006)? In other words, where are the people who are needed to generationally renew farming? As explained in the introduction, addressing this question meant going against the grain of much

A. H. Akram-Lodhi (✉)
Trent University, Peterborough, ON, Canada
e-mail: haroonakramlodhi@trentu.ca

R. Huijsmans
International Institute of Social Studies, The Hague, The Netherlands
e-mail: huijsmans@iss.nl

© The Author(s) 2024
S. Srinivasan (ed.), *Becoming A Young Farmer*, Rethinking Rural,
https://doi.org/10.1007/978-3-031-15233-7_15

research on youth and agriculture. Rather than seeking to understand youth's apparent disinterest in farming and their exodus from the countryside, the research teams focused on those youth and young adults who stayed in, returned, or relocated to rural areas and were involved in farming (often alongside various other economic activities). Thereby, the case studies presented in this book have put in the spotlight the next generation of farmers. In this concluding chapter, we draw out some important issues emerging from across the chapters and reflect on key differences. This way, we reiterate the various pathways of becoming a farmer, the main challenges experienced by these young farming women and men, and the roles that policies and organizations could play in facilitating the process of becoming a farmer.

That global farming populations are ageing is beyond dispute. In the developed world, the average age of an American farmer was 59 in 2017, up from 57 in 2007 (AGDAILY Reporters 2020); in the European Union, one-third of farm managers were 65 or older in 2018 (Eurostat 2018); and in Japan in 2015, the average age of a farmer was 67 (nippon. com 2018). Farm populations are also ageing in the global South. Based on a restricted sample of developing countries, the average age of the head of a farming household is 50; in Sub-Saharan Africa, it is 32 (Arslan 2019). What do these figures mean? Are older farmers no longer replaced by younger farmers and, thus, are we facing a gradual disappearance of smallholder farming? This is indeed what some studies suggest by pointing at youth's apparent disinterest in a farming future and their out-migration from the countryside. This line of reasoning would render the question of the future of smallholder farming a youth question. Or do farmers continue farming at an ever-increasing age, and does a new generation of farmers enter the vocation at an age when they are no longer "young," chronologically speaking? This would imply focusing on the intergenerational dynamics shaping the various delays in the generational renewal of farming. As Ben White asks: "Are young potential farm successors reluctant to start, or are they unable to start because their parents are unable or unwilling to stop?" (2020, 9). Or, do these questions simply get the story wrong by reducing farming to an exclusive occupation that fails to recognize how rural livelihoods, increasingly, are about much more than the land and straddle the rural-urban divide? This would direct our focus to changes in rural livelihoods and agrarian relations. Creating space for these different explanations led to the methodological decision to focus empirically on youth's and young adults' pathways into farming.

In order to understand young people's pathways into farming, this book adopted a comparative approach. It analysed four countries, which at first glance, might be thought to occupy very different levels of "development," as illustrated in Table 15.1. If the level of development is evaluated in terms of gross domestic product (GDP) per capita in 2020, Canada has 4 times more income per person than China, 11 times more income per person than Indonesia, and more than 20 times more income per person than India. If evaluated in terms of economic structure, Canada is more reliant on services production than China, which is more reliant on services production than India, which is slightly more reliant on services production than Indonesia. If evaluated in terms of the Human Development Index, which compresses income per person, educational achievement, and health care status into one figure, Canada has significantly greater human development than China or Indonesia, which are roughly comparable, and both of which have greater human development than India.

In agriculture, too, the countries appear at first sight to be quite different, as illustrated in Table 15.2. Canada's average farm size is well over 200 times that of India, and even more compared to Indonesia and China. Canada's farms are highly capitalized; in India and Indonesia the extent of capitalization per agricultural employee is a small fraction of that of Canada. In terms of productivity, there is also significant variability. Per unit of land, China is the most productive and almost twice as

Table 15.1 Comparative development indicators

	GDP per capita, constant 2015 US dollars	Share of services in GDP, %	Human Development Index, 2020
Canada	42,108	67	0.93
China	10,431	54	0.76
India	1798	49	0.65
Indonesia	3757	44	0.72

Sources: World Bank (n.d.), UNDP (2020)

Table 15.2 Comparative agricultural indicators

	Average farm size, hectares	Net capital stock per agricultural employee, constant 2014–2016 US dollars	Cereal production kilogrammes per hectare	Agricultural value added per agricultural worker, constant 2015 US dollars
Canada	273.4	135739	3879	113113
China	0.4	Not available	6081	5609
India	1.3	3979	3248	2076
Indonesia	0.8	5366	5227	3600

Sources: Adamopoulos and Restuccia (2014), Li et al. (2018), FAO (n.d.), World Bank (n.d.)

productive as India. However, per unit of labour, Canada's productivity is 20 times that of China and even greater than that of Indonesia and India.

Notwithstanding these differences, there are also some important similarities. First, in all four countries, agriculture constituted by far the smallest economic sector in 2017, ranging from 16 per cent of GDP in India to 2 per cent of GDP in Canada (World Bank n.d.). Second, the vast majority of farms are often described as so-called family farms. In China, India, and Indonesia as of 2015, more than 99 per cent of all farms were defined by their respective states as family farms (Araghi 1995).[1] In Canada, the definition of family farms straddles at least three management categories, which together suggest that between 73 and 97 per cent of farms might be defined as family farms (Statistics Canada 2017: Chart 7). These diverse definitions of the family farm, which are not based upon the size or scale of production, nonetheless share a common feature: those who work on the farm come from the same family, which also means that not all who are part of the family work on the farm. In fact, as pointed out in Chap. 1, it may only be one family member who does so. Despite this common form of farm management, trends in farm size diverge; in Canada and China average farm size is increasing, while in India and Indonesia, it is decreasing (Lu this volume, Adamopoulos and Restuccia 2014).

---

[1] Note that more recently China has defined a family farm as being large-scale, as explained in Chap. 5.

Third, in all four countries, rural incomes are lower than urban incomes, although the magnitude of the dispersion differs. In India and Indonesia, urban-rural income gaps are narrowing, while in China they are widening (Imai and Malaeb 2018). Recent trends in Canada are less clear, but there is no doubt that net farm incomes in Canada have been subject to a trend decline for decades (National Farmers Union 2020). With relatively low incomes, it is not surprising that in Canada, China, and India, rural farm households face a significant issue with debt, although in China, the local government often takes responsibility for the debt burdens of rural households (GOI 2021; National Farmers Union 2020; Ye et al. 2021). In this light, in all four countries, farm incomes alone are an inadequate basis by which to support a household, and so in all four countries, rural livelihoods are built by diversifying income sources out of farming into rural non-farm and off-farm employment, which can include a significant element of rural-urban migration (National Farmers Union 2020; Schenck 2018; Kumar Das and Ganesh Kumar 2018; Han and Lin 2021).[2]

Findings from the Case Studies

Pathways, Generation, and Innovation

The research teams worked with an age range of 18–45 years of age in order to capture people's trajectories of becoming a farmer, which is often a long-drawn-out process that easily takes up to a decade (Monllor 2012, 10). In addition, the research team also inquired about respondents' childhood recollections of farming, if any. In Monllor's framework, these early encounters with farm work are not part of the pathway of becoming a farmer but are important experiences nonetheless shaping people's disposition vis-à-vis a possible farming future (Huijsmans et al. 2021). Because of this sampling frame, some of the "young farmers" studied can be considered youths, while others are more appropriately referred to as (young) adults.

---

[2] Note that this source is an MDPI pay-to-publish journal and is not adequately reviewed.

The literature on generational renewal in farming distinguishes between different pathways of becoming a farmer: "continuer" and "newcomer" pathways. The introductory chapters to the country case studies make mention of such "newcomer" farmers in all four country contexts. However, only in the Canadian part of the study and in isolated cases in Indonesia are newcomers (those entering farming without a farming background) captured in the study sample. In the Manitoba study, ten such newcomers were interviewed; half were women coming from non-farming families who married men with farming backgrounds. The remaining were two couples and one single man, who all entered farming without coming from a farming family. These latter three farming households practice alternative methods of production and use direct marketing—something that is much less common among continuer farmers. Without the prospect of acquiring agricultural knowledge and resources through the family line, their trajectories into farming comprise internships on other people's farms and include an interesting case of a young couple developing a relationship with an older farmer leading to a succession plan in which the young couple would take over the older farmer's farm while also benefiting from his mentorship. The Ontario sample included a larger share of so-called newcomers: 21 (versus 28 continuers). These newcomers typically entered farming at a slightly older age than their continuer peers and had established themselves as independent farmers, on average, at the age of 27 (versus continuer farmers, on average, at the age of 24). Other notable differences were farm size and farm type. The average acreage among the newcomers was 56 acres (versus 232 acres for continuer farmers), and newcomer farmers are more commonly involved in plant growing than in livestock farming. These differences can indeed be explained in economic terms on the basis of start-up costs, but also reflect a commitment to doing farming differently from the more conventional practices commonly found among the continuer farmers.

The Canadian study further notes differences in the educational needs and social networks of newcomer farmers. Whereas continuer farmers are part of a local, if deteriorating, social network of long standing, newcomers, by way of contrast, require technical training that they had not received and an entry point into the rural community from which they were, as newcomers, often isolated. Both were very important dimensions of success in that they afforded opportunities for learning by doing. In addition, newcomer farmers are more likely than their continuer peers to be driven by the social and environmental responsibilities that are met by a working engagement with nature as well as the autonomy that less hierarchical work affords, and these are important dimensions of their aspirations to succeed in farming.

These details from the Canadian studies sites shed important light on the question of innovation among the new generation of farmers. As we detail below, the research documented novel practices that characterize these farmers as a generation, but it would be wrong to consider young farmers as more inclined to innovate. Rather than linked to age, a commitment to doing farming differently appears, amongst other things, to be related to the different kinds of pathways of becoming a farmer (i.e., continuer, newcomer) as these related to very different motivations for entering agriculture.

Across the four study countries, our study confirms the importance of migration as part of being young and growing up in rural areas (Huijsmans 2019; Ní Laoire 2000). However, our coverage of the extended age range of 18–45 years illuminates that young people's out-migration from the countryside and away from agriculture does not need to be permanent. In fact, having spent time away from farming and agricultural work is a common feature of the trajectories into farming of many of the new-generation farmers. Next to continuers and newcomers, we might speak of a third pathway into farming: returning farmers—those who have spent their childhood on a farm, leave the agricultural sector and the countryside as youth, and return some years later as young adults. The various reasons for young people to leave farming and the countryside are well-documented (in, e.g., White 2020), but how this period of migration affects their trajectories into farming has received less attention. In the Chinese study sites, returning farmers couple a dislike for the manifest class-based challenges of the hierarchical urban life they had experienced in the city with a strong recognition of the autonomous embedded entrepreneurial possibilities offered by diverse forms of rural enterprise in a rapidly growing rural economy. The attraction of rural entrepreneurial agricultural opportunities is especially strongly pronounced by men returning farmers. For women, more often it is social norms and expectations around marriage that led them to return, to work with their husbands while performing extensive unpaid care and domestic work. Even in such gendered scenarios, the urban experience could still be of benefit. For example, one of the Chinese respondents, who had worked in a restaurant in Handan city before she returned to the countryside upon marriage, reflected as follows: "We interacted with many different people and experienced many different things and have become more outward. It is also a very useful experience when we deal with other people as a farmer." Further, several of the respondents from the Indonesia study sites commented on the role of migration in supporting innovation in farming. Working and living elsewhere, they had observed new farming practices or new crops, and the money earned through migration allowed them to experiment with such practices in their home villages as well.

The country studies also make some important observations about the category of continuer farmers. Across all four countries, schooling has taken a much more prominent place in the lives of the new generation of farmers than had been the case for their parents. First, intensified and expanded periods of schooling and the social value attached to it means that gradual and informal processes of agricultural knowledge transfer from parents to children cannot necessarily be assumed, even among continuer farmers. For example, Chap. 14 reports from a re-study in the village of Kaliloro (2017–2018) that "today's teenagers are the first generation who, in many cases, have literally never set foot in their parents' rice fields," concluding that "deskilling and alienation from farming is well advanced." The other side of the coin is that among the new generation of continuer farmers, more so than a generation ago, becoming a farmer has become an actual decision. An older farmer in one of the Indian study sites recollected how he had become a farmer. He expressed this clearly: "What aspirations? My father handed farming to me and said cultivate and feed yourself! That is it. I started farming young and did not consider anything else." Second, it cannot necessarily be assumed that continuer farmers have access to land. In a context of widespread landlessness and very small landholdings, as was true for some of the Indonesian study sites in particular, continuer farmers are often sharecroppers or labourers on other people's land or would effectively work as sharecroppers on their parents' land if this land was too small to divide any further.

Regardless of pathway into farming or country of study, for the new generation of farmers, the internet plays an important role. The internet, and especially platforms such as Baidu, Facebook, WeChat, WhatsApp, and YouTube, constitutes an important source of information for tackling problems (e.g., pests) or learning more about new forms of farming (e.g., horticulture). In addition, some of the new-generation farmers also use online platforms for the marketing of their agricultural produce. This could take the form of simply checking prices, but it also enables forms of direct marketing. While offering new opportunities, the internet occasionally also creates new problems. For example, a 19-year-old farmer from one of the Chinese study sites was cheated when purchasing seeds online for a herb used in Chinese medicine: "Internet is not always a helpful assistant. I bought fake seeds online [*Baidu*] and it cost more than 3000 yuan."

Land, Money, and Markets

When I first started, I kept wanting to buy land and I just didn't have enough equity. You rent land and you can't borrow enough money to put enough inputs in the ground to grow your crop. It was a constant battle to come up with the revenues to be able to plant the crop and get established, and then build equity so that you could buy land. I would say in my experience that's the biggest barrier of getting in; it is just getting established without [already] having someone in the industry. (Woman conventional grain farmer, Manitoba)

The main barrier to becoming a young farmer is land. If you want to become a farmer, you must have land. Farming like I do, I'm really just a labourer, as I have to cultivate someone else's land. (41-year-old farmer from Kulon Progo, who helps his father, a share tenant, farming 1500m² of irrigated rice land)

The two quotes above come from very different agrarian contexts yet are surprisingly similar in content. For virtually all new-generation farmers, accessing land, let alone purchasing land, constitutes a main difficulty in becoming a farmer. In Canada, for the new generation of farmers who inherited some land, the amount of owned land does not meet the needs of the farm. In such circumstances, buying in more land is becoming increasingly difficult as the price of farmland is rising rapidly in both provinces, resulting in a problem of land affordability for women and men farmers. Credit constraints only tighten this problem. As a result, the new-generation farmers rely on renting in land, on cash lease, or increasingly on sharecropping.

In the Indian study sites, too, the biggest challenge facing young women and men farmers across the two states is declining land size, driven by families partitioning land among succeeding generations of men. Moreover, farms face increasing challenges supplementing the land that they own through rental. Processes of financialization are generating high land price increases, in light of which those who have land want to keep control of it as a hedge against risk.

In the Hebei study site (China), the principal challenges facing young women and men farmers are access to land and access to labour. In recent history, land was intergenerationally transmitted within households, but a lack of effective social protection means that older farmers increasingly rely on maintaining access to land to provide for themselves as they age. Thus, when young men farmers split from their familial home at marriage, land is a constraint. However, different from the other study countries and sites, in Hebei, the local government recognized this constraint and worked to overcome it by making interhousehold land transfers easier while at the same time supporting land consolidation to free up access to some land for young farmers recently split from their familial home. Moreover, unlike in Sichuan (China), local government provide the framework within which local brokers operate, resulting in better terms and conditions for the cucumbers produced by young women and men farmers as their products enter regional urban markets.

In Indonesia, the average farm size falls because of the subdivision of holdings arising out of the intergenerational transfer of land through inheritance. As a consequence, land holdings are becoming increasingly marginal in size. Such land transfers, when they occur, reflect prevailing social relations, particularly gender biases, with women far less likely to inherit land, and if they do inherit land, they inherit less land than men. Moreover, in Yogyakarta, rising speculative investment in land alongside growing absentee ownership has led to increasing land prices arising out of financialization. The result is especially challenging for the landless and near-landless. Across the three Indonesian case study sites, there was little chance of (aspiring) young farmers being able to access adequate amounts of land, including through purchase or rental, to commence completely independent farming. This constraint is strongly gendered; indeed, in Flores, the practice of village exogamy means that when young women marry, they move to their husband's village, assuming responsibility for managing land but without any rights over that land. While village exogamy is common in the Indian field sites too, married women stand a relatively better chance of owning land through their husbands' families than from their natal families. The land constraint is then exacerbated by the problem of low prices for farm products, generating low returns to farming and inadequate income flows, reducing the capacity to purchase land, if it were affordable.

While access to land, and associated credit markets, are essential in order to acquire the main resources for becoming a farmer, access to markets greatly affects the financial viability of farming. For example, in Sichuan, a principal challenge facing young women and men farmers is their market dependence. Commercial vegetable growing, the principal farm activity, meant that they had a lack of bargaining power with brokers, resulting in the challenge of fluctuating market prices. Efforts to overcoming the constraint of the market saw young women and men farmers shifting into more agroecological production methodologies in an effort to identify local vegetables as being safer for eating. It also required leveraging interpersonal networks, in part because both agricultural co-operatives and agricultural technology services are not designed to meet the needs of small and medium-scale young women and men farmers. Moreover, the lack of off-farm employment options meant that for those young women and men farmers whose market dependence was posing a risk to their livelihood, diversification provided the only means by which to try and overcome that risk. The most common form of diversification is into the burgeoning agro-tourism economy. In contrast, the Indian case shows that farming can be viable if there is the proximity to the urban markets that are a key determinant of farm profitability; in this light it is interesting that higher return crops are those favoured by young women and men farmers who are close to urban markets.

Commonalities and Contrasts in the Case Studies

Having reviewed the broad findings of the case studies, we can now turn in more detail to reflect upon the critical challenges facing young women and men farmers. There is a surprising number of common issues across the case studies, along with contrasts that reinforce the challenges facing young women and men farmers in Canada, China, India, and Indonesia.

The first obvious point to be made is that both aspiring and actual young women and men farmers face an agrarian crisis. In Canada, Sichuan, Madhya Pradesh, and Java, extensive price and market risks mean that being able to construct a viable livelihood capable of supporting a young farming family is a test that requires, at a minimum, an acceptance of and active participation in pluriactivity by members of the household in order to construct the foundations of a financially sustainable rural life. Indeed, the need for and extent of pluriactivity can lead to important doubts being raised about the utility of the family farm as a concept because in many situations in the case studies, it was common to find that only one household member had any significant engagement with the farm. Moreover, in many ways, the agrarian crisis drives the demographic and hence generational challenges that contemporary farming presents. The agrarian crisis is reflected in the demands that are met when trying to identify who precisely are young farmers, and in particular why, in investigating young farmers, it is necessary to expand the definition of 'young' so as to encompass the farming aspirations of young women and men that can take a very long time indeed to come to fruition.

The agrarian crisis is more than economic. It is also a gendered crisis. First, women play different but unique and specific roles across farming systems that are feminized to differing degrees. Their responsibilities for the delivery of social reproduction services means that coping with crisis generally falls upon the shoulders of women care providers within the household. Second, in the Indian context in particular, this also produces a crisis of masculinity. As women preferred to marry out of farming, young men farmers struggle in the marriage markets, thereby creating further barriers for realizing forms of social adulthood while becoming a farmer.

Second, we found youth's out-migration from the countryside to be a prevailing reality in all four country case studies—a reality that is also strongly gendered. However, the case studies show that migration can be cyclical, including in Canada. This suggests that the lines may be blurred between being a farmer and being a non-farmer, a suggestion that is once again reinforced by widespread pluriactivity, and that rather than treating farmers and non-farmers as discrete categories it may be better to approach rural lives through the continuum of a life course perspective. The case studies systematically demonstrate that becoming a young farmer in the present times is a long process.

Third, a life course perspective can be used to understand the intergenerational tensions that are identified in the case studies and that impact upon young women and men farmers' aspirations for and pathways into farming. In particular, farm succession, which governs intergenerational transmission of land, and which is important in providing the structural foundations by which young women and men can enter into farming, remains an area where open inter-familial conversations often continue to be unthinkable and, even where they do happen, remain strongly and pervasively gendered. In Canada, the lack of explicit succession planning is reflected in the exit of many young women and men from farming, to which they return later when the intergenerational transmission of land becomes more realistic. Similar, if not explicit, processes were witnessed in Indonesia. In the same vein, in Hebei province and in Indonesia, the lack of adequate social protection means that many parents are retaining land well into an advanced age, precluding access to adequate amounts of land when young men split from their familial home at marriage, and in so doing promoting out-migration. In India and Java, rising land prices due to financialization is also encouraging landowners to retain their holdings, reducing intergenerational transmission of land. In all four country case studies, the intergenerational transmission of land has strong gender dimensions, although the parameters of those dimensions are different in each case; in some instances, women are precluded from inheriting any land at all.

Farm succession demonstrates one dimension by which access to land is the most significant structural constraint facing aspiring and actual young women and men farmers in all four case study countries. The issue of access to land has more dimensions than this, however. The financialization of land is witnessed in all countries, barring China, and acts to restrict the supply of land while simultaneously driving up the price of land. Increased prices have made land less affordable both to buy and to rent, and in the Canadian and Indonesian cases has resulted in an increase in sharecropping. Restricted supply has made it more difficult to find land of an appropriate size; in the Canadian case, rental parcels are too large for small-scale young women and men farmers, while in the Indian and Indonesian cases, amounts of land available to rent are limited and are far less likely to be rented to young women in the instances where it was socially tolerated that young women might rent land. Furthermore, financial systems fail to compensate for these failures in land markets, as access to private sector finance for young men and women farmers is limited—and gendered—in all four countries. Finally, with constraints on access to land for young women and men farmers, there can be marked social differentiation in farming communities based upon scale of production, as inequalities in incomes and wealth are exacerbated.

Differentiation can also be noted in the production processes followed in all four countries, which are strongly gendered and shaped by generation. While many young women and men farmers pursue conventional production processes for commodity-oriented farming, which involves the heavy usage of externally purchased inputs, alternative production processes for organic and low-impact farming for more localized markets are also witnessed among some segments of young women and men farmers. More generally, the case studies demonstrate that young women and men farmers tend to be innovative in their decision-making, being willing to experiment in crop choices, production techniques, and marketing strategies to a greater extent than older farmers. To a degree, this differentiation also overlaps with the distinction between continuing and newcomer farmers, with continuing farmers being more likely to undertake commodity farming and newcomer farmers being more likely to undertake alternative farming. It also overlaps with the distinction between the rural born and the urban educated, with the urban educated being more likely to undertake alternative farming.

This study does not bring up explicitly the implications of climate change for (young) farmers' present and their future. It did not come up in the interviews except when farmers expressed an awareness that weather posed a risk to farming, as would be expected. In the Tamil Nadu case study, farmers identify water scarcity and drought as significant constraints. In Canada, a young continuing commodity grain farmer was explicitly aware of how their production techniques were hydrocarbon-intensive and that there was a need to restore soil fertility. Yet none of these examples were linked by young women and men farmers directly to climate change, and climate change will shape the future of farming in Canada, China, India, and Indonesia, impacting upon the agrarian crisis, migration, farm succession, and differentiation. Young farmers' views on and experiences of dealing with climate change will be important for future research.

Across the four country case studies, there is, in addition to commonalities, one very important contrast that serves to reinforce a key analytical point. In Hebei, two factors were demonstrated that served to result in young farming families opting to not only remain in farming but in fact to do so even when they believe it to be a second-best choice. The first factor is that for entrepreneurial young women and men farmers, farming remains an attractive occupation precisely because it offers the possibility of improved incomes, better livelihoods, and enhanced capabilities. This emerged as a result of the production of crops for niche markets in which prices remain favourable to farmers and possibilities for household accumulation remain. As a result, farming remains a positive choice rather than a last resort. Thus, unlike Canada, India, and Indonesia, in Hebei, farming offers a viable and indeed attractive livelihood, especially when compared to limited urban options. At the same time, unlike in Canada, India, and Indonesia, in Hebei and in Sichuan, significant state support was on offer for young farming families to manage price and market risks, again increasing the attractiveness of farming and its capacity to retain young women and men. The point here is that the contrast between China and the other case studies demonstrates a commonality: that the intergenerational renewal of farming requires farming to provide a livelihood that offers improved life chances for young women and men farmers and their families. That this may require pluriactivity is neither good nor bad, as historically smallholder farmers have always engaged in pluriactivity in order to sustain and enhance their livelihood, and many successful women and men farmers opt for pluriactivity in order to sustain rising incomes. Moreover, the intergenerational renewal of farming requires a supportive policy framework that brings with it the active support of the state to sustain young farming family livelihoods. Thus, in the Kebumen (central Java) case in Indonesia (Chap. 13), local government eased land constraints to a limited degree by actively managing the distribution of small amounts of land to young men farmers. Indeed, one could say that active state support is a necessary condition for dealing with the demographic challenges that contemporary farming faces. That support is precisely lacking in Canada, and most of Indonesia, where state agro-pessimism has resulted in only providing an accommodating policy framework for large-scale export-oriented commodity agriculture governed by world market prices. Livelihood possibilities and state support mean that the autonomy that many young women and men farmers seek can, to a degree, be realized.

Cumulatively, the case studies demonstrate that the widely articulated view that young women and men have an aversion to farming as an occupation is misleading. Rather, young women and men face economic and structural issues in their lives that significantly limit the viability of farming as a livelihood for a family, which together with the perceived attractions of urban life—shaped now by widely available social media—can lead to a decision to migrate, whether it be in Canada, China, India, or Indonesia. Whether that decision is fixed and settled, however, is not clear cut. Issues of access to land, access to training and education, and access to community and state mechanisms of social support can be seen in the case studies to have a bearing on the finality of the decision. What has the most bearing, however, as is seen in all of the country case studies, is the capacity of farming to provide a viable, ongoing, and sustainable livelihood for young women and men farmers that ensures not only intergenerational security, but also improved intergenerational livelihood possibilities and capabilities for their families. In effect, the rural work of young women and men needs to be revalorized, rather than devalorized, as it currently is in contemporary commodity farming. When farming does provide such a livelihood, young women and men farmers can realize their aspirations.

References

Adamopoulos, Tasso, and Diego Restuccia. 2014. The Size distribution of farms and international productivity differences. *American Economic Review* 104 (6): 1667–1697. https://www.sciencedirect.com/science/article/pii/S0305 750X15002703#s0050.

AGDAILY Reporters. 2020. Where are the new young farmers? The picture can be troubling. *AGDAILY*, June 29. Accessed 1 December 2021. https://www.agdaily.com/news/new-young-farmers-corteva/.

Araghi, F.A. 1995. "Global depeasantization, 1945–1990" (Appendix A). *Sociological Quarterly* 36 (2): 337–368. https://www.sciencedirect.com/science/article/pii/S0305750X15001217#s0075.

Arslan, Aslihan. 2019. How old is the average farmer in today's developing world? *International Fund for Agricultural Development blog*, July 1. Accessed 1 December 2021. https://www.ifad.org/en/web/latest/-/blog/how-old-is-the-average-farmer-in-today-s-developing-world-

Eurostat. 2018. Farming: profession with relatively few young farmers. Accessed 01 December 2021. https://ec.europa.eu/eurostat/web/products-eurostat-news/-/DDN-20180719-1.

Food and Agricultural Organization of the United Nations (FAO). 2006. *Food security*. Policy Brief, Issue 2, June. Accessed 1 June 2022. https://www.fao.org/fileadmin/templates/faoitaly/documents/pdf/pdf_Food_Security_Cocept_Note.pdf.

———. n.d. FAOSTAT selected indicators. Accessed 1 December 2021. https://www.fao.org/faostat/en/#country

Government of India (GOI), Ministry of Statistics & Programme Implementation. 2021. All India debt & investment survey NSS 77th round (January–December, 2019). Press release, September 10. Accessed 3 December 2021. https://pib.gov.in/PressReleasePage.aspx?PRID=1753935.

Han, Hongyun, and Hui Lin. 2021. Patterns of agricultural diversification in China and its policy implications for agricultural modernization. *International*

Journal of Environmental Research and Public Health 18 (9): 4978. https://doi.org/10.3390/ijerph18094978.

Huijsmans, Roy. 2019. Becoming mobile and growing up: A 'generationed' perspective on borderland mobilities, youth, and the household. *Population, Space and Place* 25 (3): 1–10. https://doi.org/10.1002/psp.2150.

Huijsmans, Roy, Aprilia Ambarwati, Charina Chazali, and M. Vijayabaskar. 2021. Farming, gender and aspirations across young people's life course: Attempting to keep things open while becoming a farmer. *European Journal of Development Research* 33 (1): 71–88. https://doi.org/10.1057/s41287-020-00302-y.

Imai, Katsushi, S., and Bilal Malaeb. 2018. *Asia's rural-urban disparity in the context of growing inequality*. International Fund for Agricultural Development Research Series 27. https://www.ifad.org/documents/38714170/40704142/27_research.pdf/86ff7619-8814-48d0-a232-fc694fcc55ce?eloutlink=imf2ifad.

Kumar Das, Varun, and A. Ganesh-Kumar. 2018. Farm size, livelihood diversification and farmer's income in India. *Decision* 45: 185–201. https://link.springer.com/article/10.1007%2Fs40622-018-0177-9.

Li, Minghao, Wendong Zhang, and Dermot J. Hayes. 2018. Can China's rural land policy reforms solve its farmland dilemma? *Agricultural Policy Review*, Winter. Center for Agricultural and Rural Development, Iowa State University. www.card.iastate.edu/ag_policy_review/article/?a=78.

Monllor, N. 2012. *Farm entry: A comparative analysis of young farmers, their pathways, attitudes, and practices in Ontario (Canada) and Catalunya (Spain)*. http://www.accesstoland.eu/IMG/pdf/monllor_farm_entry_report.

National Farmers Union. 2020. *Tackling the farm income crisis*. https://www.nfu.ca/wp-content/uploads/2020/02/Farm-Income-Crisis.25.02.pdf.

Ní Laoire, Caitríona. 2000. Conceptualising Irish rural youth migration: A biographical approach. *International Journal of Population Geography* 6 (3): 229–243. https://doi.org/10.1002/1099-1220(200005/06)6:3%3C229::aid-ijpg185%3E3.0.co;2-r.

nippon.com. 2018. Japan's farming population rapidly aging and decreasing. *nippon.com*, July 3. Accessed 1 December 2021. https://www.nippon.com/en/features/h00227/.

Schenck, Laura. 2018. *Small family farming in Indonesia—A country specific outlook*. Food and Agriculture Organization of the United Nations fact sheet. http://www.fao.org/3/I8881EN/i8881en.pdf.

Statistics Canada. 2017. A portrait of a 21st century agricultural operation. Accessed 22 June 2022. https://www150.statcan.gc.ca/n1/pub/95-640-x/2016001/article/14811-eng.htm.

United Nations Development Programme (UNDP). 2020. Human development insights. Accessed 1 December 2021. http://hdr.undp.org/en/countries.

White, Ben. 2020. *Agriculture and the generation problem.* Halifax: Fernwood Publishing.

World Bank. n.d. World development indicators. Accessed 1 December 2021. https://databank.worldbank.org/source/world-development-indicators.

Ye, Chunhui, Suwen Zheng, and Edward Gu. 2021. Positive effect of village debt on land transfer: Evidence from county-level panel data of village finance in Zhejiang Province. *PLOS ONE.* https://doi.org/10.1371/journal.pone.0255072.

Correction to: Becoming A Young Farmer

Sharada Srinivasan

Correction to:

S. Srinivasan (ed.), *Becoming A Young Farmer*, Rethinking Rural,
https://doi.org/10.1007/978-3-031-15233-7

The original version of this book has been revised. Acknowledgements has been updated with the correct funding information.

The updated version of this book can be found at
https://doi.org/10.1007/978-3-031-15233-7

Author Index[1]

---

[1] Note: Page numbers followed by 'n' refer to notes.

© The Author(s) 2024
S. Srinivasan (ed.), *Becoming A Young Farmer*, Rethinking Rural,
https://doi.org/10.1007/978-3-031-15233-7

Subject Index[1]

---

[1] Note: Page numbers followed by 'n' refer to notes.

© The Author(s) 2024
S. Srinivasan (ed.), *Becoming A Young Farmer*, Rethinking Rural,
https://doi.org/10.1007/978-3-031-15233-7